DIAGNOSTICS AND FAILURE PREVENTION IN TURBO-MACHINES

DIAGNOSTICS AND FAILURE PREVENTION IN TURBO-MACHINES

A S Rangwala
International Consultant
Turbo-Machinery

Publishing Globally
NEW AGE INTERNATIONAL (P) LIMITED, PUBLISHERS
New Delhi • Bangalore • Chennai • Cochin • Guwahati
Hyderabad • Kolkata • Lucknow • Mumbai
Visit us at **www.newagepublishers.com**

Copyright © 2015, New Age International (P) Ltd., Publishers
Published by New Age International (P) Ltd., Publishers
First Edition: 2015

All rights reserved.
No part of this book may be reproduced in any form, by photostat, microfilm, xerography, or any other means, or incorporated into any information retrieval system, electronic or mechanical, without the written permission of the publisher.

BRANCHES

- **Bangalore** 37/10, 8th Cross (Near Hanuman Temple), Azad Nagar, Chamarajpet, Bangalore-560 018
Tel.: (080) 26756823, Telefax: 26756820, **E-mail: bangalore@newagepublishers.com**
- **Chennai** 26, Damodaran Street, T. Nagar, Chennai-600 017
Tel.: (044) 24353401, Telefax: 24351463, **E-mail: chennai@newagepublishers.com**
- **Cochin** CC-39/1016, Carrier Station Road, Ernakulam South, Cochin-682 016
Tel.: (0484) 2377004, Telefax: 4051303, **E-mail: cochin@newagepublishers.com**
- **Guwahati** Hemsen Complex, Mohd. Shah Road, Paltan Bazar, Near Starline Hotel, Guwahati-781 008
Tel.: (0361) 2513881, Telefax: 2543669, **E-mail: guwahati@newagepublishers.com**
- **Hyderabad** 105, 1st Floor, Madhiray Kaveri Tower, 3-2-19, Azam Jahi Road, Near Kumar Theater Nimboliadda, Kachiguda, Hyderabad-500 027, Tel.: (040) 24652456, Telefax: 24652457
E-mail: hyderabad@newagepublishers.com
- **Kolkata** RDB Chambers (Formerly Lotus Cinema) 106A, 1st Floor, S.N. Banerjee Road, Kolkata-700 014
Tel.: (033) 22273773, Telefax: 22275247, **E-mail: kolkata@newagepublishers.com**
- **Lucknow** 16-A, Jopling Road, Lucknow-226 001
Tel.: (0522) 2209578, 4045297, Telefax: 2204098, **E-mail:lucknow@newagepublishers.com**
- **Mumbai** 142C, Victor House, Ground Floor, N.M. Joshi Marg, Lower Parel, Mumbai-400 013
Tel.: (022) 24927869, Telefax: 24915415, **E-mail: mumbai@newagepublishers.com**
- **New Delhi** 22, Golden House, Daryaganj, New Delhi-110 002
Tel.: (011) 23262368, 23262370, Telefax: 43551305, **E-mail: sales@newagepublishers.com**

ISBN: 978-81-224-3814-7

₹ 699.00

C-14-08-7895

Printed in India at Glorious Printers, Delhi.
Typeset at In-house, Delhi.

PUBLISHING GLOBALLY
NEW AGE INTERNATIONAL (P) LIMITED, PUBLISHERS
7/30 A, Daryaganj, New Delhi-110002
Visit us at **www.newagepublishers.com**

PREFACE

Failures of unimaginable proportions and consequent destruction in turbo-machinery occur under certain circumstances that are totally external to the engine's own system. A foreign object ingested into the fan of an aviation engine has the potential to bring down an entire airplane. An abrupt shedding of a large electrical load can cause a power generation turbine to speed up and cause the rotating disks to burst. Faults in the lube oil processing system can cause oil flow stoppage to the bearings, leading to rotating blades grinding against the casing.

In spite of all the advanced technologies employed in the design and manufacture of turbine engines, failures also arise from within the engine. The basic principle of air and gases flowing over the stationary and rotating airfoils in a turbine engine develops power. But the same principle is also responsible for the cause of aerodynamic interference and failure of the blades from metal fatigue due to unacceptable blade passing frequencies.

Diagnosis of faults ranging from inadequate pressure to excessive vibrations is based on detecting and measuring to ensure the parameters throughout the engine do not stray beyond their upper and lower bounds. Modern turbo-machines rely on marvelous devices, instruments and fly-by-wire control systems for smooth operation. In the event of a measured parameter going out of its prescribed limits, the central processing unit of the engine system analyzes the signals, determines appropriate remedial action and issues commands to control manipulators and actuators to rectify the situation.

Sophisticated control mechanisms notwithstanding, the operating speed of modern turbines is so high it is not possible to avoid running the engines into cataclysmic breakdowns. A failure in what may at first sight appear to be a minor component can lead to rapid deterioration and failures in a string of components of the sub-assembly. And where remedial action commands from the microprocessor-based controller are not possible, the end consequences may very well prove to be disastrous.

Automatic control theory and its methods are also put to use to improve efficiency by reducing fuel consumption, and in the process reduce exhaust gas emissions. As an example, during a take-off climb an airplane's engines go through rapid acceleration from ground idle to maximum take-off speed before the aircraft settles into a steady mode of cruise condition at a selected altitude. The microprocessor-based controller determines the ratio of air and fuel to be burned in the combustor of the turbine at each stage, and constantly adjusts the ratio to obtain optimum performance as parameters such as rotor speed, power output, compressor discharge pressure and exhaust gas temperature vary with the load demand.

This book is written to meet the needs of students in engineering colleges and practicing engineers in a large number of industries where turbo-machines are used. All of the material has been specifically tailored for college undergraduate and graduate level design engineering of rotating machine courses. Electronic spreadsheet type of calculation is also used in certain example problems to calculate control system natural frequencies, dynamic response, stability and other related design parameters. It is expected that the reader is familiar with basic to medium level calculus offered at the college undergraduate level.

The book is split into three parts. The first part focuses on failure modes in turbines and automatic control methods. The five chapters in the section look into failures in turbine components, diagnostics and control instrumentation. The second part describes various forms of turbines and discusses peculiarities in their operation. The third part looks into geometric variations, operation and control requirements. Failure modes in each type of turbo-machinery are carefully investigated to determine the root source and propagation of the failure mechanism throughout the engine. Topics as diverse as rotor vibrations, metal fatigue and combustor fires are thoroughly explored with the intent to provide the reader with skills to prevent failures in turbo-machinery.

The first chapter of part I discusses failure modes in turbine components. The second chapter delves into diagnosis and system control, the third chapter into automatic control methods, the fourth chapter into instrumentation for measuring parameters and the fifth chapter into regulation and control devices. In the second part, chapter six looks at aircraft jet engines, chapter seven describes steam turbines, chapter eight discusses power generation gas turbines and chapter nine investigates turbine applications for offshore oil platform and ship. Part III starts with a description of root cause of failures in the tenth chapter, eleventh chapter considers thermodynamics and aerodynamics pertaining to turbines, chapter twelve describes vibrations, chapter number thirteen considers steady and dynamic stresses, chapter number fourteen goes into fatigue failures and chapter number fifteen looks at thermal distortion. Chapter number sixteen considers fan and compressor operation, chapter number seventeen looks into combustion of fuel, chapter number eighteen into expansion of gases in turbine, chapter number nineteen into performance deterioration and chapter twenty into system related failures.

I gratefully remember and appreciate past students of the course on this topic who have sent in comments and reported errors, and express my hope that those who work with this treatise will do likewise. I am indebted to Mr. Mark Belloni and Dr. Fred Ehrich of General Electric Company for performing a vast amount of computational work in finite element analysis and for valuable advise on the text and layout of the book. I greatly appreciate comments provided by Dr. Ahmad Kamel, Mr. George Robinson and Dr. Raj Subbiah of Siemens-Westinghouse Power Corporation for checking the problems and reading the proof.

A.S. Rangwala

Orlando, FL

CONTENTS

Part III – Component and System Failure Modes

Part I
Diagnostics and Control Theory

1

CHAPTER

FAILURE MODES IN TURBINE COMPONENTS

1.1 INTRODUCTION

Modern turbo-machines are marvels of technology. The machines perform a plethora of services, come in all conceivable shapes and size, and are found on the land, in the air and on the high seas. Aviation engines propel passenger airplanes, fighter jets and helicopters at speeds approaching and sometimes exceeding the speed of sound with a remarkable record for safety. Industrial power generation gas and steam turbines bring electricity to provide light and comfort to the remotest corners of the world. Gas turbines derived from the lightweight characteristics of aviation engine technology produce power to drive the myriad pumps, compressors, hoists and cranes of offshore platforms to produce petroleum and natural gas. Turbochargers enhance power developed in diesel engines and at the same time burn less fuel. All the while, the machines deliver exceptional performance for power generated, fuel consumed, exhaust gases emitted, noise produced and operating life to place today's turbines in a class all their own. Suffice it to say that a world devoid of turbo-machines would be totally different form the one we live in.

The machines, however, are subject to failure from a host of situations. Conditions arising from factors external and internal to the turbine hold potential to cause situations ranging from temporary disturbances in the normally smooth operation to disastrous failures that may lead to loss of life and property. The operating speed of the turbines is very high, so events take place that may be measured in milliseconds. For instance, the time available for a pilot of a passenger airplane to take note of a possibly anomalous operation of an engine and take appropriate corrective action is so little that his action is mostly of little consequence. Note also that the pilot is in charge of operating the total aircraft, and cannot pay all of his attention to just one miscreant engine.

The root source of failures can arise from a number of factors. Inherent design fault, deficient manufacturing process, adverse environmental condition, abrupt departure from steady operation, foreign object damage, and an array of many other situations are responsible for fractures in components in the turbine. A sequence of events then may lead to a total system breakdown.

It is imperative then that a reliable method for detecting and measuring system operating parameters and manipulating and actuating devices are available to forestall a failure situation. Electronic signals from sensors, microprocessor and actuators are widely used to rectify a system parameter that is out of bounds and bring it back within its limits.

1.2 FAILURE TYPES

Aerodynamics plays a major role in turbo-machinery as air entering the inlet flows over stationary and rotating blades and makes its way to the exit of the machine. After going through the compressor section to increase pressure, air then enters the combustor section where fuel is burned. The hot compressed gases than go through expansion in the turbine section where enough power is generated to not just drive the compressor but, importantly, the driven load. The driven load may be in the form of fan of a turbofan engine, electrical generator in a power-generating unit or propeller drive shaft of a helicopter engine.

Thermodynamic considerations play a major role in cycle design of turbo-machines. Based on a certain set of assumed values, various parameters such as the annulus dimensions of axial flow compressor and blade angles are established from thermodynamic entities. If prevailing conditions result in the assumptions straying beyond limits, the likelihood of disruption in steady operation of the machine and consequent deleterious effects on the hardware increases.

Unbalanced forces and overturning moments are invariably present during operation of a turbine. Unbalance in the rotor, misalignment in the bearings or at the coupling and inadequate torque on bolts of mating components are some examples causing a state of loss of equilibrium of the loads. Unequal pressure distribution of the fluid medium around the periphery of a compressor or turbine stage will lead to a net radial thrust on the shaft.

Cyclical loads on a part can lead to fatigue failure. In areas of high stress concentration, repeated cycles of loading and unloading will lead to fatigue cracks that initiate and propagate. As the load cycles accumulate, fatigue cracks gradually spread and eventually lead to total fracture.

Thermal distortion of a component located in the hot path of the turbine may occur from elevated operating temperature. When the temperature rises from a datum level to a higher one, thermal growth will depend on the difference in the temperature levels and the coefficient of thermal expansion of the substrate material. If the component is constrained such that the thermal growth is suppressed, thermal stresses will be induced. Thermal stresses will also occur in the case where thermal gradients are present. The gradients are produced as a function of time when an engine goes through periods of startup and shutdown. Variation in temperature also occurs when a part is exposed to varying temperature levels at different surfaces.

1.3 FRACTURE DUE TO METAL FATIGUE

Fatigue failure in metal components of a turbine is the result of progressive and localized structural damage when the part material is subjected to cyclic loading. Failure takes place even though nominal maximum stress value is below ultimate tensile stress limit, and may even be below the yield stress limit of the material.

Repeated loading and unloading is the primary cause of fatigue occurring in a material. Above a certain threshold, the loads create microscopic cracks at the surface. The crack propagates till it reaches a critical size, leading to sudden fracture of the part. The shape of the component plays prime role in the fatigue life, with sharp corners leading to local stress risers and initiation points of fatigue cracks. Larger round holes and generous fillet radii are important in increasing the fatigue strength of the structure.

The process starts with dislocation movements in metals and their alloys, which eventually form persistent slip bands that nucleate short cracks. The stochastic nature of the process often displays considerable scatter, even when taking place in a controlled environment. The scatter in components with long fatigue life generally increases.

Greater fatigue damage is induced in a metal as the applied stress range increases, which leads to shorter number of available life cycles. The damage caused to a part after each load cycle accumulates, since the material is not capable of recovering even after breaks between the cycles.

A number of factors influence the fatigue life in turbine components. Running temperature, surface finish, microstructure, presence of oxidizing or some inert chemicals, residual stresses, and fretting at contact points commonly need to be closely evaluated in establishing useful life of components before retirement. Certain steel and titanium alloys have displayed a theoretical fatigue limit, below which continued loading does not lead to failure. Structural integrity is preserved and failure is ruled out if the number of load cycles is below that limit.

Research based on ultrasonic resonance experiments with frequencies between 10 and 20 kHz indicates failures occur below the theoretical fatigue limit at very high fatigue lives (10^9 to 10^{10} cycles).

High cycle fatigue strength in the range of 10^3 to 10^8 cycles can be described by stress-based parameters. A load-controlled servo-hydraulic test rig is commonly used in these tests, with frequencies of around 20 to 50 Hz. Other sorts of machines, such as resonant magnetic machines, can also be used, achieving frequencies up to 250 Hz.

Low cycle fatigue and plasticity tend to occur in tandem under 10^3 cycles in many metals and alloys. A strain-based parameter is required for fatigue life prediction in the metals. Testing is conducted with constant strain amplitudes, typically in the 0.01 and 5 Hz range.

1.4 EXCESSIVE VIBRATIONS FROM ROTATING SYSTEM

When a metal chip is loosened from the rotor due to rub occurring with a stationary component, unbalance results. Since the unbalanced mass is rotating at the rotor speed, centrifugal force 180° away from the location of the loosened chip is created. Centrifugal force is proportional to the square of the angular speed. The unbalanced force is borne by the shaft, with the load path traversing through the support bearings and eventually leading to the ground. Associated with the centrifugal force, overturning moments will also be produced. The combination of forces and moments produce stresses in the components lying in the load path. If a heavily stressed part is required to burden the additional load, the possibility of failure of that part exists.

Depending on its geometric configuration, the rotating shaft and its support bearings has its own natural frequencies of vibration. The mass distribution along the length of the rotor, shaft diameters, length segments, and material characteristics of the shaft will determine the natural frequencies. The natural frequencies are also affected by the stiffness characteristics of the support bearings.

If the operating speed regime of the turbine is away from any of the natural frequencies of vibration, and assuming the unbalance in the rotor is within permissible levels, the operation of the turbine will be relatively free of vibrations. On the other hand, when the operating speed and any of the natural frequencies of vibration are in close proximity, a condition of resonance sets in and high vibrations ensue even if the rotating unbalance is at low levels. Dynamic forces are created in mostly all load-bearing components of the rotating systems, with the forces proportional to the square of the vibration frequency. In addition to steady state stresses, if the combined equivalent stress exceeds the endurance strength of the substrate material, failure mechanism sets in, initiating at the peak stress concentration point and eventually spreading to produce total fracture.

1.5 EXTERNAL MECHANICAL SHOCK

Heavy military vehicles, such as the US Army's Abrams Main Battle Tank and the Crusader self-propelled howitzer, require an efficient hybrid propulsion system to meet large and continuous power needs. The LV100 recuperated turbine engine is well suited to provide electric power for advanced armored motor vehicles where volume and weight are a premium. Mission requirements may call for sustained high power to maintain speed even on a slope. To avoid easy detection, low emissions and noise characteristics of the vehicle are mandatory. To top it all, since the fuel tank uses valuable under the armor volume, fuel consumption over a wide range of power settings has to be low.

Aircraft engines are designed for lowest possible weight and best performance in the 80 to 100 per cent power range for fixed wing and helicopter applications. On the other hand, an aero-derivative engine adapted for military vehicular use needs to operate on rough terrain that is not conducive for the engine. For instance, the US Army's Abrams Main Battle Tank must be designed for continuous medium to high shocks from the ground. When the battle tank is traversing a ploughed field at a speed of 18 miles per hour, the vehicle and the engine are subjected to high ground shocks during field exercises. Also, the engine must be optimized in the idle to 60 per cent of maximum power output range.

Besides a good volume to shaft horsepower ratio, the turbine must contain turbo-machinery components to address a number of constraints and needs:

- Means for power augmentation to cover short term needs while maintaining speed over a rocky slope
- Components with wide operating mass flow range
- Parts designed for high thermal cycle count and shock loads
- Durability and ease of maintenance of engine system.

Naval and merchant ships often run into rough weather and high waves in the open seas. Marine gas turbines must be capable of withstanding the pitching and rolling of the ship even when in operation. Loss of power and consequent downtime under such adverse conditions can lead to considerable loss. Failure of turbine components and difficulty in replacing them would represent major disadvantages.

1.6 FOREIGN OBJECT DAMAGE

The likelihood of an aviation engine ingesting objects lying on the runway as the aircraft rolls for takeoff at many airports is high. Stones, twigs, pieces of tire rubber tread and other debris hold potential for serious trouble if ingested into the engine. Maximum takeoff thrust means tremendous suction at the inlet, and if the runway is not thoroughly maintained, the debris can be sucked into the engine and directly impact on the fan blades rotating at near maximum speed.

Instances have occurred where an aircraft dropped metal hardware pieces such as loose bolts, brackets and nuts that fell off from the aircraft as it was taking off. A subsequent flight taking off from the same runway after a short time sustained ingestion of the loose hardware lying on the ground, causing damage to the engine.

1.7 BIRD STRIKE IN AVIATION APPLICATION

Many airports throughout the world are located close to the coastline. Large birds such as the seagull and Canadian geese tend to fly high and are heavier in weight. Even airports located further inland may have bird sanctuaries located within a few miles radius from the runway. As an airplane prepares to takeoff and the engines develop maximum thrust to permit escaping the earth's gravity and climb up to altitude, the engines create a powerful suction at the inlet to the engine. If a flock of birds is passing above the runway path, chances of a bird being ingested into one of the engines are good. Multiple birds are also known to get sucked into a single engine occasionally.

Notwithstanding the turbine engine manufacturer's efforts to ensure fan blades remain intact even after one or more birds simultaneously strike the engine, partial breakage of fan blades has remained a problem in the aviation industry. The broken fan blade fragments may be large enough to cause other blades to break. Smaller fragments may even get into the high-speed compressor. The smaller airfoils spinning at the higher core speed are more susceptible to breakage. Instances have been recorded where a bird strike incident causes the entire airplane forced to crash land in the ocean or on land.

1.8 COMPRESSOR SURGE

Partial blockage arising from turbulence in selected regions of a compressor can lead to pressure surge immediately behind the blockage, which may further deteriorate into reversal of flow and stalling of the compressor. Turbulence can occur when air flowing into rotating blades is not entering at the proper angle, causing eddy currents to form at the blade's leading edge tip. Radial clearance between the rotating blades and the stator casing is also of significance. Larger than normal clearance will cause eddy currents to flow randomly in the dead space. This acts as a blockage to the flow of air coming in from the inlet.

Excessive vibrations in the engine often accompany a compressor stall, which can lead to rubbing between the rotating blade tips and the stator. Blade tips may also be torn off, both in the compressor and turbine sections. Rubbing at the seal teeth in the bearing sump and inter-stage areas will also lead to rapid loss of performance and overall deterioration of engine parts.

1.9 COAL, LIQUID AND GAS FUELS

In the aviation industry liquid fuels such as JP-5, a blended kerosene fuel, and JP-4, a naptha-based fuel, are relied on for powering the engines. Natural gas finds wider application in industrial and generation engines. Among other characteristics of flash point and freezing point, the fuel used must have the capacity to easily burn even in a lean mixture of air and fuel without leaving residual unburned hydrocarbon particles.

Lean air-fuel mixtures offer the commercially attractive proposition of burning less fuel while at the same time reducing certain harmful exhaust gas pollutants. However, the combustor must not lose the flame with the leaner mixture, since the engine would abruptly shut down. This can cause thermal fatigue damage in the hot path components.

A noteworthy trend in modern power generation gas turbines is their ability to burn both liquid and gas fuels. Due to the exigencies of market conditions and availability, liquid fuels may sometimes be required during operation, although natural gas is predominantly more cost effective. This entails provision of specially designed dual-fuel injectors for liquid and gas fuels inside the combustion chamber. When switching from one fuel to the other, it is not necessary to shut off the gas turbine, although to avoid surprises it would be advisable to reduce load on the electrical generator down to mostly zero.

Power generation plants equipped with steam turbines mostly use mineral coal for steam production in a boiler, although natural gas is also sometimes used when available in plentiful supply. The primary benefit of burning the fuel outside the turbine is absence of exposure of the hot combustion gases to the turbine airfoils. On the other hand, steam is far less corrosive on the turbine materials and also cooler than combustion gases produced inside the combustor. Still another plus is longer operating life of the rotating turbine blades running at lower temperatures. After combustion, tiny coal particles are carried away in the exhaust gas stream. A definite disadvantage for the community in its neighborhood, the particles may be removed using electrostatic precipitation method. Organic compounds are oftentimes present when the coal is obtained closer to the ground level, as opposed to mineral coal mined from far below the ground surface. This may necessitate separation of the coal from the vegetation and other organic material prior to admission to the boiler furnace.

1.10 SUSPENDED AND DISSOLVED IMPURITIES IN FUEL

Natural gas has advantages over liquid fuels. It is easier to mix with air to obtain more complete combustion, the gas can be compressed to permit transmission through pipelines in large quantities and has relatively fewer problems with contaminants. Suspended impurities such as sand and particles of iron oxide cannot be burned. But the particles may be removed in cyclone type filters, where vanes attached to a cylindrical vessel provide the gas a strong whirl. Centrifugal force on the heavy particles causes them to fly out toward the walls of the cylinder. The particles tend to be oily and gritty as opposed to loose sand, so they tend to stick to the walls. Finer suspended particles may further be removed when the gas flows through conventional filtering elements.

Besides suspended particles, liquid fuels such as benzene for boilers and gas turbines contain dissolved compounds of sodium, potassium, phosphorous, calcium and sulfur. A cumbersome process of treating the liquid chemically with other agents, sometimes in the presence of a catalyst, is necessary to produce granules, which are then washed off. Sodium and potassium compounds are thus removed without too much difficulty, as also calcium and phosphorous compounds. Sulfur, however, requires a more rigorous process. The removal of the compounds is essential to safe operation of the turbine. For instance, sulfur reacts with the chromium in the turbine blade substrate to produce sulfate compounds in the elevated temperature environment. This causes formation of a wavy pattern on the surface of the blades and vanes. The uneven surface leads to higher flow losses as opposed to the original smooth surface finish and consequent loss in power output and efficiency. The problem may be acute enough to require premature retirement of the high-cost turbine blades.

1.11 THERMAL SHOCK IMPACT

A sudden departure in sequence of events in the hot path of a gas turbine holds potential for creating thermal distress in the components. Cutoff of fuel supply to the injectors in the combustor and loss of coolant to the turbine inlet vanes are examples where the associated hardware components are exposed to thermal shock.

Thermal distortion in the form of bowed stator vanes at the turbine inlet is sometimes observed. The occurrence is mainly due to uneven temperature distribution in the gas stream, with the temperature near the inner and outer annuls walls differing from that in mid-stream.

On a system-wide basis, a gas turbine can trip in the event of loss of lubricant oil supply failure to the bearings. A notable incident occurred in a case of failure of a single-engine aircraft's lube oil pump.

The new military aircraft was undergoing acceptance trials by an Air Force pilot. The pump failure was attributed to a bearing retainer ring breaking because the groove in which the ring was seated was not machined deep enough. This caused the retainer to break and inside parts sliding axially. Subsequently the fuel supply was shut off, and the hot path components of the engine were exposed to the cold air streaming in through the inlet.

A trip in a power generation turbine is considered to have undergone four start and stop cycles because of the thermal distress experienced by the turbine blades. The steep decline in metal temperature of the blades creates thermal gradients and associated thermal stress that is equivalent to four times the level of stress produced when the engine goes through normal shutdown procedure.

1.12 LOSS OF IGNITION IN COMBUSTOR

During startup of a gas turbine the mixture of air and fuel needs to be ignited. The fuel nozzle is required to deliver a nearly stoichiometric mixture of air and fuel uniformly in the primary zone, with the fuel flow adjusted to permit engine operation from idle to full load conditions. Once a stable flame regime is obtained, the combustion of fuel is self-sustaining and proceeds continuously during normal operation.

Aviation engines sometimes lose the light in the combustor, and the engine suffers in-flight shutdown at altitude. Attempts to re-light the combustor may run into a few difficulties. For instance, combustion may not even initiate when the mass flow is too large. An unacceptably low ratio of air and fuel may be caused if the engine is accelerating and the flow of air is still building up. The defined stable flame region is also affected by pressure in the chamber. The difficulty of igniting increases as the pressure reduces, as is the case when the airplane is flying at altitude. Chemical reactions slow down as the pressure decreases, which tend to restrict the flame's stability region.

The can combustors of industrial units are provided with crossfire tubes between the cans. A certain number of combustor cans are arranged around the engine centerline along a circular path. If a single combustor can loses the flame, the crossfire tubes permit the flame to spread from the adjacent cans. The annular configuration of the combustor generally favored in aviation engines does not lend itself to a similar arrangement for quickly spreading and maintaining the flame.

Sudden loss of flame and inability to re-establish quickly will lead to thermal shock related problems in the hot path components. The impact of the shock may be intensive enough to lead to permanent distortion. Excessive excursions in the temperature of the gas stream can also result in high thermal stresses and consequent failure.

1.13 ROTATING BLADE SEPARATION

A leading example of a rotating blade separation is that of the steam turbines installed in the passenger ship Queen Elizabeth II. In November of 1968, on one of its first maiden voyages, the turbine threw out a blade. Aside from the turbine casing getting damaged, extensive damage on board the ship occurred as the blade fragment flew out and landed several hundred feet away into the water. The failure was subsequently attributed to damage due to high-cycle fatigue arising from aerodynamic interference as superheated steam made its way past the stationary stator vanes and the rotating blades attached to the rotor.

The mass of the broken blade, its center of gravity location from the shaft centerline and the operating speed squared will dictate the centrifugal force acing on the released blade fragment. If multiple blades were to fail at the same time, as often occurs in the case of a high-pressure compressor of a gas turbine, the resulting unbalance force is even greater. The resulting vibrations detected by vibration pickups in the bearings will be large enough to cause the engine to rapidly shutdown. The damage caused to the engine and its surroundings is generally heavy, and may even lead to extensive fires to totally demolish the turbine.

1.14 ABORTED AIRCRAFT TAKE OFF AND LANDING

Under certain emergency situations control towers are known to order the pilot of an aircraft to abort takeoff just as it prepares to leave the ground after the takeoff roll. The pilot is then required to pull back on the engine throttle lever in a snap motion. In preparation for takeoff the thrust developed is at its peak, with maximum airflow through the engine inlet, as well fuel flowing into the combustor. In the turbine section, elevated gas temperature will lead to thermal growth of both the casing and the rotor. In the aborted takeoff condition, abrupt shutoff of fuel to the injectors will cause a steep decline in gas and metal temperatures in the combustor and in the turbine. This causes the casing and the rotor to shrink. However, the thermal shrinkage does not occur simultaneously in the two modules, and may lead to rubbing at the tip of the rotating blades. If larger chips break off at the tip of the rotating blades, unbalance in the rotor will accumulate and cause stronger vibrations. Ground personnel likely will call for bore-scope inspection to evaluate damage sustained in the turbine after an aircraft experiences an aborted takeoff.

A similar situation, but in a reversed order, will occur in the case of an aircraft required to abort a landing just when it is approaching close to the ground. Ground personnel may even elect to take the engines off the wing if after test runs the temperature of the exhaust gases departing the last stage of the turbine is hovering near the limit.

1.15 EXTREME ENVIRONMENTAL CONDITIONS

In certain parts of the world power generation gas turbines are required to operate even when ambient temperature plunges to –40° F or less. Cold air entering the inlet and into the compressor may pose the problem of cold brittle fracture of the rotor. Since the rotor is of considerable weight, running into a few hundred thousand pounds, it is difficult to justify use of higher cost material that can withstand such low outside air temperatures. To avoid the risk, heating of the air prior to entering the intake bell-mouth may be required. If a crack in the rotor occurs in a corner fillet radius due to step change in the diameter, the potential exists for the crack to go deeper or wider as the machine accumulates operating hours.

Many other abnormalities arising from environmental and operational situations are known to take place that affect airborne, land and shipboard engines. Ingestion of ice and rainwater into the engine can force the pilot to cancel a takeoff. Strong crosswinds prevent the air from smoothly flowing into the inlet. Where an airport is located at a high altitude in a desert region, mid-day temperatures can soar to 120° F. If the aircraft is fully loaded with passengers and cargo and has a full gas tank, it may not be possible for the pilot to takeoff and leave the ground since the exhaust gas temperature limit may be breached. Spraying of water at the inlet to cool the air is not available in aviation gas turbines.

1.16 REVIEW QUESTIONS

1.16.1 Name some hazards facing machinery operating out at sea.

1.16.2 What form of corrective actions may be used for ensuring safe and trouble-free operation at sea?

1.16.3 When discussing running speeds, what sets apart an industrial power generation turbine from an aviation unit?

1.16.4 Why are hydrodynamic journal bearings not suitable for aviation engines?

1.16.5 What are the drawbacks of employing rolling element bearings in power generation turbines?

1.16.6 What location in a turbine is most suitable for measuring vibrations?

1.16.7 What role does damping play in a vibrating rotor-bearing system? Name some sources of damping in a turbine.

1.16.8 Why are the rotating blades of a steam turbine less failure-prone than those in a gas turbine?

1.16.9 Name some obstacles faced in establishing the safe operating cycles for the rotating components of a turbine.

1.16.10 What methods are used to obtain concentricity between two bolted casing sections?

1.16.11 What form of coolant is available for cooling parts in the hot path of aviation engines?

1.17 EXAMPLE PROBLEMS

Example Problem # 1.17.1: A stroboscope produces intermittent flashes of light on a machine rotating at 3450 rpm. If a point located 4.5 in. from the shaft axis is required to see the point with a blurring of less than 1/48 in., what should be the duration of the light flashes?

Solution: The point travels at circumferential speed equaling:

$$= 3450 \times 2\pi \times 4.5/60 = 1625.77 \text{ in/sec.}$$

Flash time duration multiplied by blur size will give the velocity of the point.

Thus: $t(1/48) = 1625.77,$

Hence: $t = 1/(1625.77 \times 48) = 1/78037$ second.

Example Problem # 1.17.2: A shaft spinning at 2750 rpm has an unbalance weight of 3.25 oz. at a radial distance of 25.0 in. Calculate the centrifugal force on the mass.

Solution: Centrifugal force $= W/g \times \omega^2 r = [3.25/(16 \times 386)] \times (2\pi \times 2750/60)^2 \times 25.0$

$$= 1089.93 \text{ lbs.}$$

Example Problem # 1.17.3: A simply supported beam is impacted by a free falling sphere of weight W from height h. Determine an expression for maximum momentary deflection produced in the beam.

Solution: When a load W falls a distance h on the beam, it causes a peak instantaneous deflection δ in the beam where the ball strikes. The total work done by the ball is $W(h + \delta)$. Assume a static equivalent load W_{eq} that produces an equal deflection δ at the same point on the beam, with the load applied gradually. The work done by this equivalent load is $½(W_{eq}\delta)$. If the supports at the two ends of the beam are rigid, the work done by the falling and the stationary loads are the same, or:

$$W(h + \delta) = ½(W_{eq}\delta)$$

The static deflection δ maybe expressed in terms of the equivalent static load W_{eq}. For example, if the load is applied centrally on the beam, $\delta = (W_{eq} l^3/48EI)$, where l is the length of the beam and EI represents the beam stiffness in flexure.

Example Problem # 1.17.4: If the ball in *example problem # 1.17.3* weighs 8000 lbs, produces a static deflection of 0.75 in. and a maximum bending stress of 5500 lbs/in^2 in the above problem, from what height can a ball of 4000 lb. be dropped at the same point in the beam if the maximum permissible bending stress is 22000 lb./in^2?

Solution: The required static load for a stress of 22000 lb./in^2 is:

$$W_{eq} = (22000/5500) \times 4000 \times 2 = 32000 \text{ lb.}$$

Since static stress and deflection are proportional to the load causing them, the corresponding deflection is: $\delta = (32000/8000) \times 0.75 = 3.0$ in. Substituting the values in the expression obtained from the previous example problem,

$$4000(h + 3.0) = \tfrac{1}{2}(32000 \times 3.0)$$

Hence, $h + 3.0 = 12.0$ and the required height is 9.0 in.

1.18 EXERCISE PROBLEMS

Exercise Problem # 1.18.1: Determine the natural frequency of vibration of a mass-spring system of a 4.0 lb. weight attached to a spring of 12.5 lbs/in stiffness,

Exercise Problem # 1.18.2: If a dashpot of coefficient of 2.5 in-lbs/sec is attached to the mass-spring system in *Exercise Problem* # 1.18.1, how is the frequency affected?

Exercise Problem # 1.18.3: Air at 7.5 lbs/in^2 enters an eight-stage compressor. Assuming each stage compresses the air by 22.5 per cent, calculate the pressure at the discharge end.

Exercise Problem # 1.18.4: If the temperature of air entering a 20:1 ratio compressor is 25° C and pressure is 1.01 bar, what is the temperature at the discharge end assuming γ, the ratio of specific heats at constant pressure and temperature, is 1.39 and gas constant $R = 0.287$?

REFERENCES & BIBLIOGRAPHY

Garnier, V. H., Epstein A. H., Greitzer E. M., "Rotating Stall Anticipation and Initiation in Axial Compressors", ASME paper # 90-GT-156, New York, 1990.

Hathaway, M., Gertz, J., Epstein, A. H., Strazisar, A., "Rotor Wake Characteristics of a Transonic Flow Fan", AIAA Journal 24, number 11, 1986.

Hoeft, R. F., "Heavy Duty Gas Turbine Operating and Maintenance Considerations", GER-3620B, General Electric I & PS, Schenectady, NY, 1993.

Jahnsen, W., Peters, T., Fottner, L., "Stall Inception in a 5-Stage HP Compressor with Increased Load due to Inlet Distortions", ASME paper # 99-GT-440, New York, 1999.

Kotidis, P. A., Epstein, A. H., "Unsteady Radial Transport in a Transonic Compressor Stage", ASME International Gas Turbine Conference, Brussels, 1990.

McCloskey, T. H., "Turbine Steam Path Damage: Theory and Practice, Turbine Fundamentals", Volume I, EPRI, 1999.

McCloskey, T. H., "Turbine Steam Path Damage: Theory and Practice, Damage Mechanism", Volume II, EPRI, 1999.

McDougall, N. M., Cumpsty, N. A., Hynes, T. P., "Stall Inception in Axial Compressors", ASME Journal of Turbo-machinery, volume 112, page 116 – 125, New York, 1990.

National Transportation Safety Board, "In Flight Loss of Propeller Blade – Aircraft Accident Report ", NTSB/AAR-96/06, Washington DC, 1996.

National Transportation Safety Board – Aircraft Accident Report, NTSB/AAR-90/06, Washington DC, 1990.

Ragland, T. L., "Industrial Gas Turbine Performance Up-rates: Tips, Tricks and Traps", ASME Paper # 97-GT-409, New York, 1997.

Tarabrin, A. P., Schurovsky, V. A., Bodrov, A. I., Stalder, J. P., "An Analysis of Axial Compressor Fouling and a Cleaning Method of Their Blading", ASME paper # 96-GT-363, New York, 1996.

Wilson, A. G., Freeman, C., "Stall Inception and Development in an Axial Flow Aero-engine", ASME Journal of Turbo-Machinery, volume 116, page 217 – 225, New York, 1994.

2

CHAPTER

DIAGNOSTICS AND SYSTEM CONTROL

2.1 INTRODUCTION

Diagnosis of a fault in an operating turbo-machinery chiefly relies on a system comprising a system of sensors for its detection. Depending on the nature of the error and its location within the system, multiple operating parameters in several areas may be affected. For normal operating conditions, each parameter value is required to stay within fixed upper and lower bounds. For instance, if the temperature of lubricant oil supplied to the bearing is higher than normal, it may be due to greater vibrations of the rotor, reduced flow of lubricant to the bearing, lower oil pressure or inadequate cooling of the lubricant prior to admission to the bearing sump. Clogging of the oil filter with debris also cannot be ruled out.

The task of the control system is to evaluate all of the data reported by the temperature, pressure, flow and vibration sensors and decide on a corrective action. Command instructions from the control system may be to increase the oil pump's speed or the oil cooling system's capacity. If a feedback element is used in the control loop, electronic signals from the sensors will report the usefulness of the corrective action. Special feedback elements also are able to adjust the feedback signals to further increase the gain when necessary. The complete sequence of events is reported to control room personnel so that corrective maintenance work may be carried out to prevent the system from further deterioration.

Advanced aircraft requirements for avionics and hydraulic power have changed substantially. At the same time, engine requirements for hydraulic power and oil system of advanced aircraft have increased. The capacity of the engine system to absorb and dissipate heat has decreased because engine cycle efficiencies have increased, thus decreasing burn flow at engine cruise points.

The integration is accomplished across the data bus between the aircraft's control systems and the engine's Full Authority Digital Engine Control (FADEC) system. Either system can request changes that result in the best overall thermal efficiency through a set of integrated control laws. The engine burn flow temperature is essentially maintained at a maximum permissible level to transfer as much heat as possible into the combustor and to re-circulate excess levels through heat exchangers to cool the fuel before returning it to the aircraft system.

2.2 SENSOR SIGNALS TO DETERMINE PERFORMANCE

Accuracy and speed of measuring a rapidly changing signal is of significance for effective control of the output variable. Response from the sensor's element relies on the properties of its material and construction details. The measured value is converted into an electrical, mechanical, or fluidic signal. The signal may take the form of an electrical voltage, force, displacement, pressure, or some other physical parameter. Regardless of the form, the output of the sensor depends on the value of the measured parameter.

The threshold of the measuring instrument defines the smallest increment in the measured variable that will result in a shift in the indicator output, while its sensitivity is the ratio of change in input and the corresponding shift in output.

Calibration data is mostly provided in a graphical form. The curves form a loop as the upscale and downscale curves follow different path. The loop is formed from a combination of threshold and hysteresis effects. Hysteresis defines the dependence of the value of the variable on the history of past variables and current direction of traverse along the curve.

2.3 PARAMETER MEASUREMENT

Judicious selection of the measurement instrument for a given application is essential in ensuring precision of readings and to achieve certain level of ability to repeat the data. For example, for a temperature transmitter consideration must be given to the many different element types, such as thermocouple, resistive thermal detector (RTD) or radiation pyrometer. Criteria pertaining to operating, static, and dynamic characteristics of the measuring instruments must be available prior to choosing a measurement instrument.

Some uncertainty is unavoidable in the measurement, so it is difficult to achieve the true value of the variable. Uncertainties also prevail in the measurement standards while calibrating the measurement devices. Lack of repeatability in the measurements under identical conditions may give different results. Statistical evaluation is available, but statistics cannot determine the true value of the variable. It, however, helps in understanding and confronting the uncertainties.

2.4 SIGNAL PROCESSING

The raw signal obtained from a sensor cannot be put to use. Processing of the signal obtained from the sensor calls for conditioning to convert the sensor's output signal into an electrical or pneumatic signal with the intent to use it with a controller or a display device. Typical operations performed by a signal conditioner include isolation and impedance conversion, noise reduction through filtering, amplification, analog-to-digital and digital-to-analog conversion, data sampling, and linearization.

The operational amplifier, a high-gain dc amplifier, plays the central role in most modern signal conditioners. Operational amplifiers come in a variety of IC packages. Gain, input and output impedance, and frequency response depend on external resistors and capacitors in the circuit. Behavior of the assembled circuitry can be made to fit a particular application by proper selection and placement of the resistors and capacitors.

Two inputs are needed, the 'inverting input' (v_1) and the 'non-inverting input' (v_2) voltages, which provide output terminal voltage v_{out}. Corresponding currents are: i_1, i_2, and i_{out}. The gain, A, of the operational amplifier determines the output voltage according to the expression: $v_{out} = A(v_2 - v_1)$. The open-loop response of the amplifier is a linear curve in operational region with horizontal segments at both ends to make up the saturation region. An inverting amplifier changes the sign and level of the input signal. A non-inverting amplifier circuit can increase the size of the signal, but cannot decrease it.

An integrator circuit produces an output that is proportional to the integrator of the input voltage. A differentiator circuit produces an output that is proportional to the rate of change, or derivative, of the input voltage. A differential amplifier circuit amplifies the difference between two input voltages, neither of which is zero. An instrumentation amplifier (IA) circuit is essentially a differential amplifier with high input impedance, high common mode rejection, balanced differential inputs, and gain determined by a user selected external resistor. A function generator circuit provides an output that is a non-linear function of its input. An inverting or non-inverting or differential amplifier is sometimes needed to convert the signal from a voltage to a current, or from a current to a voltage.

Strain gage, force and pressure transducers convert changes in the measured variable into small changes in the resistance of the element. A balanced or an unbalanced bridge circuit measures small changes in resistance. The current balance bridge is useful for balancing to provide a useful signal for control systems. The unbalanced Wheatstone bridge is also used for the same purpose.

Electric signal lines are prone to operate in an environment that is filled with high levels of electro-magnetic energy. The effect is to produce noise in the electric signal lines. Even with noise suppressive measures such as grounding of lines and shielding of cables, some noise is bound to creep in. Filters may be used to decrease unwanted signals in specific bands of frequencies, thus giving low-pass, band-pass and notch-pass filters.

Certain primary resistance detector elements, among them platinum, produce a linear signal calibration curve so the output goes from 0 to 100 per cent as the input varies by the same percentage. Other primary elements produce a non-linear signal calibration curve, so the signal needs to be linearized in small increments to produce a nearly ideal calibration curve.

Since physical parameters in a process are analog signals, they must be converted into digital form for processing by computers. This is a function of the data acquisition system. After processing in the computer, the digital code is converted back into analog voltage levels for input to the analog actuators, and is the job of the data distribution system. The acquisition and distribution tasks are part of the data sampling and its conversion. The data sampling process switch connects momentarily to an analog signal in a sequence of pulses separated by the sampling interval. Values of the sampled signal give an approximation of the original signal. Data converters are the interface between the analog and digital domains. The physical parameter measured by the primary sensing element is converted to produce a digital code with the aid of an analog-to-digital (*A*/*D*) converter to produce a code that accurately represents the level of the analog input voltage. The digital-to-analog (*D*/*A*) converter converts digital codes into analog voltage levels for input to the analog actuators, and is done by the data distribution system.

2.5 FIRST AND SECOND ORDER LAG

Sensors cannot provide immediate and complete response to a sudden change in the measured variable under rapidly changing conditions. The dynamic characteristics of a measuring instrument depend on its electronic features, and a period of time elapses before the total response is received. Step, ramp, and frequency responses of the measuring instrument determine its dynamic characteristics. A step change in the measured variable, such as moving a temperature probe from ice water to boiling water, produces a worst-case response, and is often used to define its dynamic characteristics. The response may be classified as over-damped, critically damped or under-damped. Response and rise time are good indicators for step response of over-damped and critically damped systems. Step response of under-damped systems is stated in terms of rise time, peak percentage of overshoot, and settling time.

In a ramp function the measured variable is gradually increased, and provides additional insight into the instrument's characteristics. The term dynamic lag indicates the time period required by the indicated value to catch up behind the input variable, while dynamic error shows difference in the variable's ideal and measured values at any time.

Frequency response and transfer function, discussed in Chapter 3, are most useful in defining the dynamic characteristics of the measuring instrument. Some measuring instruments have a more complex transfer function than the first order lag. If there are two layers of protective sheath, for instance, the two resistances and two capacitances form a two-capacity interacting system, the differential equation has a first-order and a second-order derivative term, and the system becomes a second-order lag. The frequency response of a second-order system is also more complex than that of a first-order in an under-damped system.

2.6 RESILIENCE AND RELIABILITY OF DIAGNOSTIC TOOLS

The ability of a multi-channel control to accommodate an arbitrarily malicious fault defines the resilience of the system. A resilient architecture with a fault in one of its channel prohibits the fault from propagating into other control computations. Many fault tolerant control system concepts are subject to single point failures. As an example, over the water flights of aircraft with two engines are limited in duration primarily because reliability of the engine control system becomes a critical factor subsequent to loss of one of the engines. With improvements in reliability, however, aircrafts such as Boeing 777 and 787 now are permitted to fly over the oceans. Commercial aircraft now in operation use two redundant computing channels in the control system of the engine. Such a system is deemed sufficiently reliable if in the event of loss of an entire engine control system continued flight is allowed.

Gas turbine control systems typically encompass five areas that require a systems design approach. The areas include sensors, actuators, control and computational hardware, and software and integration with the vehicle's management system. Each area requires a system design that stresses maintainability, reliability, performance, and life cycle cost. The vehicle's management system incorporates that portion of the aircraft's system that is truly flight critical, and is useful in reducing component maintenance while improving the performance of the aircraft.

2.7 EMERGENCY LEVEL

Dual radial sensors monitor the vibration and trip the unit on high vibrations. A triple-redundant system protects the gas turbine against shaft over-speed. Magnetic pickup sensors are mounted near the tooth-wheel on the rotor inside the inlet side of the bearing cavity (near the thrust and journal bearings) to monitor the speed. The sensors and the associated control components will trip the unit on over-speed. Over-speed bolts are also used to trip the unit when the speed reaches around 112.5 per cent.

The exhaust temperature of the air leaving the gas turbine is monitored with the aid of thermocouples. They have the capability to trip the unit on high exhaust temperature. The blade path temperatures are monitored by 16 thermocouples installed at about 1.3 cm (0.5 in) downstream of the last-stage blades of the turbine. The thermocouples are installed around the circumference of the turbine for monitoring the temperature difference between the combustor baskets. These thermocouples cannot be installed at the discharge of the combustors due to the high temperature, around 2500° F (1371° C). However, they can still monitor the difference in temperature between the combustor baskets due to a relationship between the location of the baskets and the location of the thermocouples. For instance, the air stream leaving a combustor flows through the turbine and is discharged from it at a 40° angle away from the combustor. Therefore, the temperature difference between two neighboring thermocouples would be representative of the temperature difference between two combustors. The concern about having a high difference in blade path temperatures is thermal cycling and fatigue. An alarm is received when the blade path temperature difference is 90° F (50° C); the unit is unloaded when the difference is 120° F (67° C); and the unit is tripped when the temperature difference is 130° F (72° C).

Acceleration detectors trip the turbine on high shaft turning speed acceleration. They are used during start-up to prevent surge in the compressor. They are also used during normal operation to prevent over-speed following load rejection, when the circuit breaker connecting the generator to the grid opens suddenly following a fault in the grid.

The pressure at the inlet to the gas turbine is below atmospheric (under vacuum). A pressure switch monitors it. The condition of the intake filters is responsible to a considerable extent for the level of vacuum. The control system trips the unit when the vacuum drops, caused by a damaged filter, or when the vacuum increases because of a clogged filter.

The unit is tripped on high turbine exhaust pressure. This trip is needed in case there is a restriction in the flow of gases downstream of the turbine, for instance in the heat recovery steam generator.

2.8 PROTECTIVE DEVICES

Protective systems in gas turbines are designed to provide protection in the event of over-speeding of the rotor, temperature rising beyond stated limits, excessive vibrations, loss of flame in the combustor, or loss of lubricating oil.

The over-speed protection relies on a transducer mounted on the accessory gearbox or on the shaft. The speed sensor will trip the unit at around 110 per cent of the operating speed. The over-temperature protection system relies on thermocouples similar to the ones used for temperature control. The flame detection system consists of at least two ultraviolet flame detectors that provide the status of the flame in the combustion cans.

Where multiple combustor cans are used, flame detectors are mounted in cans that are not equipped with igniters in order to ease propagation of the flame between cans during the ignition phase. During normal operation, a detector indicating a loss of flame in one can only annunciates an alarm in the control room. At least two detectors must indicate a loss of flame to trip the machine.

The vibration protection system normally relies on velocity transducers to provide a constant trip set point throughout the complete speed range. Two transducers are normally installed on the gas turbine with additional transducers on the driven equipment, such as the generator. Vibration monitors provide an alarm at a specified level and a trip at a higher level. Most control systems provide a warning in the event of an open-circuit, ground, or short-circuit fault.

Although the instrumentation and control system in gas turbines vary considerably between different machines, most units are protected from a number of abnormal operational modes. In the event of low lube oil pressure downstream at the oil pump or in the bearing header, the turbine trips. Excessive lube oil temperature indication from redundant resistive thermal detectors (RTD) embedded in the Babbitt (white metal) of the bearings to monitor the temperature will trip the unit. The reason for the action is that the oil viscosity drops with increased temperature, resulting in higher friction in the bearings. The elevated temperature inside the bearings will lead to bearing damage.

The shaft is prevented from moving axially in either direction at the thrust bearing. Tilted pads are installed at both sides of the shaft collar to prevent axial movement of the shaft. An oil film is established between the shaft collar and the pads at both sides of the collar. When the pads wear out, the shaft starts to move axially. This causes higher friction inside the bearing, leading to increase in the oil temperature. The RTD's trip the turbine on high thrust pad temperature. An alternative method relies on proximity probes to trip the turbine following axial movement.

2.9 ACTUATORS TO CONTROL TURBINE OPERATION

The controller interacts with the process through input signals received at selected stages of the process, and through output commands controls the process. Sensors and process data signals come form the input side. Switches, actuators, control valves and electric motors execute the command signals to provide the required output.

Actuators provide the power for executing the final element of the control system. Hydraulic and pneumatic cylinders provide linear motion, and hydraulic and pneumatic motors develop large torque to produce rotational motion. In manufacturing operations, electric motors are often used as actuators in a servo control system to position the work piece and to locate the cutting tool. Variable speed electric motors are also used to drive pumps for controlling the flow rate and pressure of the fluid.

Hydraulic and pneumatic cylinders are the preferred choice for producing linear motion from economic and maintenance considerations. Hydraulic actuators are better suited for slower, more precise, and heavier load applications. Pneumatic actuators operate at lower pressure than the hydraulic ones, and are useful for lighter loads and where faster response is necessary. The speed of the hydraulic cylinders is limited by the maximum flow rate of oil received from the pump.

2.10 BYPASS MECHANISM TO AVOID COMPRESSOR SURGE

When the speed of an aviation turbine engine is low during startup and when power output is reduced, the amount of air required to burn the fuel is also restricted. Provision must be available to bypass some of the air as it enters the inlet to the compressor. Variable blocker doors are used to reduce the amount of air flowing into the compressor inlet of aviation engines. Excess air is then directed towards the fan duct. A series of blocker doors are placed around the circumference on the inner section of the fan frame, between the splitter end of the high-pressure compressor casing and the inlet guide vanes. Opening and closing of the blocker doors is controlled through pneumatic actuators.

Aviation and aero-derived engines also have to contend with the large speed band in which the high-pressure core section of the engine operates. Variable stator vanes are used on the initial stages of the engine. Variable stator vanes perform the dual function of feeding the air from the stationary vanes to the leading edge of the rotating blades at the right attack angle, and to reduce the cross-sectional flow area at low running speed. The latter action prevents excessive airflow, thus avoiding the risk of pressure surge.

On the other hand, power generation gas turbines run at a fixed speed that is consistent with the frequency of the electrical power output of the generator. However, even during the startup of the turbine, as the engine accelerates to its full running speed, the danger of pressure surge is present. To avoid the risk, variable stator vanes are provided at the inlet to the compressor. The vanes are gradually opened as the engine picks up speed up to its full running speed.

2.11 ENSURING AIR-FUEL RATIO IN MIXTURE

The control system primary effectors coordinated by the FADEC consist of fuel metering valves for the main combustor and augmentor of a fighter engine. Hydraulic actuators are used to control variable geometry functions for the compression system, air bleed and cooling, and exhaust nozzle system.

The number of effectors has a considerable impact on the complicity of the FADEC and its software. Other substantial issues arise from the hydraulic system. The horsepower required to operate the vectoring nozzle constitutes a large amount. Also, the thermal impact on the heat load from the fuel combines with increased levels of heat from the aircraft system. Management of additional levels of heat requires architectural innovations. The imposed load, rate, temperature, weight limitation, and fail-safe requirements of the nozzle actuator require new materials and design features for the actuator.

The usual approach to thermal management is isolation of the engine's fuel and oil systems from the aircraft's own system. Thus, fuel and oil specifications are defined at the delivery point to the engine, and the systems are managed independently. The practice is possible because the flow delivered to the engine is sufficient to absorb the heat generated from the inefficiencies of the engine oil and fuel pumping systems, and could be managed within the engine. The same is true on the aircraft side of the fuel system.

2.12 AVAILABLE POWER OUTPUT

The power developed in the turbine's expander section not only produces power to compress air in the compressor section but also for the driven unit. Net available power at the coupling between the turbine and the electrical generator rotors is thus the difference of the total power developed and the power requirement of the compressor.

A sensor placed in the turbine measures the developed power output. Depending on the needs of the compressor and demand from the driven unit, the controller decides on the amount of fuel to be burned in the combustor. Pressure, temperature and flow rate prevailing at both the intake and discharge ends of the compressor will determine work to be done in pressurizing the air. Note that air flowing through the compressor will experience friction losses as it flows past the walls. Losses are also encountered as the airflow enters each row of stationary vanes and rotating blades due to turbulence. The losses and efficiency of the overall compressor are obtained from extensive performance tests carried out by the engine manufacturer.

2.13 MINIMIZE EXHAUST GAS POLLUTANTS

Aside from carbon dioxide (CO_2) and moisture (H_2O), which are inevitable products of combustion of fuel, three other components are present in the exhaust gases. Carbon monoxide (CO), unburned hydrocarbons and oxides of nitrogen (NO_x) are insidious chemicals that impact human health, vegetation and other forms of life negatively. The challenge to alleviate generation of smoke and oxides of nitrogen (NO_x, most notably nitrogen oxide and nitrogen dioxide) lies in the nature of combustion processes that lead to their formation.

Substantial reduction in the levels of carbon monoxide and unburned hydrocarbons are obtained by increasing the combustion temperature in the combustor of gas turbines. But the increased temperature also accelerates formation of oxides of nitrogen, and therein lies the crux of the problem. Increasing the temperature inside the combustor also offers the advantage of increasing the net power available in the turbine. So the focus of turbine manufacturers has been on methods to reduce formation of NO_x.

Among the many methods where research has focused on, the concept of staged combustion attempts to achieve the multiple objectives by using two or more separate zones. Each zone optimizes certain features of the combustion process. In the lightly loaded primary first zone, high combustion efficiency and minimized CO and unburned hydrocarbon production are achieved through adjustment of the actual fuel ratio and the stoichiometric fuel ratio. The primary zone at the upstream end of the combustor provides the temperature increase needed at low power settings up to idle speed. At higher power requirements, it acts as a pilot heat source for the main combustion region downstream in the axial staging version or at different radial distance from the pilot in the radial staging style. The second and subsequent zones are supplied with the fuel-air mixture suitably adjusted for all zones at varying levels of power setting. Staged combustion is extensively used in industrial turbines using gas fuels, achieving acceptable levels of pollutant emission.

2.14 DETERMINING EXHAUST GAS TEMPERATURE IN GAS TURBINE

The temperature of gases exiting from the turbine exhaust is one of primary significance in determining the overall health of a turbine installation. After accumulation of a number of operating cycles and hours, the exhaust gas temperature is heavily relied upon when ground personnel decide to take an engine off the wing of an aircraft or to schedule shutdown for an industrial power generation turbine.

A number of temperature measurement probes are located on the struts of the rear frame of the turbine. Temperature of the gases will necessarily be different closer to the outer annulus wall than that near the inner wall, and also from in between the two walls. An algorithm code is used that calculates the more representative temperature of the gas stream from the readings obtained from all of the temperature probes placed on the frame struts. To avoid the dynamic, or kinetic velocity, effects of the rushing gas stream headlong, the temperature probes are located on the downstream edge of the struts without facing the stream directly. Lead wires from the sensors need to be securely fastened in metal clad sheath to prevent low frequency noise commonly associated with improperly restrained wires.

2.15 PRESSURE AND TEMPERATURE AT COMPRESSOR DISCHARGE

Overall operating conditions prevailing in an axial flow compressor will predominate the pressure and temperature of air at the discharge end. For a given set of pressure, temperature and flow velocity at the intake end, the compressor performance curves will dictate the level of compression that may be achieved at the discharge end when the unit operates at a given speed. If the settings on the variable stator vanes in the initial stages of the compressor of an aviation engine, for instance, are not accurate, the effects will be felt at the exit end in the form of reduced pressure. Temperature reading will also be affected.

Special attention is paid to the pressure and temperature gauge readings. Engine manufacturers accumulate large data blocks from performance tests carried out at various speed points on several engines. If the pressure and temperature data consistently deviates from set standards, a closer investigation of the situation may be required.

2.16 SPEED REGULATION IN POWER GENERATION TURBINE

Controlling the running speed of an electrical power generation unit, be it gas or steam driven, arises from the need for the electrical signal of the generator to be operating at a fixed speed. Where electrical power is required to be 60 Hz, the rotor running speed of the turbine and generator train is 3600 rpm. In many parts of the world 50 Hz electrical frequency is favored, in which case the shaft speed is 3000 rpm.

A speed control governor is required for this purpose. Of even greater significance is operation of the turbine during its startup. Acceleration must be at a controlled pace due to the considerable size and weight of most turbine-generator rotors. If the acceleration rate exceeds set limits, startup of the turbine rotor is aborted, leading to an over-speed trip. The nozzle valves admitting the flow of steam to a steam turbine will close in the event, or the fuel admission valve in the combustor of a gas turbine will be shut. If the trip action is not successful, the rotating train may continue to keep accelerating. Inertia loads of the massive rotors from the runaway acceleration will be large enough to cause the system to destroy itself, perhaps even catastrophically. Turbine-generator sets are high cost items and difficult to replace, so special attention must be paid to all aspects of their operation.

2.17 LUBRICATION REQUIREMENTS FOR ROLLING ELEMENT BEARING

Rolling element bearings, either ball or roller type, have a number of parts that require a coat of oil film to perform their stated task: outer and inner raceways, ball or rollers and the cage to keep the elements in their relative angular position. In comparison, hydrodynamic journal or sleeve bearings have the shaft journal and the outer fixed bearing housing. The annular space between the journal and the housing can be easily provided with lubricant oil at one end of the bearing with oil fed through a groove, and flowing out at the other end.

The complexity of rolling element bearings precludes such a simple circulation method. Aviation engines need a continuous supply of the lubricant oil in order to carry away any heat built up and to flush out metal chips, carbon and other debris accumulated in the bearing sump, while providing load carrying capacity due to the viscous characteristics of the lubricant. Within the enclosed bearing sump, the oil is injected through a nozzle tip with finely drilled holes that creates a mist of oil. The controlled mist coats all of the parts of the bearing and provides them with an oil film of proper thickness. The diameter of the oil droplets is measured in microns. Once the oil has fulfilled its task, it drops and collects in the base of the sump. The suction side of the oil pump then picks up and drains the lubricant back into the oil reservoir.

2.18 REVIEW QUESTIONS

2.18.1 Explain the method required to convert an accelerometer's signal in *g*'s to velocity and displacement.

2.18.2 Why does calibration data from a measuring instrument follow two different paths when the signal is increasing and when decreasing?

2.18.3 Name the functions that are useful in evaluating the dynamic characteristics of a measuring instrument.

2.18.4 How is step response used to determine the damping level in a dynamic system?

2.18.5 What are the primary reasons for maintaining control of rotor speed?

2.18.6 Why is it essential to maintain temperature of bearing oil within upper and lower limits?
2.18.7 Under what circumstances is a rolling element bearing preferred to a journal bearing?
2.18.8 What does rising temperature of gases exiting from the turbine indicate?
2.18.9 What is staged combustion? What are its merits and demerits?
2.18.10 How is air-fuel ratio adjusted as the combustion process proceeds?
2.18.11 Explain the need for pressure and temperature probes at the compressor's discharge end.

2.19 EXAMPLE PROBLEMS

Example Problem # 2.19.1: The following parameters are prescribed for an aircraft engine:

Airflow rate entering inlet:	124 lbs/sec.
Velocity of air at inlet:	0
Velocity of gas at exhaust:	2439 ft/sec.
Exhaust nozzle area:	487 in^2
Pressure at exhaust nozzle:	3.78 lbs/in^2
Ambient pressure:	0 lbs/in^2

Neglecting the effects of fuel, calculate the engine's thrust.

Solution: Thrust $f_n = w_a (v_2 - v_1)/g + (p_j - p_{atm})a_j$

$= 124 \times (2439 - 0)/32.2 + (3.78 - 0) \times 487$

$= 11233.26$ lbs.

Example Problem # 2.19.2: A ball weighing 23 lbs is shot upwards at a speed of 38 ft/second from a height of 78.4 ft above the ground. Neglecting the resistance of air, calculate the time required to fall back to the ground.

Solution: Let v_1 denote initial velocity, v_2 velocity at maximum height, and v_3 velocity when the object reaches the ground. Since the object has an initial speed, it will travel upwards with the gravity pull reducing its upward speed. When the speed reduces to zero, the acceleration due to gravity will force it to come down. Assume t_1 seconds is required to reach the maximum height of x_1 ft, and t_2 seconds is required to travel $(x_1 + 78.4)$ ft.

Now $v_2 = v_1 + gt_1$ and $v_3 = v_2 + gt_2$.

Velocity at highest point is zero. Hence, $0 = -38 + 32.2t_1$ or, $t_1 = 1.18$ seconds.

Also: $x_1 = -38t_1 + (32.2 \times t_1^2)/2 = -38 \times 1.18 + 16.1 \times (1.18)^2 = -22.42$ ft upwards.

Then: $(x_1 + 78.4) = 55.98$ ft.

Also: $(x_1 + 78.4) = -0t_2 + (32.2 \times t_2^2)/2 = 55.98$, or $t_2 = 1.86$ seconds.

Thus, total time required to fall back to the ground $= t_1 + t_2 = 3.045$ seconds.

Example Problem # 2.19.3: Calculate the work done by a force of 547.85 lbs acting on an object as it is moved through 11.5 in.

Solution: The product of force and distance gives work done.

Hence: Work = 547.85 × 11.5 = 6300.28 in-lb.

Example Problem # 2.19.4: Determine the rate of doing the work when a force of 55.5 lb. acts on moving a body through 33.75 in. in 4.75 seconds in horsepower.

Solution: The rate of doing work = (force × distance)/time

= (55.5 × 33.75)/(4.75 × 12) = 32.86 ft-lbs/sec.

Corresponding horsepower = 32.86/550 = 0.060

Example Problem # 2.19.5: What is the potential energy and kinetic energy of an object weighing 3.575 lb. shot at 18.5 ft/second from a height of 45 ft above the ground at the start?

Solution: Potential energy = wh and kinetic energy = $mv^2/2$

At start potential energy = 3.575 × 45 = 160.88 ft-lbs

and kinetic energy = (3.575/32.2) × $18.5^2/2$ = 76.00 ft-lbs

2.20 EXERCISE PROBLEMS

Exercise Problem # 2.20.1: Calculate the time required by 1.25 HP to move a weight of 775 lb. through 110 in.

Exercise Problem # 2.20.2: Calculate the work done by a force of 5.85 lbs acting on an object as it is moved through 2.2.5 ft.

Exercise Problem # 2.20.3: Calculate the air flow rate of an engine developing 5800 lbs thrust if velocity of air at inlet is 0, velocity of gas at exhaust is 240 ft/sec., exhaust nozzle area is 218 in^2, pressure at exhaust nozzle is 4.5 lbs/in^2, and ambient pressure is 0 lbs/in^2.

Exercise Problem # 2.20.4: What is the potential energy of an object weighing 5.85 lbs shot at 34 ft/second from a height of 43 ft above the ground at ground level?

Exercise Problem # 2.20.5: What is the potential energy and kinetic energy of an object weighing 6.35 lbs shot at 29 ft/second from a height of 39 ft above the ground at maximum height?

REFERENCES & BIBLIOGRAPHY

Bateson, R. N. "Introduction to Control System Technology", Prentice-Hall, Clifton, NJ, 1999.

Boyce, M., "Handbook for Cogeneration and Combined Cycle Power Plants", ASME Press, New York, 2002.

Carstens, J. R., "Automatic Control Systems and Components", Prentice-Hall, Clifton, NJ, 1990.

DeRoy, B. E., "Automatic Control Theory", John Wiley & Sons, New York, 1996.

Funakubo, H., "Actuators for Control", Gordon and Breach Science Publishers, New York, 1991.

Kiameh, P., "Power Generation Handbook", McGraw Hill, New York, 2002.

Miller, R. W., "Servo-mechanism Devices and Fundamentals" Reston Publishing Co., Reston, VA. 1977.

Patrick, D. R., Fardo, S. W., "Industrial Process Control System", Prentice-Hall, Clifton, NJ, 1979.

Rangwala A. S., "Turbo-machinery Dynamics: Design and Operation", McGraw Hill, New York, 2005.

Sante, D. P., "Automatic Control System Technology", Prentice-Hall, Clifton, NJ, 1980.

3

CHAPTER

AUTOMATIC CONTROL THEORY

3.1 INTRODUCTION

Turbo-machines operate at high speeds, with pressures, temperatures and flow rates shifting rapidly throughout the engine. Ensuring smooth operation of the unit requires means to control it's characteristics without delay. Even with constant attention from the airliner's pilot or the operator, the speed at which the processes occur is so fast and the time to react to a possible emergency situation so little that human action is of no consequence. Automatic detection of a parameter that has strayed out of bounds is required to detect and measure its value. Location of the fault and determining suitable course of action are essential if the errant signal is to be prevented into a deteriorating situation. On the other hand, if the fault were due to failure of a component or sub-assembly and there is no recourse to a corrective action to rectify the problem, the control system may even require complete shutdown of the engine in order to prevent total system disintegration.

The automatic control system must have the capacity to sense and measure operating parameters throughout the engine's system, detect a fault, decide on the appropriate action to forestall and execute necessary commands. The full sequence of detection, making decision and execution must occur in milliseconds. To add to the complexity, multiple causes may be resulting in the abnormal condition. For instance, an increase in the temperature of lubricant oil returning from a bearing sump could be due to a leak in the oil circuit, defect within the bearing, failure of the oil cooling system or from imposition of excessive load on the rotating system. Prompt corrective action is necessary since the elevated temperature can lead to excessive thermal expansion of components and seize the bearing, forcing a shutdown of the engine.

3.2 CONTROL SYSTEM GOALS

The task of a control system is chiefly to obtain stable values of the variable around a set limit, irrespective of command settings for the turbine system. The system ideally must respond to a change before an error is detected. Shifts in load, speed and ambient conditions will initially trigger a variation in the controlled variable. The controller must act to generate a change in the manipulated variable. The sequence of events is completed when the shift in the manipulated variable drives the controlled variable towards the preset limit.

The goal of the control system may be based on meeting the set value. Achieving a nearly perfect operation where the objective is defined in terms of response to the change thus would meet the objective. But a residual error that returns to a steady and non-changing value may persist in other systems. The magnitude of the residual error after the system settles out is of special interest. The time required to obtain minimized residual error points to another goal for the control system. Simultaneous minimizing of the maximum value of the error, settling time, and residual error may thus be considered the primary focus of any control system.

But the objectives as defined are inherently in conflict with each other. Modifying the controller to improve its gain will bring about decreased error in the output error. But another consequence of increased gain will be to increase the settling time as the output signal oscillates about the target value. Compensation between the twin needs of sensitivity and stability is thus required in the control mechanism parameters to achieve a more desired outcome. Hence, the rate at which the error is eliminated and the settling time must be acceptable enough to ensure the controlled variable does not loiter too far in both positive and negative directions as it attempts to zero in on the target value.

3.3 BENEFITS OF CONTROL SYSTEM

Modern turbo-machines rely on fly-by-wire control systems. An array of mechanical links, switches, relays and valves do not need to be physically actuated when an airliner's pilot or operator of an engine installation notices excessive vibrations or smoke emitting from the turbine exhaust. Instead, the operator instructs a computer of his intentions at the start of flight or operation. Among other pertinent information, the instructions may be in the form of direction of flight path segments, prevailing environmental conditions, altitude and coordinates of destination of an aircraft. For a power generation unit the instructions may be in the form of fuel characteristics, expected load demand, and rate of imposition of load. The central processing unit of the computer automatically reverts to the necessary protocol and issues appropriate command instructions.

Heavy reliance on computerized control of the turbo-machinery is inevitable. Maximum power and performance rely on operation of the engine in the neighborhood of aerodynamically unstable limits that if exceeded hold the possibility of causing failures in the form of compressor stall, start a fire or set up excessive vibrations. For all practical purposes control through conventional methods in combination with operator diligence is virtually impossible.

Significant improvements in efficiency of the turbine system are achievable using automatic control theory. Timely and appropriate adjustments in air-fuel ration, for example, will reduce fuel consumption, which in the process will also reduce exhaust gas emissions. During the takeoff climb an airplane's engines go through rapid acceleration from ground idle to maximum takeoff speed before the aircraft settles into a steady mode of cruise condition at the selected altitude.

Based on the throttle lever movement and position along with other relevant bits of information, the microprocessor-based controller determines the ratio of air and fuel to be burned in the combustor of the turbine. Frequent adjustment of this ratio is made to obtain optimum performance as parameters such as rotor speed, power output, compressor discharge pressure, and exhaust gas temperature vary with load demand. Thus, the pilot can pay his full attention to navigating and controlling the overall airplane, and does not have to tinker with the air-fuel ratio. Oxygen sensors measuring the content of oxygen in the exhaust path also play a deciding role in establishing the amount of fuel to be consumed for a given set of conditions in the turbine.

3.4 EVOLUTION OF MECHANISMS

Hard-wired panels of relay logic, motor starters, fluid actuators, and solenoid valves were used in the early stages of development of controlling turbo-machines. In the absence of miniaturization vacuum tubes were used in servo controls. Variable speed drives required a dc generator to drive a dc motor. Punched paper tapes provided storage of command instructions, numerically controlled machines using digital technology arriving on the scene in the 1950's. Programming was carried out off-line. Programmable controllers with re-programmable computer logic were developed around 1970. This was followed by the microprocessor in the 1980's, which took advantage of semiconductor switching elements and memory systems.

Sensor technology, manipulation devices and microelectronics have made rapid advances. Coupled with materials technology and miniaturization of parts, the state of the art in automatic control of turbine engine has made strides that were hitherto unimaginable a few decades earlier. Sensors to measure operating parameters and actuators to develop force through a combination of magnetic and electro-dynamic forces are indispensable in modern turbo-machines.

The concepts of physics have been applied and extended to meet the demands of a production-line machine to carry out assembly of miniature components in an electrical or electronic subassembly; machining of electro-hydraulic valves and servomechanisms is also carried out to obtain exacting specifications. The application of the theoretical concepts extends to position fixing, solid-state printed circuit boards and measuring devices. The principles of advanced manufacturing techniques have spread into a number of diverse areas as their cost versus performance ratio is improved by reducing the number of configuring elements.

3.5 OPEN-LOOP CONTROL

Controlling the operation of a turbo-machine using an open loop implies reliance on observed calibrated data obtained previously from an established procedure or from analytical predictions. Comparison of the actual output with the desired value is absent.

Fig. 3.1 Open Loop Control System — Needle Valve Setting

The open-loop control system may be understood with the aid of operation of a needle valve depicted in Figure 3.1. Test data provides the calibration curve from measurements of the flow rate through the valve for various valve openings. Multiple calibration curves are obtained for different values of pressure drop across the valve.

The convenience of easy installation and little hardware requirements define the benefits of an open-loop controller. Measurement of the output signal is not essential, as also subsequent need for corrective action based on the difference between the output and the desired target value. But if unexpected disturbances in the system cause spurious output signals that stray out of bounds, manual adjustment of the mechanism may be necessary to limit the error.

3.6 FEEDBACK ELEMENT FOR CLOSED-LOOP CONTROL

Opposed to the open-loop control method is the principle of feedback. The intent is to rely on obtaining the difference between actual and desired values of the output. The output signal is sent back to modify the input signal before again entering the controller block. Besides measuring the controlled variable, the error, meaning difference in desired and measured values, is computed to modify and determine a new control action.

The control action manipulates some other variable to reduce the error. In Figure 3.2, reference S is the input signal to the mechanism and C is the output variable. The transfer function defining the ratio of output and input signals, C/S, can be readily determined from the block diagram as follows.

Error E = (reference input) – (primary feedback) = $S - C_m$

Controlled variable C = (error, E) × (forward transfer function, G) = EG

Measured control variable C_m = (controlled variable) × (feedback transfer function)
= CH

Now: $C = (S - C_m) \times G = SG - CHG$,or: $C(1 + GH) = SG$

Hence closed-loop transfer function $C/S = G/(1 + GH)$

If a coolant were made to flow over a hot surface of a body, the temperature of the object may be used to increase or decrease the rate at which the amount of coolant is made to bring down its temperature to a specified level (Figure 3.2). This is the basis of 'feedback', and control systems using feedback are referred to as 'closed loop-control' systems. The corrective action ceases if the feedback is disconnected from the circuit, and the block diagram loop is then open.

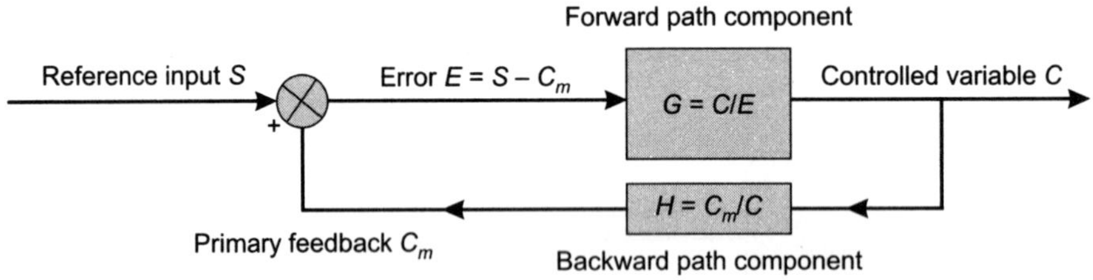

Fig. 3.2 Closed Loop System with Feedback

3.7 GAIN AND FREQUENCY RESPONSE

If a linear component is subjected to an input signal of sinusoidal waveform, the output signal will also be sinusoidal, as illustrated in Figure 3.3. The ratio of the magnitude of the output and input signals provides the 'gain' of the component. Thus, if the amplitude of the output signal is B and that of the input signal A, then gain $G = B/A$. The input signal becomes zero when $\omega t + \alpha = 0$, or $t = -\alpha/\omega s$. Similarly, the output signal approaches zero value when $\omega t + \alpha = 0$, or $t = -\beta/\omega s$. Hence, the time difference between the input and output signals is $t = (-\alpha - [-\beta])/\omega = (\beta - \alpha)/\omega$s. The phase difference of the component is defined as the difference in the phase of the output and input signals, or $\alpha - \beta$.

Complex numbers in the polar form are commonly used in control theory calculations for convenience. In complex notation the input signal is $A\angle\alpha$ and the output signal is $B\angle\beta$. The transfer function is obtained by dividing the output signal by the input signal, or $G = [B\angle\beta]/[A\angle\alpha] = (B/A)\angle(\beta - \alpha)$. Thus, the transfer function of the component is provided by the gain B/A and by the phase angle difference of the output and input signals.

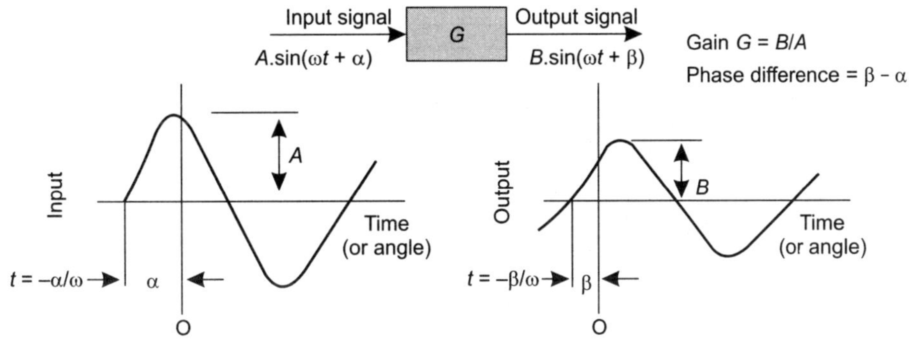

Fig. 3.3 Gain and Phase Difference of Linear Component Subjected to Sinusoidal Signal

The gain and phase difference jointly provide the 'frequency response' of the component at the given frequency. As an illustration, the gain of a certain control system may be 0.875 and the phase difference is 37° at a frequency of 20 Hz.

3.8 APPLICATION TO MECHANICAL VIBRATING ELEMENT

An important element and representative of an automatic control system is the mechanical vibrating system. The simplest mechanical vibrating system consists of a spring for storing potential energy, mass and inertia for kinetic energy, and damper for dissipating mechanical energy. The vibration process alternatively converts energy between its potential and kinetic forms. In its general sense, the vibration is a periodic motion that repeats itself in all its details after a certain interval of time called the period of vibration. Some energy must be replaced in each cycle of vibration from an external source to maintain the vibration.

An electric motor mounted on a flexible but mass-less beam supported at its ends is an example of a vibrating system, with the single degree of freedom vibration system schematically represented in Figure 3.4. The electric motor is of a certain mass m. A mass is a rigid body; so the force acting on the mass according to Newton's second law of motion is proportional to its acceleration, or md^2x/dt^2. The elastic beam has its own stiffness k, the constant of proportionality that relates the force in the spring with its deflection, or $f = kx$. Mass m is suspended by the spring from a fixed location. The ideal spring has no mass, and so the forces acting at its opposite ends are equal and opposite. Damping or energy-dissipating forces arise from friction at the end supports and from the molecules rubbing against each other within the beam. In a viscous damper, the applied force is proportional to the relative velocity of its connecting points. The damping coefficient c, provides the proportionality constant, and is the characteristic parameter of the damper. The ideal damper also is without mass, so the force at one end is equal and opposite of the force at the other end. The damper is also placed between the mass and the fixed wall. The damper does not transmit force to the mass when at rest, but as soon as it moves, it exerts a damping force of cdx/dt.

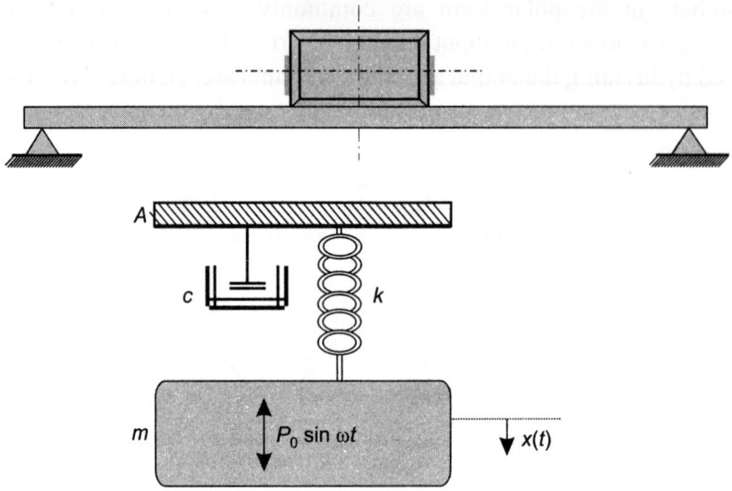

Fig. 3.4 Single Degree of Freedom Vibrating System

Free vibrations are initiated when the mass is displaced from its position at rest and then released. In the absence of external excitation the forces acting on the system may be expressed as:

$$md^2x/dt^2 + cdx/dt + kx = 0 \tag{3.1}$$

When external forces are not acting on the system, the free motion is suppressed by the energy dissipating damping mechanism. For sustained motion an external alternating force of the type $P_o\sin(\omega t)$ is required, where P_o is the amplitude and ω is the frequency. The governing differential equation of motion then takes the form:

$$md^2x/dt^2 + cdx/dt + kx = P_o \sin \omega t \tag{3.2}$$

3.9 RESONANT FREQUENCY

The role of 'resonant frequency' cannot be overemphasized in automatic control systems theory. A resonant frequency is one where the mechanical vibrating system acquires a portion of its operating energy through a more or less self-sustaining feedback loop or circuit. The single degree of freedom vibration model provides a good example. In equation (3.1), if the damping coefficient is increased, the maximum displacement x will be reduced, while if the damping is reduced (meaning low damping), the amplitude of motion will increase. In the limiting case of no damping, the equation of motion reduces to:

$$md^2x/dt^2 + kx = 0 \tag{3.3}$$

The most general solution of this equation is:

$$x = C_1 sin\sqrt{\frac{k}{m}}t + C_2 cos\sqrt{\frac{k}{m}}t \tag{3.4}$$

where C_1 and C_2 are arbitrary constants. Physically, the result as it stands is indefinite, since the constants can take any value. This is because the problem itself was not fully defined. If it is specified that the mass is pulled out of its equilibrium position to $x = x_0$, then released without initial velocity, two conditions are specified:

At $t = 0$: $x = x_0$ and $dx/dt = 0$.

The first condition gives $C_2 = x_0$. For the second condition, (3.4) must be differentiated, then get $C_1 = 0$. Substitution into (3.4) leads to the specific solution:

$$x = x_0 cos\sqrt{\frac{k}{m}}t \qquad (3.5)$$

This represents an un-damped vibration. Denoting the period of vibration by T, then:

$$T = 2\pi\sqrt{\frac{m}{k}} \qquad (3.6)$$

Customarily, $(k/m)^{1/2}$ is denoted by ω_n, called the natural circular frequency. The natural frequency f_n is given by:

$$f_n = \frac{1}{T} = \frac{1}{2\pi}\sqrt{\frac{k}{m}} = \frac{\omega_n}{2\pi} \qquad (3.7)$$

measured in cycles per second. If the negligibly damped system is subjected to the external exciting force of the type $P_0.sin\ \omega t$, and if the exciting force frequency ω coincides with the natural frequency ω_n, then a condition of resonance prevails. The amplitude of motion then theoretically reaches infinite without damping even if P_0 is small in magnitude, and the natural frequency of vibration then also becomes the resonant frequency.

3.10 DAMPING SOURCES

Free vibrations without damping persist indefinitely, which cannot happen; all free vibrations eliminate with time. The term viscous damping is associated with the expression *cdx*/*dt* since it represents the conditions of damping due to the viscosity of oil in a dashpot. Other types of damping exist, but their solution cannot be determined with as much ease.

Friction occurs due to relative motion between two mating components. Since the frictional force acts to oppose the motion of the moving part, it is a source of damping. The last stage of steam turbine blades is shrouded at the tip. Shrouds are also provided at one-third and two-third span of the blades to reduce twisting. The shrouds take many different forms. Protrusions at the blade tip essentially lean, and thus rub, against each other to reduce twisting motion and associated torsional shear stress in the airfoils. Rubbing also takes place in the dovetail region where the blade is affixed to the wheel.

Some other well-known sources of damping are frictional forces in hydrodynamic journal bearings and power loss due to wind-age. Shear forces are set up in the lubricant filling up the annular space between the rotating journal and the fixed bearing housing. Some of the oil clings to the rotating shaft journal and some other sticks to the fixed housing surface. Oil between the two surfaces experiences varying levels of shear and oppose the motion, which leads to friction losses to act as a source of damping. A somewhat similar mechanism is set up when the rotor spins inside the turbine casing. The presence of air in the compressor and burned gases in the turbine section offers resistance to the motion of the airfoils to extract energy, while slowing it down. Flow of air past the annular walls also results in friction losses and contributes to increased damping.

On the other hand, rolling element bearings are inherently free of damping, as is borne out by their alternate name of anti-friction bearings.

Mathematically difficult to quantify the magnitude of the opposing forces, the associated damping coefficient likewise cannot be determined accurately. Perhaps the ideal method is to measure the amount of thermal energy input in the form of fuel burned inside the combustor of a turbine and the mechanical output at the coupling end on the driven unit side. The difference between the two provides an estimate of losses and the relevant damping coefficient.

3.11 STABILITY OF OPERATION

Damping increases losses, but at the same time improves stability of operation in turbo-machinery. In the single degree of freedom governing equation 3.1, it can be shown that when $(c/2m)^2 > k/m$ the damping c is large. Smaller values of c pertain to more practical cases when the unit is operating in a stable steady state condition. The damping value c at which this transition to $(c/2m)^2 < k/m$ occurs is called the critical damping c_c, expressed as $c_c = 2m\sqrt{\frac{k}{m}} = 2\sqrt{mk} = 2m\omega_n$.

If the damping is less than critical damping, the motion of the vibrating mass consists of two factors, the first a decreasing exponential and the second a sine wave. The combined result is an exponentially decreasing sine wave, lying in the space between the exponential curves on both sides of the phase angle axis, shown in Figure 3.5. The smaller the damping constant c, the flatter will be the exponential curve and the more cycles it will take for the vibrations to be eliminated. Hence, the conclusion may be drawn that the system is not stable.

The rate of this dying down of the vibrations is of interest, and can be calculated by considering two consecutive maxima of the curve, *A-B*, *B-C*, and so on. During the time interval between any two such maxima, the amplitude of vibration diminishes from:

$$e^{-\frac{c}{2m}t} \text{ to } e^{-\frac{c}{2m}\left\{t+\frac{2\pi}{q}\right\}} \text{ where } q = \sqrt{\frac{k}{m} - \frac{c^2}{4m^2}} \tag{3.8}$$

The latter of these two expressions is equal to the first one multiplied by the constant factor $e^{-\frac{\pi.c}{mq}}$. This factor is the same for any two consecutive maxima, and is independent of the amplitude or of time. Thus, the amplitudes decrease in a geometric series. The logarithm of this expression is commonly referred to as logarithmic decrement, and is used to indicate stability of the vibrating mechanism. Note that frequency of vibration diminishes with increasing damping.

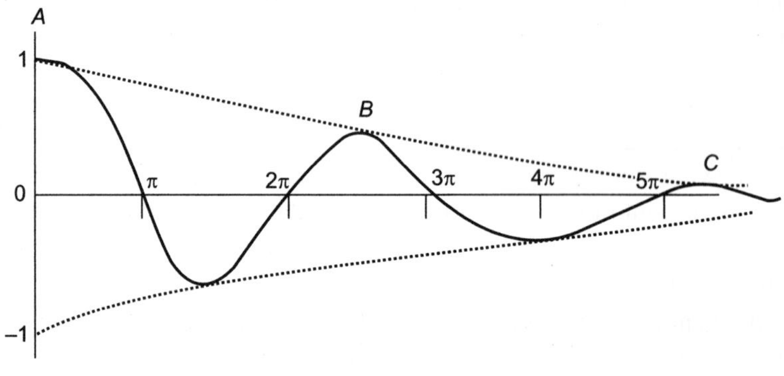

Fig. 3.5 Free Vibration of a System with $c/c_c < 1$

3.12 TRANSIENT RESPONSE

The forced vibration equation of motion in equation 3.2 yields homogeneous solution where the right hand side equation is set to zero and a particular solution. The latter is obtained from the form of the forcing function. Thus, a harmonic forcing function of the type shown in equation 3.2 will yield particular solution of the form $x = x_0\sin(\omega t - \varphi)$. Amplitude x_0 and phase angle φ are obtained from Newton's law requiring the sum of all the forces to be zero at all times, implying that the sum of the vertical and horizontal components must also be zero. Thus:

$$x_0 = \frac{P_0 / k}{\sqrt{\left\{1 - \frac{\omega^2}{\omega_n^2}\right\}^2 + \left\{2\frac{c}{c_c}\frac{\omega}{\omega_n}\right\}^2}} \tag{3.9}$$

$$\tan(\varphi) = \frac{2\frac{c}{c_c}\frac{\omega}{\omega_n}}{1 - (\omega^2 / \omega_n^2)} \tag{3.10}$$

The amplitude of motion and the phase angles are expressed in terms of ratios of frequency and damping, where c_c is the critical damping. P_0/k may be interpreted as the static deflection of the spring due to P_0. These relations are plotted in Figure 3.6. The amplitude diagram contains a family of curves, one for each value of damping. All curves lie below the one for zero damping, since damping reduces the amplitude of forced vibration. The maxima of the different curves do not occur any more at $\omega/\omega_n = 1$ but at a smaller frequency. In fact, three different frequencies need to be distinguished, all of which coincide for $c = 0$: = un-damped natural frequency $\omega_n = \sqrt{(k/m)}$, damped natural frequency $q = \sqrt{\frac{k}{m} - \left\{\frac{c}{2m}\right\}^2}$, and the frequency ω of the forcing function of equation 3.2, or the resonant frequency.

The phase angle diagram is also of interest when evaluating the dynamic characteristics of a vibrating system. For no damping, it was seen that the force and displacement are in phase below resonance and 180° out of phase above resonance. So the phase angle shows a discontinuous jump at the resonance point. For damping values different from zero the other curves in the phase angle diagram are plotted. In general, damping tends to smooth out the sharpness of the un-damped diagrams for both amplitude and phase.

The observations about the shift in phase angle are directly applicable to a rotating shaft. When the shaft is accelerating in speed and spinning in a certain direction, the bent shaft is also whirling about the unbent axis in the same direction till the shaft speed reaches the first natural frequency of vibration. The bend in the shaft and consequent whirl may occur due to unbalance in the shaft, misalignment in the bearings or some external exciting force. Further acceleration in spin speed causes the whirling motion in the opposite direction to the direction of spin. If now the spinning shaft is further accelerated till the speed reaches the second natural frequency of vibration, the whirl direction once again reverses, and the spin and whirl take place in the same direction. Still further increments in the spin speed will lead to alternate changes in the spin and whirl directions as the speed reaches third and higher natural frequencies of vibration.

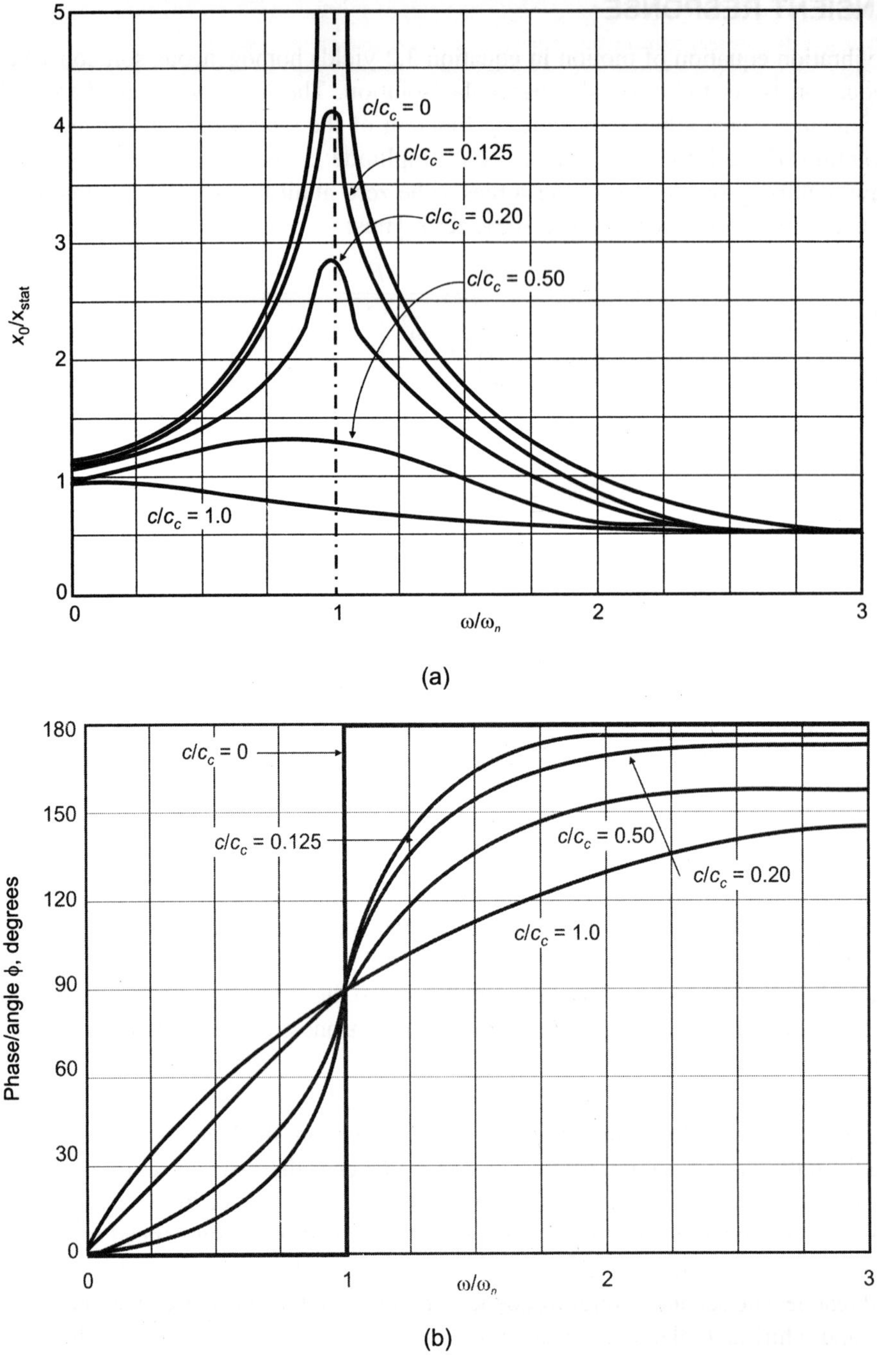

Fig. 3.6 (a) Amplitudes of Forced Vibrations, (b) Phase Angle between Force and Displacement as a Function of Frequency for Various Damping Values for Single-Degree-of-Freedom Vibration Model

Returning to the assertion pertaining to the particular solution, the general solution consists of the damped free vibration superposed on the forced vibration. After a short time the damped free vibration disappears and the forced vibration alone persists. For this reason the forced vibration may also be referred to as sustained vibration and the free vibration becomes transient vibration. The values of the constants C_1 and C_2 depend on the conditions at the start and can be calculated from these conditions using the analytical process described earlier. It is possible to describe the whole process by physical reasoning. For example, consider a mass suspended by a spring and acted upon by an external harmonic force of frequency eight times as slow as the natural frequency of the system. The mass is clamped while the external force is acting. Suddenly the clamp is withdrawn. Assume the damping in the system is such that the free vibration decreases by 10 per cent for each cycle. Assume also that the clamp is released when the forced vibration has its maximum amplitude. From the initial conditions it follows that at the instant of release the mass has no deflection and no velocity. The prescribed forced vibration starts with $x = x_0$ and $dx/dt = 0$. These two conditions can only be satisfied by starting a free vibration with $x = -x_0$ and $dx/dt = 0$. Then the combined motion will start at zero with zero velocity. Figure 3.7(a) shows the free vibration, 3.7(b) shows the forced vibration and 3.7(c) the combined motion.

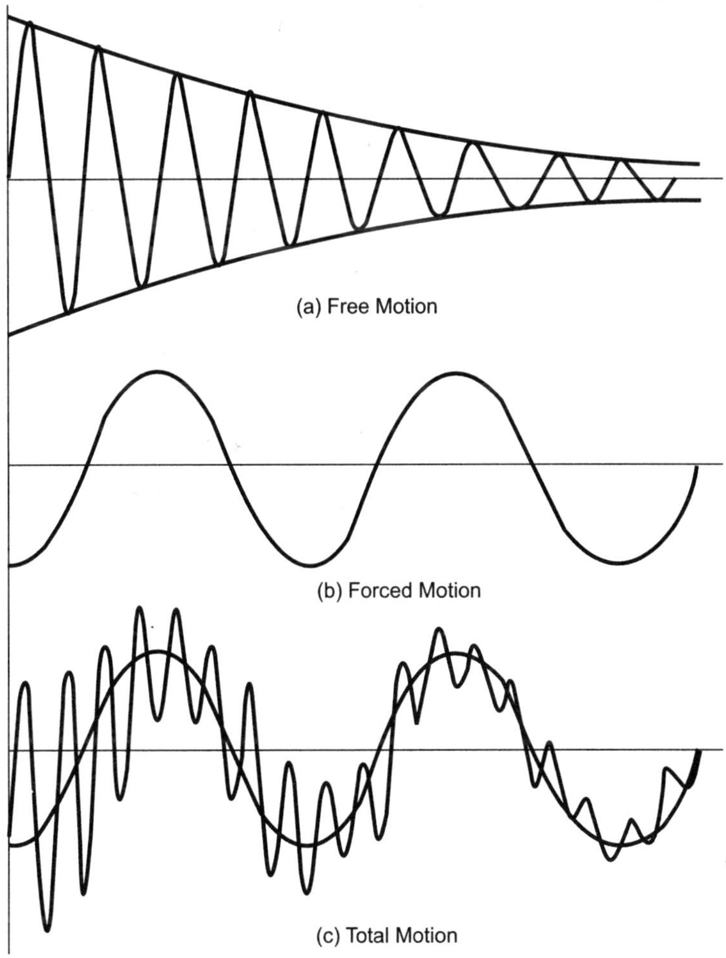

Fig. 3.7 Starting Transient Motion of a Forced Vibration

Note that the transient vibration disappears quickly and that the maximum amplitude at the start is nearly twice as great as the sustained final amplitude. If the difference between free and forced frequencies is small and damping is also small, the free and the forced vibration vectors would retain nearly the same relative motion during one revolution, the included angle between them changing only slightly. The vectors can then be added geometrically, and during one revolution of the two vectors the motion will practically be a sine wave of frequency $\omega_1 \approx \omega_2$ and amplitude the arithmetic sum of the two amplitudes. After a large number of cycles, the relative position of the vectors varies, and the magnitude of the sum vectors also changes. The resulting motion can then be described as an approximate sine wave with frequency ω_1 and an amplitude varying slowly between the sum and the difference of the amplitudes of the free and forced vibrations. This phenomenon is known as beats. The free vibrations will be eliminated after a while due to energy dissipation from the damping. In order to have sustained beats it is necessary to have two sustained, or forced, sources of vibration.

3.13 BLOCK DIAGRAM ALGEBRA

Block diagram schematics are useful in explaining the mechanics of a process and flow of information within an automatic control system. But a mathematically rigorous approach exists to simplify the system by modifying and moving the blocks. Each block representing a component has its own transfer function, which must work with other transfer functions in a systematic manner to generate the overall desired output.

Arbitrary swapping of the components is not permitted. If the position of a block were to be exchanged with another block, the system would be rendered inoperable. Combining or swapping of the component blocks for the sake of simplification is permissible only if the output and input results consistently remain unaffected.

Some basic rules to be followed in simplifying or modifying the block diagram are provided below in Table 3.1.

Table 3.1. Block Diagram Simplification

- Replacing two cascading blocks with a single block.

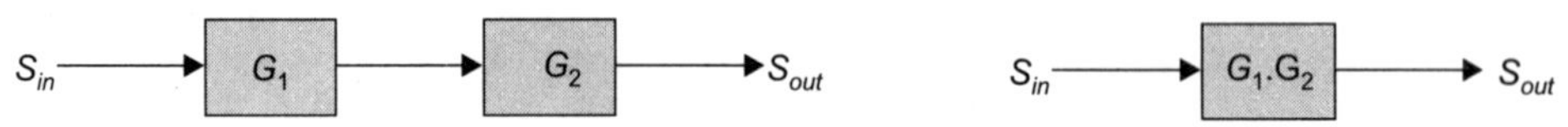

- Eliminating forward loop.

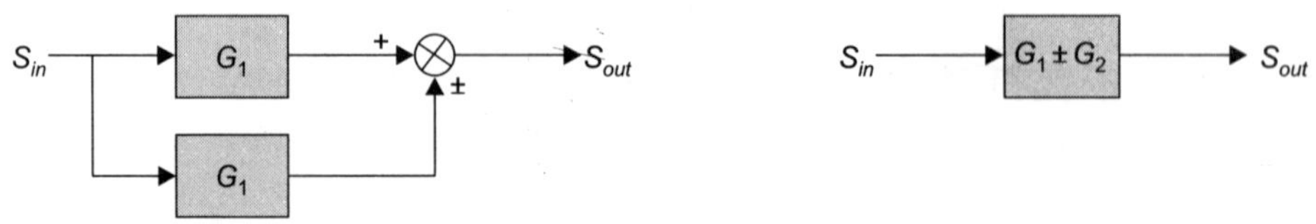

- Moving takeoff point from input to output side.

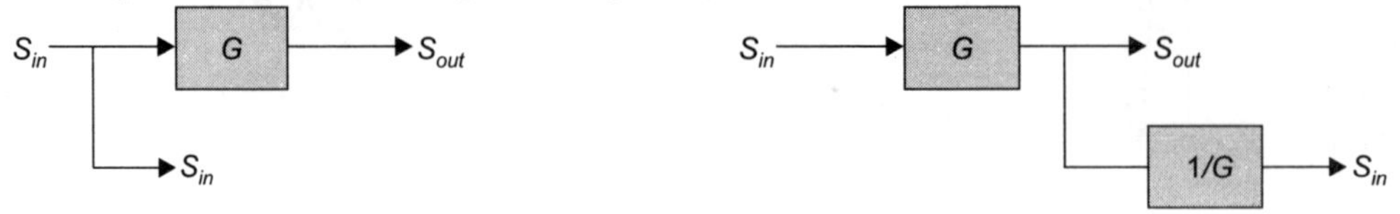

- Moving takeoff point to a block ahead.

- Moving summation point after a block.

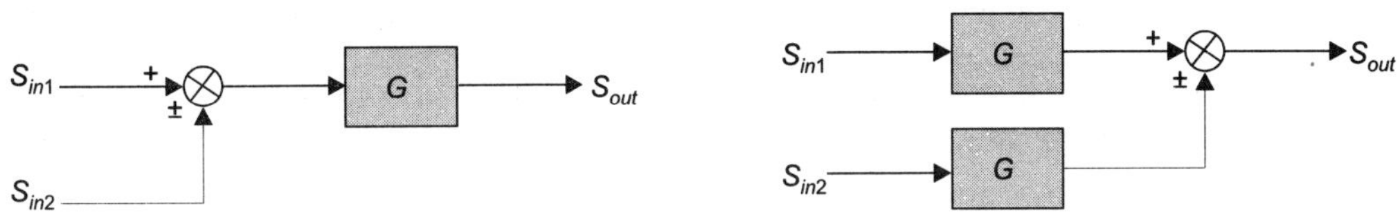

- Moving summation point from output to input block.

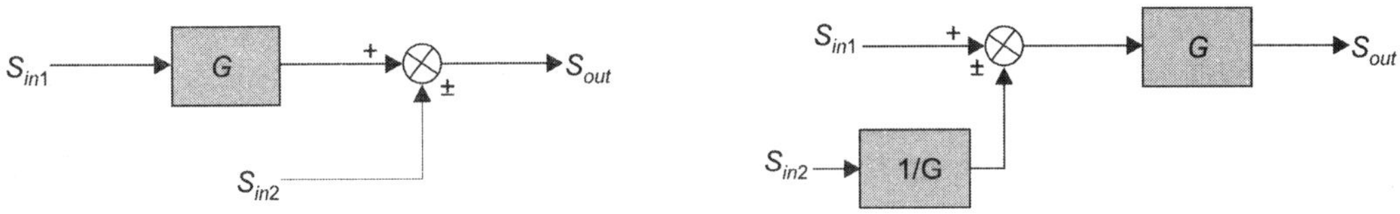

- Eliminating a feedback loop.

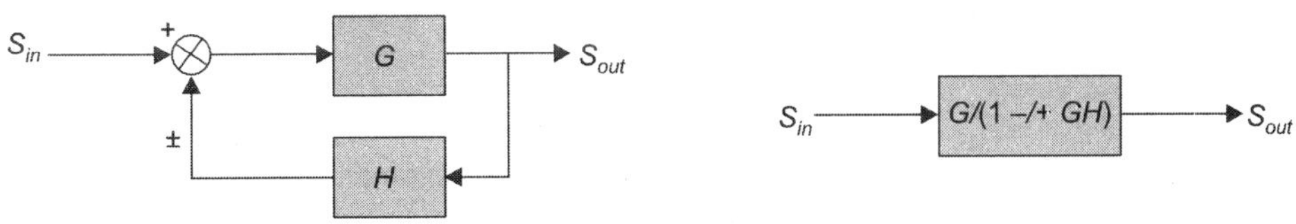

- Switching blocks.

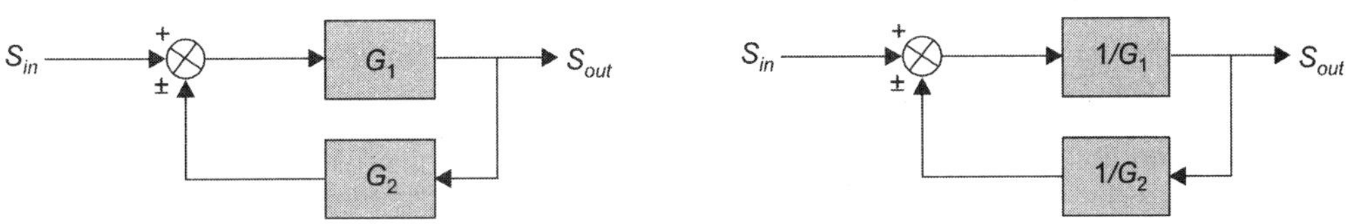

3.14 LAPLACE TRANSFORM METHOD

Laplace transforms find extensive usage in control systems theory for description, evaluation, and design. The mathematical equations for describing the time related connection between the input and the output parameters require derivative and integral functions. The Laplace transform theory offers the distinct advantage of converting a differential/integral equation into an algebraic function of frequency, which inherently is an easier process than operational calculus. The process effectively takes the problem from the time domain to the frequency domain.

The frequency-domain equations also establish the relation between the output (or the frequency response) and input parameters, the ratio of the two also defining the transfer function. After solving the algebraic equation, an inverse Laplace transformation yields the output as a function of time. Most control system evaluation and design problems avoid the inverse Laplace transformation by working directly with the algebraic equations. But the inverse transformation aids in understanding the situation and for answering stability related issues.

Note that the Laplace transformation is only one of a number of useful transforms. For instance, the logarithmic transformation converts numbers into their logarithms, thus replacing multiplication and division by the easier addition and subtraction. Other related methods such as the Fourier transform also find widespread usage.

A useful aspect of a component's characteristics is its response to sinusoidal inputs of more than one frequency. Knowledge of gain and phase difference of the component due to different frequencies is oftentimes of special interest. The analysis can be accomplished by letting the transfer function of the component to be a function of the frequency parameter s, where s has the unit of $(\text{second})^{-1}$. If the input is sinusoidal of frequency ω radians/second, and j is the imaginary operator ($\sqrt{-1}$), then the frequency parameter in the transfer function $s = j\omega$, yielding a complex number with magnitude M and angle θ, or $M\angle\theta$. Magnitude M provides the gain of the component and angle θ is the phase difference of the component.

Split into two categories, the functional Laplace transform relates to particular functions such as $\sin(\omega t)$ or $e^{-\alpha t}$, while operational transforms are concerned with the result of an operation involving one or more functions. The notation $\mathcal{L}\{f(t)\}$ and $F(s)$ are commonly used to indicate the Laplace transform of f(t)., which is defined by the expression:

$$\mathcal{L}\{f(t)] = F(s) = \int f(t)e^{-st}dt \quad (3.11)$$

where the upper and lower limits of integration are 0 and ∞. As an illustration, the

Laplace transform of a constant K is:

$$F(s) = \int K.e^{-st}dt = -(K/s)(e^{-s\infty} - e^{-0}) = -(K/s)(0 - 1) = K/s \quad (3.12)$$

When the Laplace transform of two functions acting independently are known, then operational transforms provide a means to obtain the Laplace transform of the two functions added together, or the derivative or integral of a single function. Table 3.2 provides Laplace transform pairs of commonly encountered functions in control systems theory. The tabulation aids in finding the Laplace transform of a time-domain function or the inverse Laplace transform of a frequency-domain function.

The inverse Laplace transform converts a frequency-domain function of s into a time-domain function of t. If the frequency domain function is available in the tabulation of Laplace transform pairs, the inverse transform is the associated time-domain function in the table. But the frequency-domain expressions in control system analyses mostly occur as a ratio of polynomials made up of integer powers of s, so the task is more complex than a simple conversion. For instance, consider a function with both polynomials factored in the form:

$$X(s) = 5[(s + 3)(s + 7)]/[s(s + 2)(s + 4)] \quad (3.13)$$

If the ratio of polynomials is split into a sum of terms where each term is in the Laplace transform table, the task of obtaining the inverse transform is facilitated, and indeed, the 'method of partial fractions' does precisely the same. The above polynomial may be expanded using partial fractions to assume the form:

Table 3.2. Laplace Transform Pairs

Time domain, $f(t)$, $t > 0$	Frequency domain, $F(s)$
1. K	K/s
2. Kt	K/s^2
3. Ke^{-at}	$K/(s + a)$
4. Kte^{-at}	$K/(s + a)^2$
5. $K.\sin(\omega t)$	$K\omega/(s^2 + \omega^2)$
6. $K.\cos(\omega t)$	$Ks/(s^2 + \omega^2)$
7. $Ke^{-at}.\sin(\omega t)$	$K\omega/[(s + a)^2 + \omega^2]$
8. $Ke^{-at}.\cos(\omega t)$	$K(s + a)/[(s + a)^2 + \omega^2]$
9. $K(1 - e^{-t/\tau})$	$K/[s(\tau s + 1)]$
10. $2Ke^{-at}.\cos(bt + \theta)$	$\{K\angle\theta/[s + a - jb]\} + \{K\angle-\theta/[s + a + jb]\}$
11. $K.f(t)$	$K.F(s)$
12. $f_1(t) + f_2(t) - f_3(t)$	$F_1(s) + F_2(s) - F_3(s)$
13. $df(t)/dt$	$sF(s) - f(0)$
14. $d^2f(t)/dt^2$	$s^2F(s) - sf(0) - df(0)/dt$
15. $d^3f(t)/dt^3$	$s^3F(s) - s^2f(0) - sdf(0)/dt - d^2f(0)/dt^2$
16. $\int f(t)dt$	$F(s)/s$
17. $f(t - a)$	$e^{-as}.F(s)$

Note: K represents a numerical constant.

$$X(s) = 5[(s + 3)(s + 7)]/[s(s + 2)(s + 4)] = K_1/s + K_2/(s + 2) + K_3/(s + 4) \quad (3.14)$$

where K_1, K_2, and K_3 are constants to be determined. In general, a term is present on the right side for each root of the polynomial in the denominator of the left side. Multiple roots for factors such as $(s + 2)^n$ will have a term for each power of the factor from 1 to n. Thus:

$$Y(s) = 8(s + 1)/[(s + 2)^2] = K_1/(s + 2) + K_2/[(s + 2)^2] \quad (3.15)$$

Complex roots are commonly encountered, and will occur in conjugate pairs. The corresponding constants in the numerator will also be complex conjugates. If $j = \sqrt{-1}$, K is a complex constant and K^* is its complex conjugate, then:

$$Z(s) = 7.5/(s^2 + 2s + 5) = K/(s + 1 - j2) + K^*/(s + 1 + j2) \quad (3.16)$$

A two-step approach is useful in determining each distinct non-multiple real or complex root. To determine the value of the constants, for instance K_2 in equation (3.14), multiply both sides of the equation by the denominator of K_2, $(s + 2)$. Next, replace s on both sides by the root of the factor that was multiplied (or, replace s by –2). The remaining terms on the right side will be zero, leaving only the constant term K_2. The left side reduces to a numerical value, that of the constant K_2. The process is repeated for each term on the right side of the equation. In the case of complex roots, the reduction of the left side will be a complex term, as may be expected. Additional steps are needed in the event of multiple roots. The inverse transformation completes the execution. The procedure is illustrated in example problems # 1.8 to 1.10.

3.15 REVIEW QUESTIONS

3.15.1 Define gain for open and closed-loop control systems.

3.15.2 How do newer control methods for turbines differ from those used 50 years back?

3.15.3 How do the latest fly-by-wire control systems reduce exhaust gas emissions?

3.15.4 Why does hunting, or speed oscillations, occur in a power generation gas turbine?

3.15.5 Name some devices commonly used to execute command decisions.

3.15.6 Describe the methods that prove to be useful in simplifying block diagrams for a control system.

3.15.7 Why do free vibrations die out? What is required to maintain vibrations in a system for prolonged periods?

3.15.8 What is time delay? How does it affect stability of a control system?

3.15.9 Explain the terms natural frequency, resonant frequency and damped frequency in a mechanical vibrating system. How may the same terms be used in an electrical circuit?

3.15.10 Why does mechanical energy dissipated in a vibrating system not affect the principle of energy conservation?

3.15.11 What are beats? Under what circumstance do they manifest in vibration systems?

3.16 EXAMPLE PROBLEMS

Example Problem # 3.16.1: The input signal to a vibration measurement instrumentation is 0.45 kHz sinusoidal signal with a peak amplitude of 0.013 V and a phase angle of 37°. The output has peak amplitude of 8.5×10^{-3} ins. and phase angle of 127°. Calculate the gain, phase angle, and transfer function of the instrument.

Solution:

$$\text{Gain} = \text{output/input} = 8.5 \times 10^{-3} \text{ in./0.013 V}$$
$$= 0.654 \text{ in./V}$$
$$\text{Phase difference} = (\text{output angle}) - (\text{input angle}) = (127 - 37)^\circ$$
$$= 90^\circ$$
$$\text{Transfer function} = 0.654\angle 90^\circ \text{ in./V}$$

Example Problem # 3.16.2: A spring of stiffness 2.35 lb./in has one end fixed and the other end attached to a mass weight of 1.39 lb. Assuming no damping or external forces acting on it, write the expression describing the free motion of the mass.

Solution:

$$\text{The term } \sqrt{(k/m)} = \sqrt{[(2.35 \text{ lb./in})/(1.39 \text{ lb.}/386 \text{ in/sec}^2)]}$$
$$= 25.55 \text{ radian/sec.}$$

Using equation 3.4, the expression for the free motion of the mass is:

$$x = C_1 \sin 25.55t + C_2 \cos 25.55t$$

Example Problem # 3.16.3: A single degree of freedom vibrating system has a 57.35 lb. weight, spring stiffness $k = 124.75$ lb./in. and damping $c = 0.125$ in./sec. The system is subjected to a sinusoidal force of amplitude 47.25 lb. and of frequency 18 radians/sec. Determine (*i*) critical damping, (*ii*) damped natural frequency, (*iii*) amplitude of motion and corresponding phase angle.

Solution: (*i*) From section 3.11 critical damping:

$$c_c = 2\sqrt{(mk)} = 2 \times \sqrt{[(57.35 \text{ lb.}/386 \text{ in/sec}^2) \times (124.75 \text{ lb./in})]} = 8.61 \text{ lb-in./sec.}$$

(*ii*) And natural frequency:

$$\omega_n = \sqrt{(k/m)} = \sqrt{[(124.75 \text{ lb./in})/(57.35 \text{ lb.}/386 \text{ in/sec}^2)]} = 28.98 \text{ radian/sec.}$$

Damped natural frequency:

$$q = \omega_n \times \sqrt{[1 - (c/c_c)^2]} = 28.98 \times \sqrt{[1 - (0.125/8.61)^2]} = 28.97 \text{ radian/sec}$$

(*iii*) Amplitude of motion:

$$x_0 = (P_0/k)/\{[1 - (\omega/\omega_n)^2]^2 + [2(c/c_c \times \omega/\omega_n)^2]\}^{1/2}$$
$$= (47.25/124.75)/\{[1 - (18/28.98)^2]^2 + [2 \times (0.125 \times 18/28.98 \times 8.61)^2]\}^{1/2}$$
$$= 0.6166 \text{ in.}$$

Corresponding phase angle:

$$\tan(\varphi) = 2(c/c_c \times \omega/\omega_n)/[1 - (\omega/\omega_n)^2] = 0.02937, \text{ or } \varphi = 1.68°$$

Example Problem # 3.16.4: Using the block simplification technique, reduce the control system shown below into an equivalent single block.

Solution: The variables may be expressed as: $E = S - T$

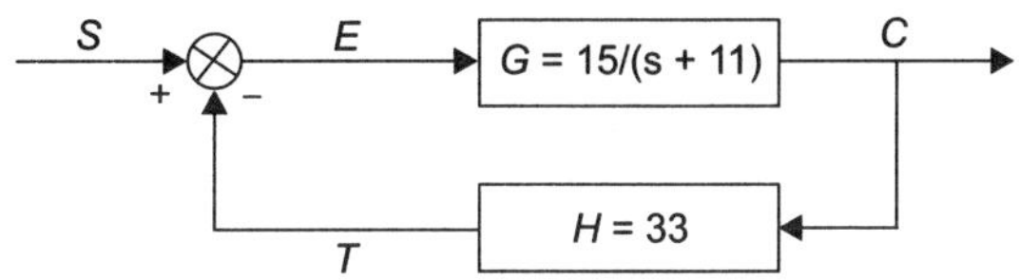

$$C = GE$$
Hence: $C = G(S - T)$
$$T = CH$$
And: $C = G(S - CH)$
Simplifying: $C = GS - GCH$
Or: $C/S = G/(1 + GH)$

Substituting values of G and H: $C/S = [15/(s + 11)]/[1 + \{15 \times 33\}/(s + 11)]$

The single block representation of the control system is given below.

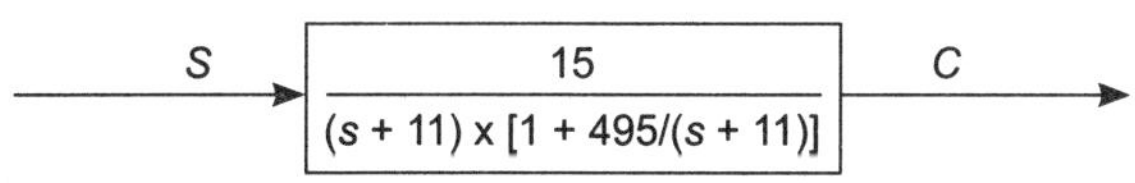

Example Problem # 3.16.5: Determine the period of vibration and the natural circular frequency of vibration for a mass of 47.5 lb. and spring of stiffness 28.75 lb./in.

Solution: The period of vibration T is given in equation 3.7:

$$T = 2\pi \times \sqrt{(m/k)} = 2\pi \times \sqrt{[(47.5 \text{ lb.}/386 \text{ in/sec}^2)/(28.75 \text{ lb./in})]}$$
$$= 0.411 \text{ sec.}$$

The natural circular frequency $f_n = 1/T = 1/0.411 = 2.43$ cycles/sec.

3.17 EXERCISE PROBLEMS

Exercise Problem # 3.17.1: A vibrating system has a mass weight of 2.8 lb., spring coefficient 3.25 lb./in and damping constant 0.1 lb-in/sec. Determine the damped natural frequency of vibration.

Exercise Problem # 3.17.2: A spring of stiffness 4.7 lbs/in has one end fixed and the other end attached to a mass weight of 2.78 lb. Calculate the critical damping.

Exercise Problem # 3.17.3: The input signal to an electrical circuit instrument is 325 Hz sinusoidal signal with a peak amplitude of 0.143 V and a phase angle of 129°. The output has peak amplitude of 7.5×10^{-3} A and phase angle of 38°. Calculate the gain and phase angle of the instrument.

Exercise Problem # 3.17.4: Determine the gain of a closed-loop system where forward loop $G = 25/(s + 32)$ and feedback loop $H = 27.5$.

Exercise Problem # 3.17.5: Calculate spring stiffness if the natural frequency of vibration of a mass-spring system is 42.5Hz and mass weight is 18.35 lbs.

REFERENCES & BIBLIOGRAPHY

Bateson, R. N., "Introduction to Control System Technology", Prentice-Hall, Clifton, NJ, 1999.

Carstens, J. R., "Automatic Control Systems and Components", Prentice-Hall, Clifton, NJ, 1990.

DeRoy, B. E., "Automatic Control Theory", John Wiley & Sons, New York, 1996.

Foxboro Company, "Compensated Mass Flow Control Loop", Bulletin # C-40A, Foxboro, MA, November, 1985.

Miller, R. W., "Servo-mechanism Devices and Fundamentals" Reston Publishing Co., Reston, VA. 1977.

Olesten, N. O., "Numerical Control", John Wiley & Sons, New York, 1970.

Patrick, D. R., Fardo, S. W., "Industrial Process Control System", Prentice-Hall, Clifton, NJ, 1979.

Rangwala A. S., "Turbo-Machinery Dynamics: Design and Operation", McGraw Hill, New York, 2005.

Sante, D. P., "Automatic Control System Technology", Prentice-Hall, Clifton, NJ, 1980.

4

CHAPTER

INSTRUMENTATION FOR PARAMETER MEASUREMENT

4.1 INTRODUCTION

Considerable differences exist in the principles, design, performance, application suitability and cost of sensors. Fifty different technologies have been identified to measure pressure, temperature and other parameters. Hundreds of companies worldwide are engaged in the business of manufacturing sensors. Numerous ways are available to derive the parameters in turbine operation. Where parameters are rapidly changing, such as dynamic pressures, the piezo-electric procedure is widely employed.

Analog pressure sensors may be split into two basic categories. The force collector types are electronic pressure sensors that generally use a diaphragm, piston, bourdon tube, or bellows to measure strain (or deflection) due to applied force (pressure) over an area.

Electronic sensors for turbo-machinery use properties such as density, speed of sound, mass and index of refraction to infer operating parameters. The capacitive principle relies on a diaphragm and pressure cavity to create a variable capacitor to detect strain due to applied pressure. Metal, ceramic, and silicon diaphragms are widely used. A number of these technologies are useful when the pressures are low. The electro-magnetic method measures the displacement of a diaphragm to indicate changes in inductance (reluctance), 'linear variable differential transformer' (LVDT), Hall-Effect, or by eddy current principle.

As sensors and actuators become more complex they provide support for various modes of operation and interfacing. Some applications require additional fault-tolerance and distributed computing. Such high-level functionality can be achieved by adding an embedded micro-controller to the classical sensor/actuator, which increases the ability to cope with complexity at a fair price. A smart transducer, for instance, is an analog or digital transducer or actuator combined with a processing unit and a communication interface.

In the machine vision field, a single compact unit that combines the imaging and image processing function is often referred to as a smart sensor. They are often made using CMOS, VLSI technology and may contain MEMS devices to lower cost. They may provide full digital outputs for easier interface or they may provide quasi-digital outputs like pulse width modulation.

4.2 SENSORS AND TRANSDUCERS

A sensor forms the complete assembly required to detect and communicate a particular event in a turbine process. A transducer, on the other hand, is the element within the assembly that detects the event. In special cases, a sensor and a transducer have the same characteristics, and the two are identical.

A transducer converts an input signal into an output in order to exploit it and accomplish the sensing mission. As an illustration, a pressure sensor may use a diaphragm and/or strain gauge to indicate pressure differential across the diaphragm, but the complete sensor will additionally consist of a display element and the electronics required to energize and condition the output of the detecting transducer, as well as the elements required to house and drive the display feature. A compass is a simple sensor of magnetic north, wherein the magnetic element in the compass is the transducer or northern direction indicator. The needle, housing and compass face comprise the remainder of the north sensor. Still another example is that of a bi-metallic spring element that can both detect temperature change; if a pointer is attached to the bi-metallic spring, it would constitute the entire sensor.

4.3 SENSOR CHARACTERISTICS

In selecting the measuring instrument for an application, certain questions must be addressed. The questions pertain to: (*i*) identification of measured variable, (*ii*) accuracy required at various levels of input, (*iii*) nature of item being measured, (*iv*) upper and lower range limits, (*v*) over-range, and (*vi*) occurrence of dynamic changes (rate of change and frequency) in the measured variable. Questions concerning the sensor must also be answered: (*i*) primary sensor identification, (*ii*) effect of primary sensor on measured variable, (*iii*) intent of output, (*iv*) analog or digital signal, (*v*) output signal type, (*vi*) provision for signal conditioning, (*vii*) transmitter operation, (*viii*) environmental conditions, (*ix*) thermal sensitivity, and (*x*) economic considerations.

When the measured variable is continuously varying in its upper and lower range limits, the measuring device sometimes changes the signal in discrete steps. The 'threshold' of the measuring instrument defines the smallest increment in the measured variable that will result in a shift in the indicator output, while its 'sensitivity' is the ratio of change in input and the corresponding shift in output. Environmental effects of a measurement instrument include ambient temperature and pressure, fluid temperature and pressure, mounting method and location, and electro-magnetic effects. Operating conditions dictate the environment to which a measurement instrument is subjected.

Static characteristics of a measurement instrument at room conditions describe its precision when the measured value is constant or changing slowly. The instrument output's 'accuracy' provides the degree of conformity to the ideal value of the measured variable as determined by an established standard. 'Error' is defined by the difference in the measured and the ideal values. 'Repeatability' establishes the dispersion, or standard deviation, of a measurement instrument. A repeatable but inaccurate instrument will produce a tight pattern that is not centered at the ideal value, the offset being attributable to 'bias' or 'systemic error'. Thus, an automatic controller that is repeatable but inaccurate may still be useful if adjustment to compensate for bias is made.

The 'dynamic characteristics' of a measuring instrument are of interest under rapidly changing conditions. A case in point is the acoustic pressure oscillations in the combustion chamber of gas turbines. Liquid fuel from both the reciprocating or rotary vane and screw type fuel pumps does not come out in a continuous stream. The fuel is injected in the form of pulses, with pulse frequency dependent on the speed of the pump. The pulse frequency of a single-acting reciprocating pump will be the same as the operating speed, while a double-acting reciprocating pump will be twice the operating speed. For vane and screw pumps the pulsation frequency depends on the running speed and the number of vanes or screws.

As the fuel ignites, it sets up heat release waves in the axial and circumferential directions of the combustor. The heat waves in turn set up pressure waves in the gases. This leads to dynamic stresses in the combustor liners. The primary task of the liners is to contain the hot products of combustion. Also, the liners must contend with thermal growth as the engine goes from cold condition at start to hot running condition, and so are lightly restrained at selected stop points. Hence, it is essential to impose minimal gas pressure load on the liners. This is borne out by the limit of 2.5 lb./in^2 peak-to-peak dynamic pressure imposed in the combustion zone. Consequently, dynamic pressure gauges are needed to ensure the limit is not exceeded with dynamic pressure gauges.

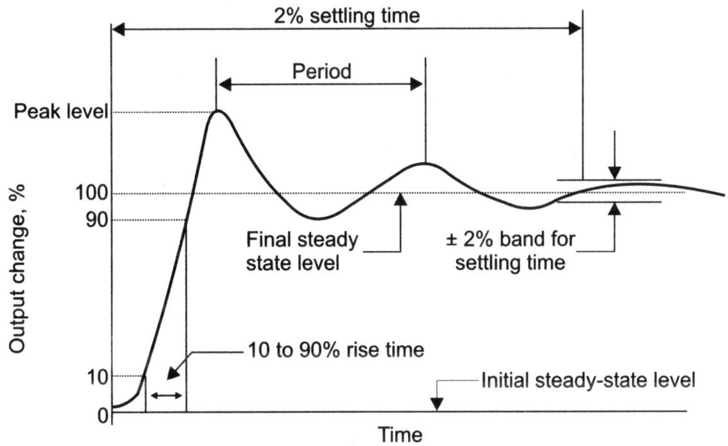

Fig. 4.1 Step Response Curve for Under-damped Measuring Instrument

Generally, sensors do not provide immediate and complete response to a sudden change in the measured variable. Depending on the electronic characteristics of the instrument, a certain period of time elapses before the total response is received. Step, ramp, and frequency responses of the measuring instrument are needed to determine its dynamic characteristics. A step change in the measured variable, such as moving a temperature probe from ice water to boiling water, produces a worst-case response, and is often used to define its dynamic characteristics. The response may be classified as over-damped, critically damped or under-damped. Response and rise time are good indicators for step response over-damped and critically damped systems. Step response of under-damped systems is stated in terms of rise time, peak percentage of overshoot, and settling time (Figure 4.1).

4.4 DISCRETE AND CONTINUOUS MEASUREMENT

Most common sensors and transducers for motion and sequence control are on-off switches. Actuated by a number of system parameters, the switches respond by completing or breaking the control circuit, or provide an input to the programmable logic controller.

Electro-mechanical limit switches are the most common type, but inductive, capacitive, light sensitive, and optical types are also used. A mechanical actuated lever or pushbutton triggers the electro-mechanical switch. Depending on the function required to be performed, the switch may be normally in the open or closed position. Switching characteristics of a particular switch design must take into consideration actuating force, pre-travel, acceptable over-travel, movement distance, actuating point, releasing position, and repeatability (*ref*: Balluff, 1987).

More sophisticated than the mechanical type, inductive and capacitive proximity switches and photoelectric sensors accomplish non-contact discrete position sensing. Inductive probes are activated by the presence of a metal in the sensing zone, and have four functional sections – sensing head, oscillator, detector, and output amplifier (*ref*: Balluff, 1988). An inductive alternating radio frequency field is developed near the sensing surface by the resonant circuit coil of the oscillator. Energy is absorbed when a metal piece traverses the radio frequency field, which causes the oscillating motion to stop.

If a metal is not available for a target, capacitive proximity switches may be used. The device has the same four functional units of the inductive sensor, but the oscillator does not operate continuously. When an object approaches the active surface of the sensor, capacitance in the region increases. When a preset value is exceeded, the oscillator starts to produce a signal that activates a trigger. The control circuit then uses the output signal.

Unlike discrete sensors to detect individual events, continuous sensors are active throughout the operation for monitoring a parameter and producing an output to represent the parameter.

Devices for continuous sensing include flow parameter sensors and vibratory characteristic transducers. The output from the measuring instruments may be displayed to indicate the operating condition of a machine or a process plant, or may be used as feedback and control signals.

4.5 PRESSURE SENSING TECHNOLOGY

The piezo-resistive strain gauge uses the piezo-resistive effect of bonded or formed strain gauges to detect strain due to applied pressure. Some common piezo-resistive materials are quartz, silicon (mono-crystalline), poly-silicon thin film, bonded metal foil, thick film, and sputtered thin film. The strain gauges are connected to form a Wheatstone bridge circuit to maximize the output of the sensor and to reduce sensitivity to errors. This is the most commonly employed sensing technology for general-purpose pressure measurement. These technologies are suited to measure absolute, gauge, vacuum, and differential pressures. The method is also widely employed for measuring rapidly changing dynamic pressures.

The optical technique calls for physical change of an optical fiber to detect strain due to applied pressure. A common example of this type utilizes Fiber Bragg Gratings method. The technique finds favor in challenging applications where the measurement may be highly remote, under high temperature, or may benefit from systems that are inherently immune to electromagnetic interference. Another analogous technique utilizes an elastic film constructed in layers that can change reflected wavelengths according to the applied pressure or strain.

The potentiometric method is based on motion of a wiper along a resistive mechanism to detect the strain caused by applied pressure.

Changes in resonant frequency in a sensing mechanism may be used to measure stress, or changes in gas density, caused by applied pressure. This technology may be used in conjunction with a force collector, such as those in the category mentioned above. Alternatively, resonant technology may be employed by exposing the resonating element itself to the media, whereby the resonant frequency is dependent upon the density of the media. Sensors have be made out of vibrating wire, vibrating cylinders, quartz, and silicon MEMS. This technology is relied upon to provide stable readings over time.

Changes in thermal conductivity of a gas due to density changes permit measurement of pressure. A common example of this type is the Pirani gauge.

The process of ionization measures the flow of charged gas particles (ions) as they vary due to density changes, which provides the method to measure pressure. Common examples are the hot and cold cathode gauges.

4.6 GAS PRESSURE SENSOR

The pressure sensor measures pressure of gases or liquids. Pressure is an expression of the force required to stop a fluid from expanding, stated in terms of force per unit area. A pressure sensor usually acts as a transducer; and generates an electrical signal as a function of the imposed pressure.

Pressure sensors are used to control and monitor in a variety of situations. The sensors may also be used to indirectly measure other variables such as force, fluid or gas flow, speed, water level, and altitude. Pressure sensors are also referred to as pressure transducers, pressure transmitters, pressure senders, pressure indicators, piezometers, and manometers.

Pressure sensors are also used to measure in a dynamic mode for capturing rapidly changing values in pressure. Commonly manufactured out of piezoelectric materials such as quartz, in a gas turbine the dynamic sensors are capable of measuring compression of air, combustor and turbine gas pressures.

Pressure switches rely on pressure sensors, such as those found in optical devices and cameras and function in a binary (on/off) manner. When pressure is applied to a sensor, it acts to complete or break an electrical circuit.

Pressure sensors can be classified in terms of pressure and temperature range of operation, but more importantly the type of pressure they measure. Pressure sensors are variously named according to their purpose, but the same technology may be used under different names.

An absolute pressure sensor measures the pressure relative to perfect vacuum, while a gauge pressure sensor measures the pressure relative to atmospheric pressure. Thus, when a tire pressure gauge indicates zero, it is indicating ambient pressure.

Vacuum pressure sensors are used for measuring pressures below atmospheric pressure, and show the difference between the low pressure in question and atmospheric pressure (or, negative gauge pressure). It may also be used to measure a low pressure relative to perfect vacuum, thus providing absolute pressure.

A sealed pressure sensor is similar to a gauge pressure sensor except that it measures pressure relative to some fixed pressure rather than the ambient atmospheric pressure. Note that ambient pressure varies according to location and to weather conditions.

4.7 OTHER APPLICATIONS OF PRESSURE GAUGE

In conjunction with the venturi effect, pressure sensors may be used to measure flow. Differential pressure is measured between two segments of a venturi tube that have a different aperture. The pressure difference between the two segments is directly proportional to the flow rate through the venturi tube. A low-pressure sensor is almost always required as the pressure difference is relatively small.

A pressure sensor may also be used to calculate the level of a fluid. This technique is commonly employed to measure the depth of a submerged object, or level of contents in a storage tank. For most practical purposes, fluid level is directly proportional to pressure. In the case of fresh water where the contents are under atmospheric pressure, 1 lb./in^2 = 27.7 in H_2O, and 1 Pa = 9.81 mm H_2O. The basic equation for such a measurement is:

$$P = \rho g h$$

where P = pressure, ρ = density of the fluid, g = gravity constant and h = height of fluid column above the pressure sensor

4.8 FLUID FLOW RATE

A differential pressure sensor measures the difference between two pressures, one connected to each side of the sensor. Differential pressure sensors are used to measure flow rates by measuring the change in pressure across a restriction. Many other characteristics, such as pressure drops across oil filters or air filters may be obtained with a differential pressure sensor. Fluid levels may be obtained from comparison of the pressure above and below the liquid.

Technically speaking most pressure sensors are differential pressure sensors, since a gauge pressure sensor is merely a differential pressure sensor in which one side is open to the ambient atmosphere. Detecting leakage in a system can be accomplished with a pressure sensor when used to measure decay of pressure due to system leak. This is commonly done either through comparison with a known leak using differential pressure, or by means of utilizing the pressure sensor to measure pressure change over time.

4.9 ALTITUDE SENSING

When an aircraft is flying at a certain height, it is imperative to know its altitude. Pressure, density and temperature of the air entering the inlet to the engine vary with the altitude. The altitude sensor makes use of the relationship between changes in pressure relative to the altitude. This relationship is governed by the following equation:

$$h = (1 - (P/P_{ref})^{0.100284}) \times 145366.45\text{ft}$$

This equation is calibrated for an altimeter, up to 36,090 feet (11,000 m). Outside that range, an error will be introduced which can be calculated differently for each pressure sensor. These error calculations will factor in the error introduced by the change in temperature as the altitude increases.

Barometric pressure sensors can have an altitude resolution of less than 1 meter, which is significantly better than the Global Positioning System (GPS), or about 20 meters altitude resolution. In navigation applications, altimeters are used to distinguish between stacked road levels for vehicles.

4.10 POSITION AND DISPLACEMENT

Linear positioning and feedback plays a major role in manufacturing and quality inspection checks of turbo-machinery components. The 'Linear Variable Differential Transformer' (LVDT) is noted for sensing position accurately, and works with a fluid power cylinder. The 'linear potentiometer', or 'pot', has a tightly packed coil of wire wound around a supporting core. The terminals of the coil are connected to a known voltage potential. Another lead is connected to a brush that moves with the cylinder rod along the coils. The brush taps a voltage that is proportional to its distance along the coil. Location of the movable contacting brush on the core determines the magnitude and phase of the signal in the secondary windings of the LVDT. Depending on the core's relative position, the secondary signal is in phase for one direction of travel and 180° out of phase in the opposite direction. The voltage output at any point is expressed by $V = ED/L$, where V is the voltage output, E is source voltage, D is distance from the zero end, and L is total length of potentiometer, ins.

A 'magneto-strictive linear displacement transducer' (MLDT) consists of a non- magnetic stainless steel probe welded to an enclosed base. The torsional strain is detected by the electronics. The time elapsed between the emission of the pulse and receipt of the strain signal provides a precise indication of the position of the magnetic ring, and hence of the cylinder rod.

Tracking the position of rotating elements is required in many machines. Rotary potentiometers, resolvers, encoders, synchros, and 'rotary variable differential transformer' (RVDT) are commonly used for the purpose. As in a linear pot, a rotary potentiometer consists of a tightly packed coil with a brush. The output voltage, proportional to the angle of rotation, may be mathematically expressed: $V = E\theta/\theta_T$, where V is the voltage output, E is source voltage, θ is angle of rotation from the zero end and θ_T is total rotation of the potentiometer.

4.11 VELOCITY, ACCELERATION AND APPLIED FORCE

Velocity and acceleration can be computed from the output of linear and rotary position transducers by including the time rate of change of displacement, or time derivatives, calculations in the computer software.

Angular velocity may also be computed with digital devices producing a pulsed output. Typical of these are magnetic pickups that sense the passing of indicators on a rotating disk, and electro-optical units that detect pulses of light through slits in a rotating disk. Photo-reflective devices are also used to detect passing of a strip of reflective tape on a rotating shaft.

Analog measurement of shaft speed may be obtained with a tachometer-generator. The output from the generator is mostly linearly proportional to the rotor speed up to its maximum, or saturated, speed.

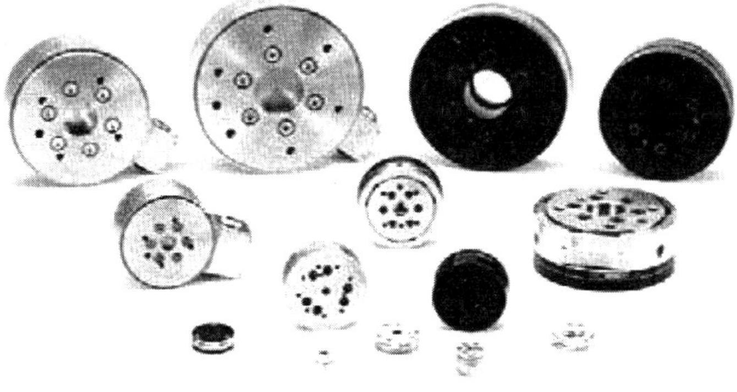

Fig. 4.2 Force Sensors (*Courtesy*: ATI Industrial Automation)

Newton's second law of motion does not imply that there are no forces acting on a body if it is not accelerating, only that there is no net unbalanced force. Two methods for force measurement rely on producing a measurable balancing force, the null balance and the displacement method (Figure 4.2). A beam balance is an example of the null balance method, where the unknown mass is placed in one pan and a calibrated mass in the other pan until the beam is balanced. A spring balance is an example of the displacement method. The unknown mass is placed in a pan supported from a calibrated spring. The stretch in the spring balances the force of gravity on the unknown mass. The spring stretch provides a measure of the mass.

The 'strain gauge load cell' is a displacement type of force sensor. If an unknown force acts on an elastic member of length L, the bar is stretched to length $L + \Delta L$. The strain ε is $\Delta L/L$. A strain gauge converts the elastic strain into an electrical resistance, based on the premise that the resistance of a wire changes when it is stretched, or strained. Bonded strain gauges are cemented directly on an elastic member at the point where the strain is to be measured. Un-bonded strain gauges measure the total displacement of an elastic member, with a mechanical linkage stretching a strain wire. Change in resistance of the strain wire gives a measure of the displacement. The gauge factor, G, of a strain gauge is given by: $G = (\Delta R/R)/(\Delta L/L)$, where ΔR = change in resistance, Ω, R = resistance, Ω, ΔL = change in length, and L = length of strain gauge. Gauge factor is usually between 2 and 4, length L ranges from 0.5 cm to 4 cm, and resistance R ranges from 50 to 5000 Ω. Stress is defined form Hooke's law for elastic materials: $E = \sigma/\varepsilon$, where E = modulus of elasticity, N/m^2, σ = stress, N/m^2, and ε = strain, m/m.

4.12 TORQUE SENSOR

A torque sensor, also called torque transducer or torquemeter, is a device for measuring and recording the torque on a rotating system of a turbine engine (Figure 4.3). In all turbo-machinery applications, knowledge of the torque developed is essential. Power output of the turbine, a function of torque and speed, must match the operator's demand setting requirement.

Torque sensors commonly use strain gauges applied on the surface of a rotating shaft. A means to power the strain gauge bridge is necessary. Static torque is relatively easy to measure. Dynamic torque in a rotating shaft, on the other hand, is not easy to measure, since it generally requires transfer of an electrical or magnetic signal from the shaft being measured to a static system outside the machine system.

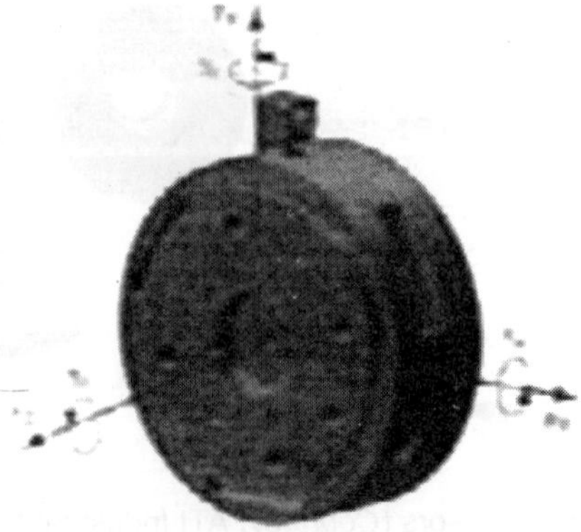

Fig. 4.3 Torque Sensor (*Courtesy*: ATI Industrial Automation)

One way to achieve this is to condition the shaft or a member attached to the shaft with a slip-ring arrangement, where a carbon brush contacts a series of permanent magnetic domains. This is similar to a commutator for electric motors. The magnetic characteristics of these domains will vary according to the applied torque, and thus can be measured using non-contacting sensors.

Signal from the rotating shaft can also be received using slip rings, wireless telemetry, or rotary transformers. Newer torque transducer designs add signal conditioning electronics and an A/D converter to the rotating shaft. Stator electronics then read the digital signals and convert those signals to a high-level analog output signal, such as +/–10VDC.

A more recent development is the use of SAW devices attached to the shaft and remotely interrogated. The strain on these tiny devices as the shaft flexes can be read remotely and output without the need for attached electronics on the shaft. The probable first use in volume will be in the automotive field as, of May 2009, Schott announced it has a SAW sensor package viable for in vehicle uses.

Still another way to measure torque is through of measurement of twist angle or the phase shift. The angle of twist resulting from applied torque is measured using two angular position sensors and measuring the phase angle between them.

4.13 TEMPERATURE PYROMETER

Temperature plays an important role in controlling the performance and operation of turbo-machinery. Temperature readings may be transparent to the operator, but the actual fluid temperature is of interest for determining performance. Thus, temperature sensors are required to provide a continuous readout. For instance, if the temperature of oil returning to the oil tank from the sump of a turbine bearing exceeds a certain level, the oil loses its viscosity and hence its capacity to provide hydrodynamic action. The system may then trigger a valve to divert the flow of oil through a cooler. On the other hand, if the oil is too cold in cold climes, the sensor may turn on a fluid heater. Either response may turn on an indicator light, or annunciator, on the control panel to inform the operator about the action taken.

Thermocouples, thermistors, and Resistance Temperature Detectors (RTD) are commonly used for measuring temperature. A thermocouple consists of two wires made from dissimilar metals joined at the ends. The difference in temperature between the two junctions creates an electric potential due to the 'thermoelectric' or the 'Seebeck' effect. Measuring the electrical potential between the junctions is interpreted as a temperature differential, and if the temperature at one junction is known, the temperature at the other junction can be established. Response to the temperature differential is non-linear, requiring a compensation of the temperature to be included as a feature in the signal conditioning process. Commercially available thermocouples have lead wires encased in a protective sheath.

'Thermistors' have a large change in electrical resistance in response to a change in temperature, although the variation is considerably non-linear. The associated software must include a lookup table in the memory to convert the observed resistance to the right temperature level. The devices are mostly used to measure temperatures over a limited range, and are restricted to temperature alarm and switching circuit applications.

A RTD works on the principle that electrical resistance of a metal wire changes with temperature. Virtually any metal shows this property, but nickel, tungsten, copper, and platinum are more commonly used. Although not as sensitive as thermistors, the change in resistance due to temperature is more linear. The feature makes the RTD more convenient from the perspective of signal conditioning.

Platinum RTDs are more widely used because of their linear characteristics over a wider temperature range. The metal wire may be shaped as a flat strip, disk, cylinder, or other configuration to accommodate installation and application needs. The element must be protected from stresses, physical damage, and hostile environment, since that would change its resistance. Many types of electronic circuits accept the RTD signal. A common one uses the RTD as a leg of the Wheatstone bridge.

4.14 REVIEW QUESTIONS

4.14.1 How is the capacitive principle used to detect strain?

4.14.2 How does a sensor differ from a transducer?

4.14.3 Define the threshold, sensitivity, error and repeatability of a measuring instrument.

4.14.4 Can an on-off switch be used to during an operation to produce an output to represent a process parameter?

4.14.5 Mention the preferred technique when the measurement may be highly remote, under high temperature, or is immune to electromagnetic interference.

4.14.6 Explain the ionization method to measure pressure.

4.14.7 How does an altitude sensor function?

4.14.8 Discuss the differences between measurement of static and dynamic torque in a shaft.

4.14.9 Describe the analog and digital methods of measuring the speed of a spinning shaft.

4.14.10 Explain the principles on which thermocouples, thermistors, and resistance temperature detectors operate.

4.14.11 How do un-bonded and bonded strain gauges differ, and how do they function?

4.15 EXAMPLE PROBLEMS

Example Problem # 4.15.1: A strain gauge force sensor for a cantilever beam has the following data:

Beam: material = steel, $E = 2 \times 10^{11}$ N/m^2

Maximum allowable stress = 2.75×10^8 N/m^2

Dimensions: $b = 2.0$ cm, $h = 0.5$ cm, $L = 12.5$ cm

Strain gauge: Gauge factor = 2, nominal resistance = 150 Ω

Determine the maximum force that may be measured with a strain gauge force sensor.

Solution: The maximum force that can be accepted by the cantilever beam may be determined from the maximum stress developed:

$$f_{max} = (S_{max} \times bh^2)/6L$$
$$= (2.75 \times 10^8 \times 0.02 \times 0.005^2)/(6 \times 0.125)$$
$$= 183.33 \text{ N.}$$

Example Problem # 4.15.2: A pneumatic force transmitter has an effective area of 3.75 in^2. If the air supply varies between 4.25 and 25.5 lbs/in^2, determine the range of force of the transmitter.

Solution: Expression for force is: $f = (25.5 - 4.25)A$

$$= (25.5 - 4.25) \times 3.75$$

$$= 79.69 \text{ lb.}$$

Example Problem # 4.15.3: A turbine flow meter ($V = KN$), where N = pulse count has a value of K = 17.4 cm^3 per pulse. Determine the volume of liquid transferred for a pulse count of 193.

Solution: For a turbine flow meter, $V = KN$, where N = pulse count.

Then: $V = 17.4 \times 193$

$$= 3358.2 \text{ cm}^3$$

Example Problem # 4.15.4: A linear pot is required to have an output of 21.7 volts for a positioning circuit when its output shaft moves through 1.85 in. for a positioning circuit. Find the transfer function value of the pot.

Solution: The pot's transfer function = (21.7 volts)/(1.85 in.)

$$= 11.73 \text{ volts/in.}$$

Example Problem # 4.15.5: A fluid of density ρ = 1128 kg/m^3 is to be stored in an open tank. A static pressure measurement probe is placed 0.885 m below the top of the tank. Determine the range of a pressure transducer for 100 per cent output when the tank is full.

Solution: Expression for static pressure $p = \rho gh$

Maximum pressure $p_{max} = 1128 \times 9.81 \times 0.885$

$$= 9793.13 \text{ Pa, or } 97.93 \text{ kPa.}$$

4.16 EXERCISE PROBLEMS

Exercise Problem # 4.16.1: A displacement float level transducer has a float mass of 2.85 kg, area of float is 48 cm^2, length of float is 1.95 m, and liquid in the tank is water. If the equation for force on the float is given by $f = Mg - \rho gAh$, find the minimum and maximum values of force on the float.

Exercise Problem # 4.16.2: A bellows pressure element has an effective area = 12.25 in^2, and a spring rate = 28 lbs/in. A parallel spring has rate = 82.5 lb./in. Determine the pressure range of the sensor if the motion in the bellows is restricted to 0.895 in.

Exercise Problem # 4.16.3: Calculate the density of a fluid stored in an open tank if a static pressure measurement probe placed 1.95 ft below the top of the tank and the maximum pressure transducer reading is 14280 lb./in^2 when the tank is full.

Exercise Problem # 4.16.4: A hydraulic cylinder is required to exert a force of 48500 lbs. Calculate the piston area if oil pressure is 3250 lb./in^2.

Exercise Problem # 4.16.5: Determine the value $(v_2 - v_1)$ that will saturate an operational amplifier of gain $A = 10^6$ and supply voltages are –25 V and +25 V, where the amplifier saturates at 80 per cent of supply voltages, or about –20 V and +20 V.

REFERENCES & BIBLIOGRAPHY

Balluff, Inc., "Electro-mechanical Single and Double Limit Switches with Accessories", Catalogue # 103E, Issue # 105, Florence, KY, 1987.

Balluff, Inc., "Non-Contact Switching", Catalogue # 321E, Edition 8806, Florence, KY, 1988.

Banner Engineering Corporation, "Handbook of Photo-Electric Sensing", 2nd edition, Minneapolis, MN, 1993.

Bateson, R. N., "Introduction to Control System Technology", Prentice-Hall, Clifton, NJ, 1999.

Carstens, J. R., "Automatic Control Systems and Components", Prentice-Hall, Clifton, NJ, 1990.

CEC Instruments, "Pressure Transducer Handbook", Pasadena, CA, 1989.

Funakubo, H., "Actuators for Control", Gordon and Breach Science Publishers, New York, 1991.

Norvelle, F. B., "Electro-Hydraulic Control Systems", Prentice Hall, Upper Saddle River, NJ, 2000.

Parr, E. A., "Industrial Control Handbook", Volume I, Industrial Press, Inc., New York, 1987.

Ryoji, O., "Intelligent Sensor Technology", John Wiley, New York, 1992.

5

CHAPTER

REGULATION AND CONTROL DEVICES

5.1 INTRODUCTION

Operational control of modern turbo-machinery depends on pressure, temperature, vibrations and other parameters from input signals at selected locations. The signals are analyzed to determine the need for corrective action and appropriate control commands to ensure optimum performance. Switches, control valves, actuators and electric motors execute the command signals to provide the required output.

Switches make or break the connection in a circuit, either mechanically with an actuator or electro-mechanically with a solenoid. A set of switches may be assembled in a control relay to execute multiple functions. Time-delay relays permit switching action at the right instant. For example, the relay for switching a three-phase electric motor may be in the form of a motor starter. Solid-state components are used extensively.

Actuators provide the power for executing the final control element. Hydraulic and pneumatic cylinders provide linear motion, and hydraulic and pneumatic motors develop large torque to produce rotational motion. Electric motors are often used as actuators in a servo control system, for instance, in numerically controlled machining centers to position the work piece and to locate the cutting tool. Variable speed electric motors are also used to drive pumps for controlling the flow rate and pressure of the fluid. Hydraulic actuators are better suited for slower, more precise, and heavier load applications. Pneumatic actuators operate at lower pressure than the hydraulic ones, and are useful for lighter loads and where faster response is necessary.

The solenoid valve is the simplest type of electrically operated valve, and has been used over a large span of time with relatively few changes to the original concept. Electro-hydraulic proportional control valves (EHPV) represent a step up in the technology, moving up from the relatively less demanding electrical switching to the more sophisticated operation by electronics.

Servo system components offer the ultimate control capability, with continuous monitoring, feedback provision and corrective action to improve efficiency, accuracy, and repeatability. Applications for servo valves for turbo-machines include operation of variable stator vanes of an axial flow compressor and control of rotor speed for power generation turbines. Controlling the rate of flow of fuel to the injectors in the combustion system to ensure development of required power setting and managing the speed of the rotor to maintain air pressure at the compressor's discharge are some other examples where servo system components play a primary role.

The turbo-machinery industry has played a leading role in originating and refining the programmable logic controller, or PLC. Design criteria for the first logic controllers were initially developed in the 1950's and 1960's. The objective was to eliminate replacing the inflexible and hard-wired relay panels with a computer that can survive in the industrial environment. Initially the units were little more than elaborate switches, but their capabilities and sophistication grew rapidly.

From its modest beginnings, the modern PLC has developed into far greater than a substitute for relay logic panels. Individualized modules that execute a host of functions such as logic, timing, counting, sequencing, and proportional-integrating-derivative (PID) control are included in the PLC's. The computers have the added advantage of carrying out arithmetical calculations, analyze data, and exchange data with other programmable controllers and host computers.

5.2 ANALOG AND DIGITAL CONTROL METHODS

Electronic 'analog control' calls for an error detector and a control mode unit. An analog controller requires a single operational amplifier as a function generator and some resistors and capacitors to implement the transfer function of the desired control mode or combination of modes.

The analog proportional controller uses three resistors to form an inverting amplifier. The circuit has two inputs, an error, and an output offset. The proportional gain is obtained by dividing the feedback resistor by the error input resistor. The offset resistor must equal the feedback resistor to satisfy the time domain equation.

The proportional plus integral (PI) controller needs two resistors and a capacitor to implement the transfer function. The gain is obtained by dividing the feedback resistor with the input resistor. The integral action rate is equal to the reciprocal of the product of the input resistor and the capacitor.

The proportional plus derivative (PD) controller uses four resistors and a capacitor. The circuit is a proportional controller with a parallel combination of a resistor and a capacitor placed in series with the input resistor.

A digital controller measures the controlled variable after specific intervals of time, called the sampling time, Δt. The measurement is converted to a binary number for input to a digital computer. The computer subtracts each reading of the sample from the set point value to obtain a set of error samples. Microprocessor based digital controllers are used for advanced tuning, multi-variable control and expert systems. Digital controllers in the closed-loop usually implement the proportional plus integral plus derivative (PID), PI or PD modes.

The integral mode sometimes presents computational difficulties that can lead to unsatisfactory results. When the value of $PI\Delta t$ is less than 1, it is convenient to work with the reciprocal of $PI\Delta t$, which could be stored in the computer as an integer. If the value of $PI\Delta t$ is very small, the computer may ignore relatively large errors because of insufficient resolution. The change would then be lost unless a special provision is made to include the change in the calculations for the next sample. The end result is an error offset that the integral mode is unable to eliminate. One solution is to increase the precision by increasing the word length in the computer.

The derivative mode can also present computational difficulties that can lead to unsatisfactory results. A slowly changing signal, for instance, results in a fluctuating derivative action mode. The derivative mode estimates the rate of change of the error, de/dt. Since Δt is fixed by the sampling rate, attention is focused on the change in the error. To obtain a better estimate of Δe, estimators of previous samples may be used to smoothen the derivative term and thus improve the estimate.

5.3 SERVO VALVES

By definition, a servo-mechanism implies return flow of information, or feedback, from the machine system. Numerous servo-mechanisms have been invented since the mid-1700s. Many were associated with raining steam in a boiler for controlling water level, water and steam flow, steam pressure and speed of machinery. Technological advances of the twentieth century in electrical power generation units, aircraft engines, and automobile turbo-chargers spurred greater advances in servo technology.

Servo valves are engaged in nearly all aspects of fluid power system operations, the most common application being for motor speed control. The valves provide direction and flow control with a sliding spool that is positioned by a torque motor. The torque motor is a low-power electro-magnetic component assembled from permanent magnets, two pole pieces, an armature, and two coils, and is operated by low-voltage DC power. The coils may be connected in series, parallel, or in the more commonly used push-pull arrangement. Increasing the voltage input to one coil and reducing the voltage to the other coil by the same amount causes the armature to rotate, with its position determined by differential torque. The current input is expressed as a difference between the two coils, ΔI, and the control power P is expressed by the equation: $P = (\Delta I)^2 R$, where ΔI = differential current and R = resistance of one coil. If the coils are connected in parallel, the control power is: $P = (I_p)^2 R/2$, where I_p = total current through the circuit. For series connection, the control power is: $P = (I_s)^2(2R)$, where I_s is current in the series circuit.

The spool and the bore in which it travels are precision machined. Circumferential grooves are machined into the land areas to reduce transverse loading. The width of the lands relative to the width of the ports in the valve bore provides the lap. The most commonly used is the zero overlap spool, where the land width matches the port width. When the spool is centered, there is no flow through the valve. The valve is useful for closed-loop position, speed and force control applications because of its precise metering characteristics about the neutral position. Overlapped spools have the advantage of providing lower leakage flow in the neutral position, but the dead band compromises the precision about the neutral position. Under-lapped spools provide rapid response to commands about the neutral position, but have non-linear flow characteristics around the region.

5.4 ELECTRO-HYDRAULIC PROPORTIONAL CONTROL VALVE

As the name suggests, the electro-hydraulic proportional control valve (EHPV) produces an output that is proportional to an electronic input. The valve may be placed between the solenoid valve and the servo valve in performance, cost, and complexity. Maximum frequency response is around 10 Hz, which is faster compared with the frequency capability of the solenoid valve, but considerably slower compared with that of the servo valve. The design of the actuating device, the proportional solenoid, allows the spool to be stopped at intermediate positions at the ends of the solenoid stroke. The associated electronics controlling the spool position offers a high degree of flexibility compared with the solenoid valve.

The performance can be improved and applicability extended by including a spool position feedback loop in the circuit. External feedback systems, higher frequency response, better performance in hysteresis, dead-band, threshold and other features can also be included in proportional control valves. The enhanced performance and control features blur the distinction between servo valves and proportional control valves, but they increase the cost.

Aside from performance and control characteristics, physical features such as manufacturing process and design define the proportional control valve. For instance, proportional control valves require a proportional solenoid, whereas servo valves are operated by torque motor. The spool of a proportional control valve is mostly machine produced, while servo valve spools need substantial manual lapping and finishing due to the tighter clearances and tolerances.

5.5 ADMISSION, EXTRACTION AND NON-RETURN VALVES FOR STEAM TURBINE

Besides the traditional stop and control valves, oftentimes steam turbines have additional ports to admit or extract steam. In addition to admission at the inlet, steam turbines may admit steam at lower pressure locations in the downstream sections. Additional steam at lower pressures is commonly called for in steel mills, paper mills and combined cycle plants provided with triple pressure steam generators.

Extraction turbines have multiple locations to extract steam for facility services at various pressures and flows. Steam is removed at many stages to match with the facility's pressure and flow requirements. The extractions may be automatically or manually controlled. Extraction for feed-water heating requires control valves to regulate the steam flow and to maintain the extraction steam pressure at a constant level. The valves are hydraulically opened and shut through spring-loaded mechanism. But the valves are not leak-tight, and may pass 5% steam flow in the closed position.

Non-return valves (NRV), also called check or clack valves, are needed downstream of the extraction connections to the turbine. The function of the valves is to permit flow of extraction steam in the outgoing direction and restrict backward flow into the turbine when turbine extraction pressure is lower than the lines it feeds. The valves are designed to be spring-loaded shut when there is no extraction pressure. Air or hydraulically assisted actuator is also provided to close the valve when the system is pressurized. Malfunctioning of extraction NRV's have been noted to be a primary cause of over-speed when the turbine is being shutdown.

5.6 STOP, TRIP AND THROTTLE AND INTERCEPT VALVES

Important to any turbine is the ability to start and stop the machine under normal control as well as in emergency conditions. Quick shutdown is normally accomplished by either the main steam stop valves or the trip and throttle valves placed at the inlet piping to the steam turbine, or on the turbine shell. The valves must be leak-tight since leakage may keep the turbine turning at low speed after shutdown. An even greater concern is over-speed since the valve did not close completely after a shutdown or trip.

High-pressure system fluid or lube system oil is commonly used for the actuators to open the valves. In case of loss of oil pressure, spring-loaded closure of the valves provides fail-safe method. Smaller steam turbines may have manually controlled valve with a large hand wheel to stop and to start the unit. Hand operated valves are sometimes mounted in the nozzle inlet to manually increase steam flow.

Steam turbines with reheat direct the steam back to a boiler superheater section after going through the high-pressure section. Additional valves are then installed between the high-pressure and the subsequent sections. Reheat stop valves provide leak-tight protection, but an intercept valve is installed in series with the reheat stop valve to avoid over-speeding. The valves also open with oil pressure and spring-loaded shut if the oil pressure is lost under trip or over-speed conditions.

Contaminants in the steam, wear of mating valve parts, or damaged valve seats can cause sticking or leaking of these valves in service. To ensure over-speed protection the valves must be regularly checked.

5.7 SPEED GOVERNOR AND CONTROL VALVE

Control valves are provided on the turbine shell to regulate the flow of steam to the turbine for starting, adjusting power output, and maintaining speed control with the turbine governor system. A single inlet valve with separate actuator, cam lift inlet valve assemblies, and bar lift inlet valve assemblies are available. The valve assemblies are normally mounted onto a steam chest integral to the shell or bolted to the shell.

The cam lift valve arrangement utilizes cams, bearings and bushings mounted on a camshaft to regulate the position of each valve. A hydraulic servomotor drives a rack and pinion connection to the camshaft to indicate the position determined by the governor. The bar lift valve method has a hydraulic cylinder to lift all of the valves attached to the bar together. Collars on each valve stem are set at different heights. The opening sequencing for steam admission during starting and load adjustments is changed. To avoid the possibility of the valves sticking in open or closed position and thus losing control of the operation of the turbine, the valves must be regularly inspected and cycled.

5.8 OVER-SPEED PROTECTION AND TRIP LOGIC

The most destructive event for a turbine arises from over-speeding of the rotor. When hoop direction stresses arising from the centrifugal forces acting in the rotating disks exceed the burst capability of the disk substrate, the disk, blades and shaft are catastrophically damaged. Centrifugal forces and stresses increase with the square of speed, so if rotor speed is 110 per cent of the normal, centrifugal stresses will climb up to 121 per cent from the normal value. Though the occurrences are not frequent, the need to maintain the speed within safe limits cannot be overemphasized.

Turbo-machines use a mechanical, electronic or a combination for protection from over-speed. A spring-loaded piston mounted in the turbine shaft at the front of the turbine is used in the mechanical device. When the speed approaches 10 per cent above the normal running speed, the piston pops out and hits an oil dump valve lever and depressurizes the oil supply to the stop, trip and throttle and intercept valves. The action causes all stop and intercept valves to close spontaneously. Some mechanical systems also have a flywheel ball governor driven through the turbine shaft. Any shift in the governor position leads to change in oil pressure to the turbine control valve servomotor or actuator. Under over-speed conditions, the flywheel governor connects with the oil dump valve lever to close the steam stop valve.

A number of magnetic speed pickups are installed on the turbine shaft in the electronic system of over-speed protection. The turbine control system and its software logic electronically cause the oil dump valve to depressurize the oil system and close all stop and intercept valves. Among some electronic versions the primary and emergency backup systems operate independently. Others have test switches to test the primary system for proper operation without actually tripping the turbine. For most turbines the over-speed protection system orders the turbine control valves to close as well. Since control valves are not leak-tight by design and their closure rate is much slower than that of stop and intercept valves, they are not relied upon to provide any over-speed protection.

Aside from the over-speed protection system, the trip logic used in the control system to open the circuit breaker associated with the turbine's generator also has an effect on the quality of the protection. Two trip schemes are available, sequential and simultaneous tripping. In sequential tripping the steam turbine trips first, followed with opening of the generator circuit breaker as the turbine speed and decaying power gradually cause the generator reverse power relay to open the breaker. Larger steam turbines operating at high inlet pressures and temperatures depend on the sequential method due to the need to dissipate the energy in the turbine before opening the breaker to minimize the over-speed level as the unit shuts down.

Simultaneous tripping is utilized when both the turbine stop or trip and throttle valve and the generator circuit breaker are opened at the same time, regardless of whether the turbine or generator protection system initiated the trip. This type system is utilized successfully on small to medium size steam turbines where the steam pressures and temperatures are low and there is little steam volume in the turbine to cause an increase in speed on shutdown. Note that all over-speed and trip protection systems must be regularly tested through simulation and actual tests of the overall system.

5.9 SERVO VALVE, PUMP AND MOTOR

The servo valve is useful for closed-loop position, speed, and force control applications because of its precise metering characteristics. Single-stage servo valves are used where the flow requirement is low, under five gallons per minute. The valve has a sliding spool that directly connects to the torque motor armature. The force of the torque motor and the stroke of the spool dictate its flow capacity. A rod connects the torque motor's armature and the spool.

Multiple stages deliver higher flow rates. The second and third stages have a pilot-operated sliding spool. A sleeve in the pilot stage provides internal feedback mechanism of the valve. Two pressure sources are available, p_c for piloting the main spool, and p_s for the system. The larger end of the spool has twice the area of the smaller end to obtain static balance when the spool is in the neutral position.

Counterclockwise movement of the torque motor armature pushes the pilot spool to the left, which initiates an internal feedback mechanism. Flow to the larger end terminates, with the stopping point determined by the deflection of the torque motor. De-energizing the torque motor returns the pilot spool to its neutral position, and the main spool is pushed back to the right. A similar but opposite sequence takes place when the torque motor armature is deflected clockwise.

Servo controllers are built inside servo pumps and servo motors to provide variable displacement. The servo assembly includes a servo valve, a hydraulic cylinder, and an internal feedback unit.

Depending on the size and application, servo pumps can handle maximum flow rate of 500 gallons per minute and pressure of 5000 lbs/in^2. The piston pumps may be single or double acting, and may be arranged as either inline or radial units. Servo pumps are mostly used to control speed of large hydraulic motors for adjusting pump displacement in response to feedback signals from the motor.

5.10 DC, INDUCTION, SYNCHRONOUS AND SERVO MOTORS

DC motors ranging from fractional to several hundred horsepower are versatile drives for accurate control of speed over a wide range, and are capable of operating in the reverse direction. The high torque to inertia ratio gives them quick response to control signals, and deliver over three times their rated torque for short periods. DC motors also have the unique capability for dynamic and regenerative braking, momentarily turning into a DC generator due to the inertia of the load during braking.

The armature and field windings may be connected in series, shunt, compound, and separately excited configurations. Series connection provides the highest starting torque and the largest no-load speed of the four connections. The shunt motor has a comparatively lower starting torque and no-load speed. The compound motor expectedly has speed and torque characteristics that lie between those of the series and the shunt connections. The separately excited version is a special case of the **shunt** motor, allowing independent control of the armature and field voltages.

The need to make electrical contact between the armature winding and the brush and commutator assembly of the DC servomotor is a considerable drawback. In a brush-less motor the rotor is a permanent magnet, the winding is in the stator, and a solid-state circuit replaces the brush and commutator assembly.

Stepping motors are brush-less DC motors used for converting digital pulses into fixed incremental motion. Pulse inputs may be provided from a computer or a programmable controller. The control unit converts the pulses into step motion of the motor shaft through power transistors. The motor's stator has an even number of equally spaced poles, or teeth, each with an electrical coil. The rotor has equally spaced external teeth with a small gap between the rotor and stator teeth. The number of teeth on the rotor and the stator determine the size of the step, also called the stepping angle.

AC motors dominate control system applications that require a single operating speed. Single-phase motors are used where power requirement is limited to 20 hp. Poly-phase motors extend the range to substantially higher horsepower ratings. For variable speed applications, improvements in solid state switching components and integrated circuits have made variable speed AC drives competitive with DC motors. Split into single-phase and poly-phase groups, and further sub-divided into synchronous and induction categories, gives a total of four major groups.

Single-phase squirrel cage induction motor, the most widely used type, does not have brushes to wear out or to generate sparks. The simple construction of the motor has low initial and maintenance costs. Resembling a squirrel cage, the rotor winding is in the form of copper or aluminum conductors fitted in slots in the rotor and connected at the ends by continuous rings. The stator has one or two windings that are externally excited. The rotor winding is shorted, and obtains energy through electro-magnetic induction. The induction motor functions as a rotating transformer, in which the stator winding is the primary and the rotor winding is the secondary.

Unlike a DC motor, the stator winding's magnetic field is not stationary. The poles of the field alternate between two positions, in effect rotating about the rotor's axis. The speed at which the field of the induction motor rotates is called the synchronous speed. A motor with a two-pole stator has the synchronous speed equal to the line frequency of 60 Hz, or 3600 rpm. For a four-pole motor the synchronous speed is one-half the line frequency, or 1800 rpm. The rotor, however, runs at less than this speed, and the difference between the synchronous and actual speeds is referred to as slip. The torque developed by the motor is proportional to the slip, and even at no-load some slip occurs to overcome wind resistance and friction losses.

The torque produced in both directions of rotation is equal, so a single stator winding induction motor cannot start by itself. Physically pushing the rotor in one direction with a starting device causes the rotor to gain speed. A second stator winding rotated 90° form the main winding and called the start winding provides the starting impetus, and is the preferred method for starting a single-phase induction motor. A speed sensitive switch disconnects the start circuit at a preset speed.

Poly-phase induction motors use three phases, and are either of the squirrel cage or of the wound rotor type. In a wound rotor the conductors are connected in series to form three windings, with the windings joined at one end in a Y connection. The other set of windings is connected to a slip ring, with one slip ring for each winding. Three brushes are in contact with the three slip rings to permit external resistances to be connected in series with the windings. The other end of the resistors is connected together in a Y connection. The resistors limit the rotor current during startup when the inertia load is large. If external resistors are missing the wound motor becomes a squirrel cage motor.

A two-pole three-phase induction motor's stator has three windings located 120° apart. The windings may be connected in either a delta or a Y connection. During each sixth of a cycle, the poles move to a new location rotated through 60°. The rotation of the field poles in six steps is smoother than the two-step rotation of the single-phase motor. The three-phase induction motor is self-starting, eliminating the need for a starting device. A four-pole three-phase motor has six windings, located at 60° intervals. Each coil uses stator slots spaced 90° apart.

Synchronous motors are distinguished by their precise constant speed of operation. Ranging in size from fractional to several thousand horsepower, the motors are designed for single-phase and poly-phase operation. The stator winding of synchronous motors is mostly identical to that of the corresponding single or poly-phase induction motors. Magnetic poles in the rotor lock into step with the rotating poles in the stator. The rotor may not be excited or may be excited by direct current.

Two methods are used for starting a synchronous motor. A separate prime mover starts the motor and brings the motor to nearly synchronous speed, and disconnecting at the proper speed. The other method calls for a squirrel cage winding to be added to the rotor. The motor starts as an induction motor and accelerates to near speed. At synchronous speed the current in the squirrel cage conductor reduces to zero.

The lighter weight and inertia and lower inductance of the armature permit DC servomotors to respond quickly to command signals for a shift in position or speed. Similar to servo pumps in many ways, servomotors can alter the displacement in response to command and feedback signals. The AC servomotor is a modified two-phase reversible induction motor for servo operations. Amplified AC error signal, which is either in or out of phase, is applied to the control field. A constant AC voltage is applied to the reference field through a 90° phase shifting network.

5.11 PNEUMATIC ACTUATORS

Pneumatic systems are more suitable for light and medium loads. Two-stage reciprocating compressors are capable of delivering air above 1000 kPa, and the air may be readily stored in pressure tanks for regular production operation or for emergency operations during power failures.

Pneumatic actuators lend themselves to easier control of power and operation with pressure and flow control valves and directional valves in response to load demand. The compressible nature of air makes it difficult to obtain good accuracy in controlling speed, and operating efficiency is also relatively low. Their performance, however, may be improved by adding proportional valves in the system.

Pneumatic cylinders have a considerably simple construction, and are widely used for converting the energy of the compressed air into mechanical reciprocating form of motion. Cylinder designs vary with operational mode (single acting, double acting, or differential) and form of piston (piston or ram).

When the pneumatic cylinder is operating at no load, the volume of air flowing in and out of the cylinder is the same. At low speed the inertia force from the moving parts is low, and the force acting on both sides of the piston will equal the applied load. When the operating speed is high, the pressure inside the cylinder increases to the supply line pressure within the starting time to overcome the applied load, while the discharge pressure is close to atmospheric level. The acceleration rate depends on the pressure differential, and is given by: $a = g/W\,[\mu P\pi D^2/4 - F_c]$, where a = acceleration, m/s^2, g = acceleration due to gravity, m/s^2, W = weight of load, N, μ = thrust coefficient of cylinder, P = air supply pressure, Pa, D = cylinder bore, m, and F_c = actual load on cylinder, N.

Pneumatic oscillating actuators are employed where the angle of rotation of the shaft is limited. As opposed to the pneumatic cylinder, the energy of the compressed air is converted into reciprocating motion in rotary actuators. The conversion mechanisms may be broadly split into vane and piston types. In the vane type, the energy of the compressed air acts on the vanes and directly transmits the power to the shaft to produce torque. The surface area of the vane, number of vanes and applied air pressure determine the output. Single, double, and triple vane oscillators are commonly used. The angle of oscillations increases, but the torque developed increases with the number of vanes. Complete shielding of the airflow in vane oscillators is difficult to obtain, and some leakage is inevitable. But the units are compact and are reliable.

Inside the piston of oscillating actuators, the linear motion of the pneumatic cylinder's piston is converted into rotary motion through screws or gears. With the rack and pinion arrangement, turning the pinion that is directly attached to the piston develops the required torque. Internal cushioning may be built in the cylinder. A small increase in friction is encountered due to the lateral thrust on the pinion and the piston, but the efficiency is still between 80 and 90 per cent. In screw type of oscillators the linear motion of the piston is converted into rotary motion through a screw mounted on the output shaft. The piston is provided with a device to prevent it from rotating. The rotary angle increases with the stroke, and may exceed 360°. Efficiency of the actuator is about 80 per cent due to the large friction losses in the anti-rotation device.

5.12 AIR MOTOR

The displacement type of air motor is designed for general industrial use, where only pressure energy of the compressed air is used. Where high levels of rotation are the norm, velocity type of air motors are more suitable. Both the pressure energy and the velocity of the flowing air are put to use in velocity type of air motors.

One form of construction calls for vanes inserted in radial rotor slots, and the rotor is placed in an eccentric casing. Compressed air enters the region with reduced radial space and exits at the larger clearance end. The vanes slide inside the slots either by springs or by air pressure. The gear type of air motor employs double helical gears for expansion of the air to provide a low-speed, high-torque unit.

In piston type of motors, compressed air acts on the piston face to provide power through an inclined plate, cam and crankshaft. Each piston has a connecting rod that links it to a crankshaft. The pistons operate through sequential supply and discharge of compressed air with the aid of a rotary distribution valve that is connected to the crankshaft. The reciprocating components are dynamically balanced and the outputs of the cylinders overlap, so vibrations are relatively low. Provision must be made for lubrication. Switching the connections to the supply and discharge ports and adjusting the distribution valve timing reverses the rotation direction. The axial piston type of air motor generally has between 4 and 6 cylinders arranged concentrically around the output shaft. An inclined wobble plate is connected to the piston rod, which rotates the output shaft through a gear mechanism.

5.13 HYDRAULIC MANIPULATION

The pressure energy of hydraulic fluid acting on a piston is converted into mechanical energy in hydraulic cylinders. Greater actuation force compared with pneumatic devices may be derived either externally through the rod, or by keeping the rod fixed, through the cylinder. Single-acting cylinders are used for loads in a single direction, with a retracting spring returning the piston back to the start position at the end of the cycle. Double-acting cylinders are for two directional loads. Telescopic cylinders are used for larger strokes in proportion to the axial length.

In the design of hydraulic cylinders, sealing and lubrication of the sliding parts, strength of the components in relation to the internal pressure, capacity to absorb shock, and rust prevention must be considered. Non-contacting seals or hydrostatic bearings may be needed to reduce coulomb friction in units where high-pressure characteristics are called for.

Thrust of the hydraulic cylinder, $F_c = \Delta p A_c \lambda$, where Δp = pressure differential, A_c = internal cylinder area and λ = efficiency of hydraulic cylinder varies with operating pressure, construction features, friction characteristics of packing seals and size.

Seamless steel tubes are frequently used to meet pressure requirements of hydraulic cylinders. Internal diameter of the cylinders is stipulated in ISO 3320. Wall thickness, t, is calculated from the equation: $t = pD/(2S)$, where p = nominal pressure, MPa, D = inner diameter, $S = \sigma/5$, N/mm^2, and σ = minimum material tensile strength, N/mm^2. To prevent wear on the cylinder's inner surface, the piston is made from cast iron or copper alloys. Packing is mostly used to obtain piston sealing. Besides tensile and compression loads, buckling capacity of the piston rod must be determined. The rod must maintain a good rubbing surface.

The consequences of extremely high working pressures in hydraulic cylinders are wear and scuffing in the piston or the rod. Imposition of bending loads in the rod is chiefly responsible for a number of failures, hence aligning the direction of the external reacting force with the axis of the cylinder is of significance. Furthermore, reacting loads on the mount bolts must avoid shear. Leakage of oil may be prevented by ensuring absence of scratch marks on the sliding surfaces and maintaining purity of the oil. Non-metallic seals must be periodically checked and replaced.

It may not be possible to avoid mixing of air in hydraulic circuits, due to which air can also get into the cylinder. Accumulation of air leads to jerking motion, sticking, slipping, and vibrations. Air must be continuously purged during operation to prevent the undesirable effects. Since air bubbles move upwards due to buoyancy, provision of a port where they collect helps in purging the air.

Hydraulic motors are required to produce large torque in a control element or load. The motors may be divided into four common types: piston motor, gear motor, vane motor and turbine motor. Gear and vane motors are more suitable for small to medium displacement, while piston motors are for larger displacement capacity.

External gear motors have a set of gears housed inside a two-piece casing. Fluid entering the space between the meshing gears causes the two gears to rotate in opposite directions. The gears transport the oil from the inlet to the outlet port. One of the gears delivers the power output. Internal gear hydraulic motors have an internal gear with circular teeth meshing with an external gear having one more tooth. Keeping the internal gear stationary permits the external gear to walk around the internal gear. The design provides combined features of a hydraulic motor and a speed reducer.

Torque is developed in a vane motor from the action of pressurized oil on a number of radial sliding vanes fitted in slots on the rotor. The vanes are forced out towards the inner surface of the casing by spring action and by centrifugal force as the rotor spins at high speed. Friction and wear related problems occur mostly at vane tips. Instability may be encountered at low speeds, and the motors also suffer from low starting torque.

Radial piston motors have the pistons arranged around a drive shaft. Fluid entering the cylinder forces the piston against an eccentric thrust ring attached to the shaft, which causes the rotor to spin. In axial piston motors, the axes of the cylinders run parallel with the cylinder axis. A wobble plate attached to the rotor is made to rotate by the normal component of the reaction thrust that is produced from the motion of the piston.

Turbine motors have a low starting torque, but operate at high speeds. The motor has a number of rotating blade sets attached to the rotor, separated by rows of stationary blades. Fluid entering the motor flows through the canted turbine blades and is deflected, which imparts a tangential force that spins the rotor. The stationary blades redirect the flow of the fluid towards the next row of rotating blades.

5.14 REVIEW QUESTIONS

5.14.1 Name a few automatic control devices used for turbines. How do they affect their performance?

5.14.2 What role do valves play when a turbine runs into over-speed condition?

5.14.3 Describe a multi-stage servo valve. How does it function?

5.14.4 What is a servo controller, and what is it used for?

5.14.5 What is a commutator? What type of motors requires it?

5.14.6 Mention the characteristics and some methods of starting a synchronous motor.

5.14.7 How can oscillating motion be obtained with a linear acting actuator?

5.14.8 How does air inside a hydraulic actuator affect its operation, and how may the problem be remedied?

5.14.9 Discuss the merits of a hydraulic motor with pistons and compare with a turbine motor.

5.14.10 How does a turbine exceeding its normal running speed pose a problem?

5.14.11 How does an analog control system differ form a digital one?

5.15 EXAMPLE PROBLEMS

Example Problem # 5.15.1: The resistance of coils connected in a push-pull circuit of a torque motor is 38 Ω, and rated at 379 mA. Find the voltage of each coil when the armature of the torque motor is centered and the maximum value of ΔI.

Solution: The maximum voltage in the coils is:

$$E_{max} = I \times R = 379 \text{ mA} \times 38\ \Omega$$
$$= 14.402 \text{ V}$$

The armature is centered when: $E = E_{max}/2 = 7.201 \text{ V}$

Differential current ΔI is maximum when maximum voltage is applied on one coil, and the other coil has no voltage.

Then: $\Delta I = E_{max}/R - 0 = (14.402 \text{ V})/(38\ \Omega) = 379 \text{ mA}$

Example Problem # 5.15.2: Determine the maximum control power of a motor if the resistance of the coils connected in a push-pull circuit is 133 Ω, and rated current is 873 mA.

Solution: Maximum control power is:

$$P = (\Delta I)^2 \times R = (873 \text{ mA})^2 \times (133\ \Omega)$$
$$= 101.36 \text{ W}$$

Example Problem # 5.15.3: Find the amplitude response in decibels if a valve spool moves in a sinusoidal wave form 0.195 in. in response to a 0.376 in. sinusoidal input.

Solution: Amplitude response = $20 \times \log(\text{output/input})$

$= 20 \times \log(0.195/0.376)$

$= -5.70$ dB

Example Problem # 5.15.4: A 4.175 in. diameter hydraulic cylinder with a force capability of 28335 lb. is used. What is the required working pressure of the oil?

Solution: Working pressure = $28335/[3.14159 \times 4.175^2/4.0]$

$= 2069.76$ lb./in^2.

Example Problem # 5.15.5: A servo valve is rated at 8.35 gallons per minute and 748 lb./in^2. If it is to be operated in a 925 lb./in^2 system, what is the adjusted flow rate for optimum power transfer?

Solution: Optimum power transfer occurs when the pressure drop Δp across the valve is 925/3 = 308.33 lb./in^2. Then flow rate $Q = QR \times [\Delta p/\Delta pR]^{0.5}$.

$= 8.35 \times [308.33/748]^{0.5}$.

$= 5.36$ gallons per minute.

5.16 EXERCISE PROBLEMS

Exercise Problem # 5.16.1: An actuator is required to exert a force of 7700 lb. to act through a distance of 8.0 in. in 13.5 s. Calculate the required force and operating speed if the overload factor is 20 per cent.

Exercise Problem # 5.16.2: An armature controlled DC motor has friction torque $T_f = 0.0185$ N-m, torque constant $K_T = 0.125$ N-m/A, maximum current $I_{max} = 3.25$ A. Find the maximum output torque T_{max}.

Exercise Problem # 5.16.3: A 2700 lb./in^2 working pressure cylinder exerts a force of 31500 lb. If the pressure is 1600 lb./in^2, what is the available force?

Exercise Problem # 5.16.4: An armature controlled DC motor produces output torque T_{max} of 0.325 N-m when maximum motor speed ω_{max} is 395 radian/s. Find the maximum output torque T_{max}.

Exercise Problem # 5.16.5: An armature controlled DC motor has maximum current $I_{max} = 8.35$ A, voltage constant $K_E = 0.115$ Volt-s/radian, maximum motor speed $\omega_{max} = 522$ radian/s, and armature resistance $R = 2.2\ \Omega$. Find the maximum armature voltage E_{max}.

REFERENCES & BIBLIOGRAPHY

Bateson, R. N. "Introduction to Control System Technology", Prentice-Hall, Clifton, NJ, 1999.

Bower, J. L., Schultheiss, P. M., "Introduction to Design of Servo-Mechanisms", John Wiley & Sons, New York, NY, 1964.

Carstens, J. R., "Automatic Control Systems and Components", Prentice-Hall, Clifton, NJ, 1990.

CEC Instruments, "Pressure Transducer Handbook", Pasadena, CA, 1989.

Funakubo, H., "Actuators for Control", Gordon and Breach Science Publishers, New York, 1991.

Herman, S. L., Alerich, W. N., "Industrial Motor Control", Delmar Publishers, Inc, Albany, NY, 1985.

Ichikawa, T., "Hydraulic Engineering", Asakura Shoten, 1979.

Kozoh, O., “Development of Multi-Cam High Torque Hydraulic Motor”, Hitachi-Hyoron, volume 56, Number 11, 1974.

Morris, N. M., “Control Engineering”, McGraw-Hill Book Co., (UK) Ltd., 1968.

Norvelle, F. B., “Electro-Hydraulic Control Systems”, Prentice Hall, Upper Saddle River, NJ, 2000.

International Organization for Standardization, ISO 4572, Geneva, 1972.

International Organization for Standardization, ISO 4406, Geneva, 1992.

Johnson, J. E., “Electro-Hydraulic Servo Systems”, Penton, IPC, Cleveland, OH, 1977.

Merritt, H. E., “Hydraulic Control Systems”, John Wiley & Sons, New York, NY, 1967.

Mannesmann Rexroth, GmbH, “The Hydraulic Trainer: Proportional and Servo Valve Technology”, Lohr am Main, Germany, 1986.

Maskrey, R. H., Thayer, W. J., “A Brief History of Electro-Hydraulic Mechanisms”, ASME Journal of Dynamic Systems Measurement and Control, New York, NY, 1978.

Reliance Motor Control, “Electro-Craft Handbook – Brush-less Servo Systems”, Racine, WI, 1992.

Shikano, M., “Application Technology on Control Use of Actuators”, Encyclopedia of Factory Automation, volume 6, 1983.

Shimazu Manufacturing Co., “Shimazu Gear Motor”, Shimazu Catalogue.

Takenaka, T., “Hydraulic Engineering”, Yokendo, 1970.

Tonyan, M. J., “Electronically Controlled Proportional Valves”, Marcel Dekker, New York, NY, 1985.

Tuji, S., “Pneumatic Engineering”, Asakura Shoten, 1973.

Vickers Inc., “Vickers Industrial Hydraulics Manual”, Rochester Hills, MI, 1992.

Zeines, B., “Automatic Control Systems”, Prentice-Hall, Inc, Englewood Cliffs, NJ, 1972.

Part II
Turbo-Machinery Types

6

CHAPTER

TURBINES FOR AIRCRAFT PROPULSION

6.1 INTRODUCTION

Turbojet, turbofan, turbo-shaft and turboprop engines drive a plethora of aircrafts ranging from huge passenger and cargo planes to helicopters and cruise missiles, come in every conceivable size and shape and fly to the remotest corners of the world. The engines have amassed a tremendously desirable record of reliability, flying billions of passengers safely and comfortably at speeds that were hitherto unheard of during the first half of the previous century. Newer gas turbine engines power the aircraft with ever reducing noise and pollution levels while burning lesser fuel, and are easily the most significant technological achievement in the history of aviation.

Turbojet engines, with their relative simplicity of design and construction, are predecessors of the later breed of turbine propulsion systems for aircrafts. Basically comprising of a multi-stage compressor to pressurize air, a combustor to burn fuel and a multi-stage turbine, the jet engine function is facilitated by an intake to channel smooth flow at the front and by an exhaust system at the rear.

The power to propel the aircraft is obtained from the reactive forces of the exhaust gases in turbojet and turbofan engines. A large percentage of airflow in the turbofan principle does not pass through the main engine, and hence comes under the classification of bypass engines.

Thermodynamic laws control the conversion of thermal energy into mechanical energy, while establishing the limit on efficiency that depends on the temperature at which heat is added to and removed from the gases. The laws of mechanics, on the other hand, determine propulsive efficiency, mostly because it is based on the conversion of two forms of mechanical energy. Defined by the ratio of thrust power delivered to the aircraft and mechanical power delivered to engine mass flow, in principle it may approach unity.

The engine that emerged as the most practical and dependable for aircraft use from before World War I until after World War II was the conventional piston, or reciprocating, engine; today many light and medium-light aircraft use this engine. The gas turbine engine has advanced to the stage where it has proved to be a superior power plant for both military and commercial aircraft for which high power and high speed are important.

6.2 MAJOR CONSIDERATIONS IN AVIATION

Aviation turbine engines generate thrust by accelerating the flow of air through the system, with energy increase accomplished in two consecutive steps. Airflow is pressurized by mechanical power in the compressor and heated in the combustor by the addition of fuel. The energized gas molecules have enough energy to turn the turbine and compressor rotor, then discharged in the exhaust nozzle to convert the remaining heat energy for further acceleration. Gas is then ejected into the atmosphere at a high velocity to generate thrust.

A unique feature of all aviation related hardware, and especially the engines, is the weight requirement. In order for an aircraft to have the capacity to lift off from the ground and to stay aloft in the air, engine system weight must be as low as physically and operationally possible. This requirement poses interesting and considerable ramifications when compared with power generation and other land based turbines. All components, whether rotating or stationary, must be thin walled to maintain weight and mass inertia properties within set limits. From engineering, manufacturing and operation aspects this calls for unique solutions, even necessitating development of brand new technologies. Because the parts have small wall thickness throughout, they lack adequate stiffness and have low natural frequencies of vibration, making them susceptible to metal fatigue. And since aircraft engines operate in a large speed range, synchronous vibrations must be determined and provided for to ensure structural integrity under many different load conditions.

A number of super alloys possessing the right combination of strength and creep resistance are used extensively. Machining of thin-walled aircraft engine parts pose a special problem, requiring extra precautions to achieve the specified dimensions while still avoiding gouging due to improper cut size, speed or feed rate. Special cutting tools to withstand the tougher machining traits of super alloys require frequent replacement due to excessive wear, even when liberal amounts of cooling fluid are dispensed.

The philosophy behind the design and configuration of bearings used in aircraft engines very well illustrates the minimum weight concept. Hydrodynamic fluid film bearings are the norm in most other turbine designs. However, they tend to be axially long in order to provide the right length to diameter ratio. If they were to be used in aircraft engines, this requirement would translate into a longer and consequently heavier engine. Specially designed rolling element bearings are the rule for aircraft power plants. Ball bearings capable of withstanding axial thrust and radial loads are used near the fan. Roller bearings are used at other locations. Three bearings are commonly required for each spool of the engine; a two-spool system would then call for six main engine bearings. When compared with equivalent power generation turbines, the weight of an aircraft engine's rotor weight is a small fraction of its industrial counterpart. This factor certainly facilitates the design of rolling element bearings when static load carrying capacity is evaluated. Dynamic loads due to rotating unbalance in engines for larger applications can be considerable, mostly because of the presence of vibration modes within the operating speed regime and because they operate at high and varying speeds.

Hydrodynamic journal bearings have a distinct advantage over rolling element bearings in that the lubricant film provides damping to dampen lateral shaft motion. Rolling element bearings inherently have a low coefficient of friction, and hence they dissipate less mechanical energy. But it also means less capacity to dampen vibratory shaft motion. In order to overcome this drawback, damping is introduced in the form of squeeze film dampers. Working in conjunction with the bearing, the damper is equipped with a squirrel cage or other form of centering device to provide the proper stiffness and damping characteristics.

6.3 MISSION REQUIREMENTS

Except for short durations during engine start up and shut down, power generation turbines operate at a near constant speed in order to permit generation of electric power at 50 or 60 Hz. An aircraft engine must go through flight idle, ground idle and cruise speed before the takeoff speed is reached. Thus, airplane engines have a large operating speed range, running into several thousand rpm.

In a two-spool construction the high pressure rotor operates independently of the low-pressure rotor. The two rotors are aerodynamically connected, without any form of mechanical coupling. Three-spool engines are also manufactured, and offer the benefits of improved aerodynamics and component matching.

6.4 FIGHTER JET ENGINE

A single-spool (or single-shaft) turbojet is the most basic form of jet engine, comparable to the simple-cycle power generation engine. Air enters through an intake, and is compressed by an axial compressor while also raising its temperature. The hot compressed air exits the compressor into a diffuser to lower the flow velocity while further raising its pressure. It then enters the combustion chamber, where it is mixed with fuel and combusted, resulting in heavily elevated temperature of the compressed gases. The very hot and compressed combustion gases then enter a turbine section, where thermal energy is converted to mechanical work, with the pressure and temperature difference between the turbine inlet and outlet increasing as the gases flow over the stationary and rotating rows of airfoils. A portion of the mechanical power developed in the turbine drives the compressor, completing the engine cycle.

The compressor and turbine of a turbo-jet engine taken together form a net-zero mechanical energy system, so all the mechanical shaft power developed in the turbine is consumed by the compressor. The net output of a turbojet is not shaft power but kinetic energy of the jet exhaust itself. Although the expansion process in the turbine reduces the pressure and temperature of the gas, there remains considerable thermal energy and pressure in the gases leaving the turbine. These energy forms are partly converted into kinetic energy by expansion to ambient pressure through a propelling nozzle, thus forming a high-velocity flow to provide reactive propulsion.

After the end of World War II, two-spool turbojets came into being to make it easier to throttle back compression systems with a high design overall pressure ratio between the combustor inlet pressure and the intake delivery pressure. The two-spool layout permits the compressor to be split into a low-pressure (LP) compressor and high-pressure (HP) compressor. Each compressor is mounted on separate co-axially running shafts, driven by its own HP and LP turbines. Aside from the geometric complexity, the two-spool turbojet operates much like a single-spool engine.

6.5 TURBINE DRIVEN PROPELLER ENGINE

Turbine driven propeller engines, also called turboprop engines, are gas-turbine engines that deliver almost all of their power to drive the propeller. The gas turbine is designed specifically for this application. The engine's exhaust gases contain little energy compared to a jet engine, and play only a minor role in the propulsion of the aircraft. Limited expansion of the gases also occurs in the propelling nozzle prior to exiting to the atmospheric. The propelling nozzle provides a relatively small proportion of the overall thrust.

Power is transmitted from the turbine output shaft to the propeller shaft through a speed reduction gear. While the power turbine may be integral with the gas generator section, many turboprops feature a free power turbine on a separate coaxial shaft. This enables the propeller to rotate freely at a near constant speed and independent of the compressor speed.

While most turbojet and turbofan engines rely on axial-flow compressors for compressing the air, turboprop engines usually have at least one centrifugal compressor stage. Centrifugal compressors have the advantage of being simple and lightweight, but at the expense of a streamlined shape.

The propellers tend to be inefficient when the tips reach or exceed the speed of sound. A reduction gearbox is placed in the driveline between the power turbine and the propeller to allow the turbine and the propeller to operate at their most efficient speed. The reduction gear may be a combination of helical gears and a planetary gear system with 4 or more planet gears walking around the sun gear. The arrangement gives the exterior of the gearbox the shape of a tear drop. While reducing the speed, the reduction gear converts the low torque output to a higher level in the same proportion as the speed is reduced.

Turboprops are popular for small or slow aircraft for short-distance commuter airliners since they yield better fuel efficiency at lower speeds. Medium military transports and patrol planes such as the C-130 Hercules and P-3 Orion also benefit from their high takeoff performance and mission endurance. Similar to reciprocating propeller engines, turboprop engines can be provided with controlled variable pitch propellers so thrust may be adjusted independently of the engine speed. The variable pitch propellers can also be used to generate negative thrust while decelerating on the runway. Additionally, in the event of an engine outage the pitch can be adjusted to a vaning or feathering pitch to minimize drag of the non-functioning propeller.

Though turboprop engines are favored on small subsonic aircraft, some aircraft outfitted with turboprops have cruising speeds in excess of 500 kt (926 km/h, 575 mph). Large military and civil aircraft, such as the Lockheed L-188 Electra and the Tupolev Tu-95, have also used turboprop power. The Airbus A400M is powered by four Europrop TP400 engines, which are the third most powerful turboprop engines ever produced, after the Kuznetsov NK-12 and Progress D-27. Commercial aircraft with turboprop engines include the Bombardier Dash 8, ATR 42, ATR 72, BAe Jetstream 31, Embraer EMB 120 Brasilia, Fairchild Swearingen Metroliner, Saab 340 and 2000, Xian MA60, Xian MA600 and Xian MA700, Fokker 27, 50 and 60.

Turboprops have superior specific fuel consumption at flight speeds below 450 mph (390 knots; 725 km/hr) because the jet velocity of the propeller and of the exhaust is relatively low. They are widely used where high-performance short-takeoff and landing (STOL) capability and efficiency at modest flight speeds are essential. Turboprop driven airliners now operate at near the same speed as small turbofan-powered aircraft but burn two-thirds of the fuel per passenger. In comparison to a turbojet flying at high altitudes, enhanced speed and better fuel efficiency, a propeller aircraft operates at a much lower ceiling.

6.6 HIGH BYPASS TURBO-FAN ENGINE

Turboprop engines have better performance characteristics at moderate flight speeds, and the turbojet engines at very high speeds. But a middling speed range exists where the turbofan engine, also referred to as the ducted fan engine, finds its place.

Modern turbofans evolved from the two-spool axial-flow turbojet engine essentially by increasing the relative size of the low-pressure (LP) compressor. A considerable amount of air exiting the unit actually bypasses the core, or gas-generator, stream, and passes through the main combustor. Civil-aviation high-bypass turbofans usually have a single large fan disk, whereas most military-aviation low-bypass turbofans such as combat and trainer aircraft applications have multi-disk compressors. This acts as a compromise between greater power-to-weight ratios, supersonic performance, and the capability of using afterburners, versus the higher fuel economy of a high-bypass design. Modern military transport turbofan engines are virtually identical to their civilian counterparts.

A turbo-shaft engine may be made up of two major parts assemblies: the 'gas generator' and the 'power section'. The gas generator consists of the compressor, combustion chambers with igniters and fuel nozzles, and one or more stages of turbine. The power section consists of additional stages of turbines, a gear reduction system, and the shaft output. The gas generator creates the hot expanding gases to drive the power section. Depending on the design, the engine accessories may be driven either by the gas generator or by the power section.

An unusual example of the turbo-shaft principle is the Pratt & Whitney F135-PW-600 engine for the STOVL F-35B. In the conventional mode it operates as a turbofan, but when powering the lifting fan, it switches partially to turbo-shaft mode to send power forward through a shaft like a turboprop and partially to turbojet mode to continue to send thrust to the rear nozzle.

The ratio of the mass-flow of air bypassing the engine core compared to the mass-flow of air passing through the core defines the bypass ratio. The engine produces thrust through a combination of these two portions working in concert. Engines using more jet thrust relative to fan thrust are known as low bypass turbofans, while those that have considerably more fan thrust than jet are known as high bypass. Most commercial aviation jet engines in use today are of the high-bypass type, but most modern military fighter engines are low-bypass.

Since most of the air flowing through a high-bypass turbofan has a low-velocity, even when combined with the much higher velocity turbine exhaust, the net average exhaust velocity is considerably lower than in a pure turbojet. Engine noise is largely a function of the exhaust gas velocity, hence turbofan engines are significantly quieter than a pure-jet of the same thrust. Other factors include turbine blade and exhaust outlet geometries, such as noise-reducing 'chevrons' seen on the Rolls-Royce Trent 1000 and General Electric GEnx engines used on the Boeing 787.

Since the efficiency of propulsion is a function of the relative airspeed of the exhaust to the surrounding air, propellers are most efficient for low speed, pure jets for high speeds, and ducted fans in the middle. Turbofans are thus the most efficient engines in the range of speeds from about 500 to 1000 km/h (310 to 620 mph), the speed at which most commercial aircraft operate. Turbofans retain an efficiency edge over pure jets at low supersonic speeds up to roughly Mach 1.6, but have also been found to be efficient when used with continuous afterburner at Mach 3 and above.

The majority of turbofans follow the same basic configuration, with a large fan at the front of the engine and a relatively small jet engine behind it. There have been a number of variations on this design, however, including rear-mounted fans that can easily be added to an existing pure-jet design, or designs that combine a low-pressure turbine and a fan stage in a single rear-mounted unit.

Economy of operation is the outstanding characteristic of high bypass turbofan engines for flight velocity of Mach 0.8, equivalent to 500 knots at 40,000 feet altitude. The configuration also offers advantageous efficiencies at high subsonic cruise conditions required for commercial airplanes, and is equally important for long range combat aircraft cruise under the sound barrier.

The turbine section converts more mechanical energy from the hot gases than is needed to drive the compressor, hence the excess energy powers the fan located upstream of the main compressor. A portion of the air entering the main intake splits from the main flow after passing through the fan to bypass the main engine, expanding in a separate diffuser to furnish its own thrust. The bypass ratio determines the split between airflow in the core engine and the fan's exhaust duct. Bypassing has the distinction of better fuel consumption characteristics over engines without this feature of comparable thrust. Lower values of bypass ratio, in the range of 0.2:1 to 1:1, are more convenient for supersonic combat airplanes. Higher values of 5:1 are more suitable for higher capacity wide body airplanes such as Boeing 747 and Airbus A300. The size of the C-5A Galaxy transport aircraft is so large that its stipulated propulsion characteristics of increased thrust at takeoff and low fuel consumption at cruise can only be satisfied by the use of four high bypass thrust engines operating at high turbine inlet temperature. Higher thrust levels are achievable, particularly during takeoff, as a consequence of acceleration of a larger mass of bypass air, since the flow through the remainder of the engine may contribute as little as 15 per cent of the combined total. Fuel is burned more efficiently by increasing the bypass ratio at the same flight speed, although the turboprop version is more economical to operate. Another benefit accrues from lower noise levels due to the low exit velocities of the propulsive jets. Figure 6.1 provides a cross sectional layout of Pratt and Whitney's 2000 series of engine.

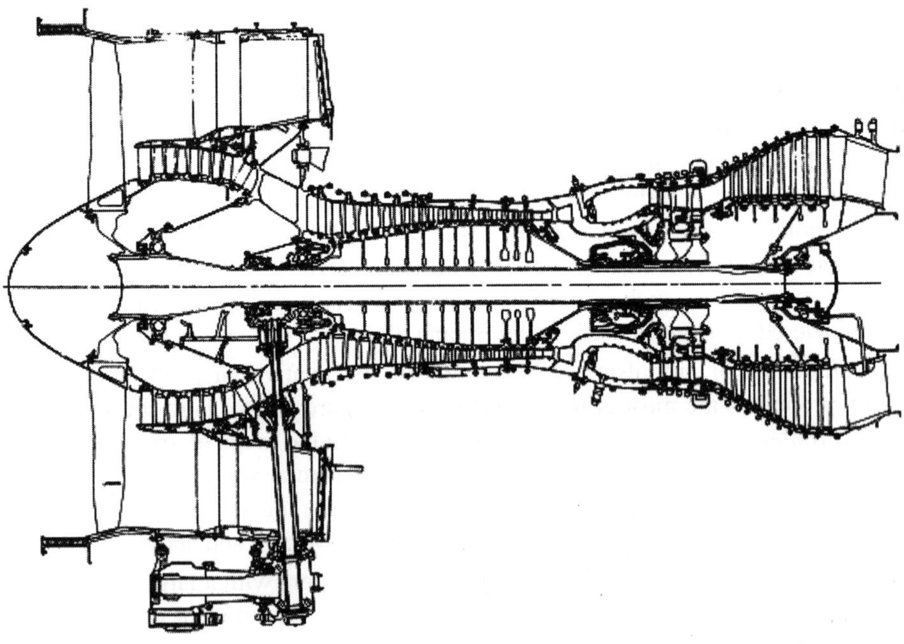

Fig. 6.1 Pratt and Whitney's 2000 Series Engine Enclosed in Nacelle

6.7 HELICOPTER ENGINE

Although jet engines, or gas turbine engines, are usually thought of as thrust, or force, producing engines, they are also used to drive rotors or propellers or for combination of thrust production and drive power

(Figure 6.2). Turbo-shaft and turbo-prop engines are basically jet engines that do not produce thrust. Power reserved for producing thrust is converted into mechanical power to drive rotors or propellers. Turbine blades can be configured so that they change thrust into large amounts of power to drive an output shaft. Alternatively, the blades can allow most of the thrust to be used as propulsion and take certain proportion to drive the compressor section of the engine.

Overhead rotating propellers produce aerodynamic thrust to permit helicopters to stay aloft in the air. Helicopters use turbo-shaft jet engines to power their rotors. The UH-1 Huey, the Bell Jet Ranger, and the AH-64 Apache are some helicopters that derive their power from gas-turbine engines. Older and some smaller helicopters have used piston engines for their power source, but most new versions rely on gas turbine engines because of the superior ratio of power and weight. From considerations of reliability and maintainability, gas turbine engines also are more economical to operate. Failure rates for gas-turbine engines tend to be low since there are fewer moving parts.

Fig. 6.2 Helicopter in Flight

Figure 6.3 shows the arrangement of modules in a T-53 turbo-shaft engine for a helicopter. Air enters the inlet at the forward end as it proceeds to the radial compressor section, then to the diffuser, combustor and turbine sections and finally to the exhaust at the aft end. A shaft from the turbine transmits power to the output shaft at the compressor end of the engine.

Power is transferred from the engine's output shaft to the helicopter's overhead propellers through the transmission and main gearbox. A combination of helical and bevel gears are employed to transfer power from the essentially horizontally mounted engine's shaft axis to the vertically oriented propeller rotor axis. The speed is reduced from the turbine speed of several thousand rpm to a few hundreds of RPM in the overhead propellers. In the process the torque is multiplied in the same proportion as the speed is slowed to an acceptable level for the helicopter's propeller system. The transmission drives the mast, which directly turns the rotors. Another shaft branches out of the transmission to directly drive the tail rotor. The tail rotor develops the right level of thrust to balance the counteracting torque produced in the helicopter's structure as the main propellers turn. The tail rotor thus provides equilibrium, and

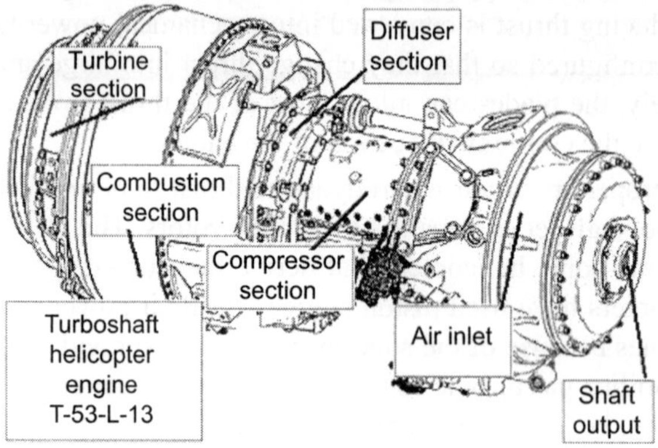

Fig. 6.3 Helicopter Turbo-shaft Engine

stabilizes the vehicle to ensure the helicopter does not go into a spin from the counteracting torque. An accessory gearbox mounted on the engine draws power to drive the fuel and oil pumps, electrical generator and fuel control mechanism for the engine itself.

6.8 TILTED ROTOR AIRCRAFT

A tilt-rotor aircraft combines the vertical lift capability of a helicopter with the speed and range of a conventional fixed-wing aircraft. The aircraft is provided with a pair or more of powered prop-rotors mounted on swiveling shafts or nacelles at the end of the fixed wing for lift and propulsion. For vertical flight, the prop-rotors are angled so the plane of rotation is horizontal, lifting in the same manner as a helicopter rotor does. As the aircraft gains speed, the prop-rotors are progressively tilted forward, with the plane of rotation eventually becoming vertical. In this mode the wing provides the lift, and the rotor provides thrust as a propeller. A tilt-rotor aircraft differs from a tilt-wing in that only the rotor pivots rather than the entire wing. This method trades off efficiency in vertical flight for efficiency in short takeoff and landing (STOL) and short takeoff and vertical landing (STOVL) operations.

The rotors can be configured to be more efficient for propulsion, for instance with twist at the root-tip. Also, it avoids a helicopter's issues of retreating blade stall, so the aircraft is capable of achieving higher speeds than a helicopter. The Bell Boeing V-22 Osprey is a multi-mission, military, tilt-rotor aircraft with both a vertical takeoff and landing (VTOL), and STOL capability (Figure 6.4). It is designed to combine the functionality of a conventional helicopter with the long-range, high-speed cruise performance of a turboprop aircraft.

The Osprey is provided with a three-bladed prop-rotor, turboprop engine and transmission nacelle mounted at the two tips of wing. Under Federal Aviation Administration classification it is a powered lift aircraft. For takeoff and landing, it typically operates as a helicopter with the nacelles vertical and rotors horizontal, although autorotation characteristics are poor partly because the rotors have low inertia. Once airborne, the nacelles rotate forward 90° in 12 seconds for horizontal flight, converting the V-22 to a more fuel efficient, higher speed turboprop aircraft. STOL rolling-takeoff and landing capability is achieved by having the nacelles tilted forward up to 45°. For compact storage and transport, the V-22's wing rotates to align front-to-back with the fuselage. The prop-rotors can also be folded. Composite materials make up 43% of the V-22's airframe. The prop-rotors blades also use composites.

Fig. 6.4 V-22 Osprey Tilt-rotor Aircraft

The V-22's two Rolls-Royce AE 1107C engines are connected by drive shafts to a common center gearbox so that one engine can power both prop-rotors in the event of an engine failure. However, if a prop-rotor gearbox fails, that prop-rotor cannot be feathered. Both engines must then be stopped before an emergency landing. Most Osprey missions will use fixed wing flight 75 per cent or more of the time, reducing wear and tear on the aircraft and reducing operational costs. This fixed wing flight is higher than typical helicopter missions. The V-22 design loses 10% of its vertical lift over a tilt-wing design when operating in helicopter mode because of airflow resistance due to the wings, but the tilt-rotor design has better short takeoff and landing performance. Wash from the rotor usually prevents usage of the starboard door in hover, and the rear ramp is used for rapelling and hoisting.

6.9 AUXILIARY POWER UNIT

The primary purpose of an aircraft auxiliary power unit (APU) is to provide power to start the main engines. Turbine engines must be accelerated to a high rotational speed in order to provide sufficient air compression for self-sustaining operation. Smaller jet engines are usually started with an electric motor, while larger engines are usually started with an air turbine motor. Before the main engines are to be turned, the APU is started, with a battery or a hydraulic accumulator. Once the APU is running, it provides power to start the aircraft's main engines. Depending on the design, APU's are capable of producing electric, pneumatic or hydraulic power.

APUs may also be used to run the main engine's accessories while they are shut down. This allows the cabin to be comfortable while the passengers are boarding before the aircraft's engines are started.

Electrical power is used to run systems for preflight checks. Some APUs are also connected to a hydraulic pump to allow the crew to operate hydraulic equipment such as flight controls or flaps prior to engine start. This function can also be used on some aircraft as a backup in flight in case of engine or hydraulic failure. Aircraft with APUs can also accept electrical and pneumatic power from ground equipment if an APU has failed or is not to be used.

APUs fitted to extended-range twin-engine operations (ETOPS) aircraft are a critical safety device, as they supply backup electricity and compressed air in place of the failed generator of the main engine. While some APUs may not be started during flight, ETOPS-compliant APUs must be capable of starting during flight at altitudes up to the aircraft's service ceiling, which may reach up to up to 43,000 ft (13,000 m) from a complete cold-soak condition. Hamilton Sundstrand's APS5000 for the Boeing 787 Dreamliner is one such APU. If the APU or its electrical generator is not available, FAA requirements do not permit the aircraft to be released for ETOPS flight, and is then forced to take a longer non-ETOPS route.

Fig. 6.5 Honeywell-APU-Airbus

APUs providing electricity at 400 Hz are smaller and lighter than their 50/60 Hz counterparts, but are costlier; the drawback being that such high frequency systems suffer from voltage drops.

The load compressor is generally a shaft-mounted compressor, and provides pneumatic power for the aircraft, though some APUs extract bleed air from the power section compressor. There are two actuated devices, the inlet guide vanes that regulate airflow to the load compressor and the surge control valve to maintain stable or surge-free operation of the turbo machine. The third section of the engine is the gearbox. The gearbox transfers power from the main shaft of the engine to an oil-cooled generator for electrical power. Within the gearbox, power is also transferred to engine accessories such as the fuel control unit, the lubrication module and the cooling fan. In addition, there is also a starter motor connected through the gear train to perform the starting function of the APU. Some APU designs use a combination starter and generator for APU starting and electrical power generation to reduce complexity. Figure 6.5 shows an APU unit on a maintenance stand.

On the Boeing 787 aircraft, the APU delivers only electricity. The absence of a pneumatic system simplifies the design, but the higher demand for electricity requires a heavier generator. Onboard solid oxide fuel cell (SOFC) APUs are being researched. Figure 6.6 shows exhaust nozzle of an APU at the tail end of an Airbus A380 airplane.

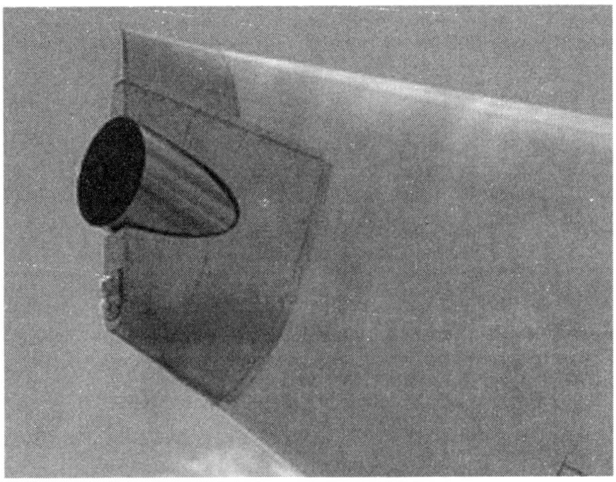

Fig. 6.6 Auxiliary Power Unit Exhaust at Tail End of Airbus A380

6.10 FAN AIRFOILS

Of particular interest in high bypass turbofan engines are the fan blades. General Electric's CF6-6 engine fan uses 38 titanium blades to generate a considerable portion of the total thrust. Weight and stiffness (and hence natural frequency) control of the blades may be achieved by the simple expediency of drilling holes in the tip. Due to the large fan diameter, the root of the blade may be operating in the subsonic regime, but tip peripheral velocity may exceed the speed of sound during maximum thrust conditions. A circular profile for the tip section of the airfoil aids in meeting the more rigorous demands of the supersonic flow, but conventional airfoil shapes are employed at the hub to serve the intended purpose.

Air drag can be limited by reducing the outer diameter of the fan, and hence a smaller hub diameter is desirable. A varying distribution of energy along the length of the blade, reaching a maximum at the tip, aids in achieving efficiency and stall margin targets. A key consideration in the control of vibrations, dynamic stresses and operating life is the blade length. Mid-span shrouds promote structural integrity of the blades, and are also advantageous in the event of a bird strike. However, the shrouds come with a penalty in efficiency, since they act as an obstruction for the airflow. Erosion from rain, hailstone and ingestion of ice are other factors to be kept in mind in the design of fan blades. Close matching of dimensions, both in the blade and in the mating disk dovetail, is of special significance because of the high centrifugal loads. The problem becomes even more acute since the blades must permit easy removal in the field. Fan disks must be designed to withstand axial, radial and tangential direction loads arising from aerodynamic and centrifugal forces on the blades. Titanium has proved to be a reliable material for the disk because of its advantageous strength and weight characteristics. The fan stage maybe followed by one to four low pressure compression stages, located just aft of the split in the air stream for bypass engines. Figure 6.7 provides details of the fan module in a high bypass turbofan engine.

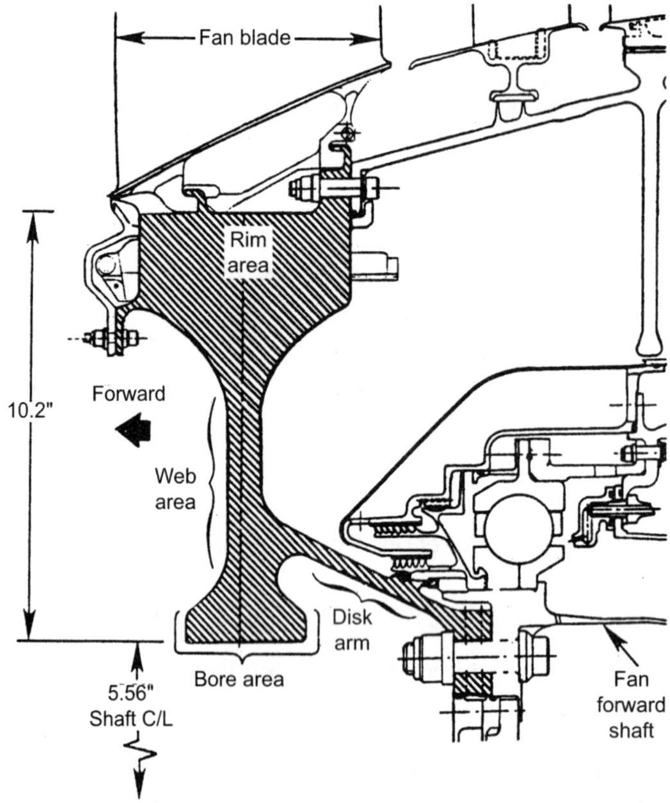

Fig. 6.7 CF6-6 Engine Fan Disk (NTSB/AAR-90/06 Report)

6.11 COMPRESSOR ROTATING BLADES AND STATIONARY VANES

Axial and radial centrifugal compressor types find wide acceptance in aircraft engines. Smaller engines can easily take advantage of the centrifugal impeller's requirement of reduced flow area at the inlet to obtain a pressure ratio as high as 5:1 in one stage, while operating at nearly 80 per cent efficiency. The rotating impeller's mechanical energy is used to create centrifugal forces in the air stream. Engines are also designed with a row of two to four axial stages followed by a centrifugal stage. Two impellers turning at a high speed, exceeding 30,000 rpm, may also be employed in an opposed arrangement to increase the flow capacity.

In a radial compressor air enters the eye of the impeller axially, then goes through a 90° rotation to exit at the periphery into the diffuser to go through another 90° turn, before discharging into the manifold. Adequate precautions are needed to ensure flow separation does not occur in the flow passage and to avoid the associated losses. The diffuser transforms the high flow velocity into pressure head through the gradually increasing cross sectional area of the passages within.

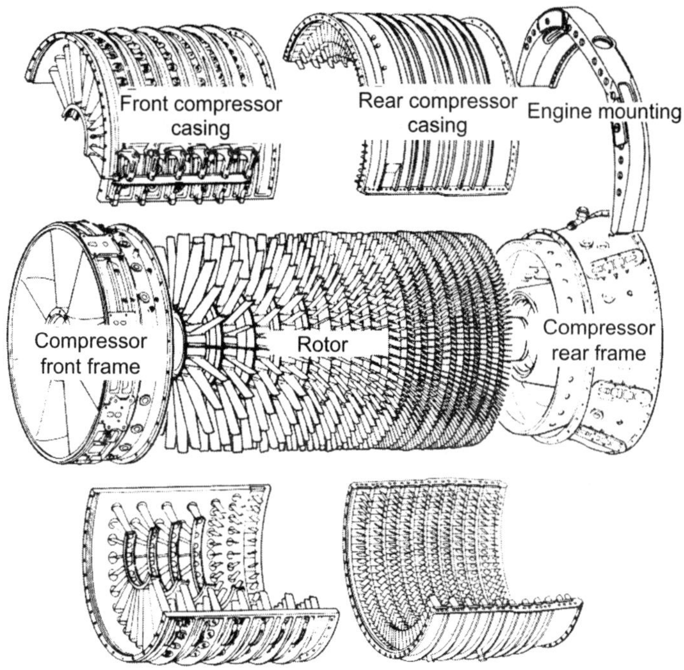

Fig. 6.8 Axial Compressor Module of General Electric J79 Engine (Hunecke, 1997)

Most engines, however, rely on the axial compressor for air pressurization to handle the large mass flow rate necessary for high levels of compression and thrust generation. Since flow direction is mostly axial, turning of flow direction is eliminated. Also, the consistent external dimensions in an axial flow machine help hold down aerodynamic drag. Axial compressors rely on the principle of lowering stream velocity to raise pressure, and hence are more susceptible to flow fluctuations.

A considerably larger number of parts go into the building of an axial compressor, thus increasing the complexity of design and manufacturing cost. Major components are support frames at the front and rear ends, an external casing with stator vanes attached to it and a bladed rotor.

The front support frame in a turbofan engine is made of eight to twelve radial struts attached at the inner radius to a hub and to a case at the outer radius. The hub provides accommodation for bearings for the fan and compressor rotors. Bypass engines are provided with an intermediate splitter ring to provide the inner flow path for air going through the fan duct. Bearing housings are attached to the frame. Since load-carrying requirements of the low-pressure rotor are high, especially in the event of a blade loss, two bearings are provided at the front end, one of them close to the fan. A housing for the high-pressure compressor rotor's bearing is attached to the aft side of the frame. Casing for the compressor also is bolted on this end. An accessories package is attached at the bottom of the engine, and is driven by a radial shaft going through a frame strut, power being taken from the compressor rotor through bevel gears.

Variable stator vanes find extensive usage in aircraft engines, primarily because of their large operating speed range. Starting with the inlet guide vanes, the next three to five stages may be equipped with vanes that may be set at a particular angle. Holes must be drilled in the casing for individual vanes, with a bushing provided to permit free rotation of the vanes. An actuating mechanism using many different parts is required for setting of the angle in the vanes for all the stages. The setup is complex, but it provides handsome returns in the form of improved efficiency.

The compressor case has a cylindrical geometry with an axial split to permit installation of the rotor. Made out of lighter weight titanium or a stainless steel, the material must be capable of maintaining radial clearance at blade tips due to thermal growth during operation. Fixed stage vanes are inserted as an assembly in T shaped grooves machined in the casing.

Since the engine's front mount in bypass engines is located at the compressor case forward flange, engine thrust forces will be reacted at this point for eventual transfer to the pylon and aircraft wing. Note also that the mount arrangement is mostly at the top of the casing, so thrust force along the engine centerline will set up bending moments in the casing. Some compressed air is nearly always bled and transported through manifolds for cabin pressurization, de-icing and other assorted tasks.

A combination of a drum and discs is the common form of rotor construction, with a shaft segment at both ends to carry the bearings. Disks may be attached at either extremity of the drum by spacer rings. Rotor blades are twisted in order to get the right flow inlet angle along the full length. Variation in magnitude of the inlet angle arises because the blade has a much lower tangential velocity at the root than at the tip, while axial flow velocity must be maintained along the full face. Blade height reduces gradually as the airflow progresses through the stages, in nearly the same proportion as the pressure rises. Blade roots may be of the dovetail shape, or of the fir tree type if the loading is considerable. Attachment of the blade to the disk must be firm, but still permit room for growth during operation. Longer blades tend to move a little in the seat, but lock under the action of centrifugal forces during operation (*ref*: Rangwala, 2004). Axial entry of the dovetail in the disk or drum grooves is common in forward stages, but circumferential insertion is favored for the latter stage blades. Blade retainers may be in the form of a key or a fitted bolt.

The rear frame of a compressor is required to transfer the compressed air stream to the combustor, and so its configuration depends on the burner system type to be used in the engine. Cross section area gradually is made to increase in the direction of the flow in order to reduce stream velocity and increase pressure. The center portion of the frame houses the gas generator bearings for absorbing axial and lateral direction loads. Frame struts must also accommodate tubes for oil lubrication and venting. If the primary engine mount is located at the compressor rear frame, as is the case in many combat aircraft, structural integrity of the frame under the action of engine operating loads is of significance.

Operation of the engine during cruise is set by the equilibrium operating line. Acceleration of the engine is more crucial, since more fuel is being injected and burnt in the combustor. Turbine inlet temperature consequently rises, so engine components downstream of the compressor are momentarily accepting less airflow. This will have the effect of raising compressor pressure ratio and discharge pressure, and can be detrimental to the compressor if the spike in generated power causes the compressor to surge. Thus, a margin is necessary between the operating and surge lines to prevent the compressor from surging, and is usually kept at 20 per cent. But a generous surge margin may not be achievable over the full operating range of the compressor. At low corrected engine operating speed steady state operating line will likely come near the surge contour. Remedial action in the form of modulated airflow by adjusting the stator vanes or by bleeding mid-stage compressor air should help alleviate the situation. Lapses in surge margin are likely to take place either at very low or high corrected speeds, which in turn are caused by very low or high temperature at compressor inlet.

6.12 ANNULAR COMBUSTOR CONFIGURATION

Aviation engines rely on annular configuration of the combustor for burning the liquid fuel since it is lighter in weight than the can type preferred in most industrial and power generation turbines. Lack of radial structural members, however, tend to reduce structural strength of the combustor inner and outer liners, or flame tubes. Made from a cobalt based alloy, the liners hold the elevated temperature product of combustion of fuel and air. Figure 6.9 provides details of one design of the annular combustor. The flange between the outer combustor and turbine casings is scalloped to reduce weight.

After going through compression, air passes through the diffuser where the gradually increasing cross-sectional flow area reduces the flow velocity while increasing the pressure. The reduced flow velocity prevents the air from rapidly flowing through to permit greater combustion of the fuel entering from the fuel feed and vaporizing tubes. Nearly complete combustion of fuel is obtained by promoting turbulence as the air enters the combustion chamber through swirler cups placed at its entry. The fuel droplets then more thoroughly mix with the air to obtain total combustion. To prevent gas temperature from rising rapidly, only some of the air is admitted between the combustor liners. The richer mixture of the fuel and air inhibits sudden temperature increase, which leads to reduced levels of NO_x formation. As the flow progresses towards the turbine inlet, secondary and dilution air holes are provided in the liners to admit more fresh air. Thus, combustion takes place in multiple regions at lower temperature levels. At the same time, enhanced combustion reactions ensure other objectionable components of combustion, carbon monoxide (CO) and unburned hydrocarbon particles, are kept to a minimum.

The liners are supported lightly at support points in radial, tangential and axial directions in order to maintain thermal stresses within limits. The liners have flexibility to permit thermal growth axially and radially as the engine is started from the cold to the maximum thrust condition. The liners require thermal barrier coating for protection from the main hot gas stream. Further protection is obtained through a layer of cooler air that circulates around the outer surfaces of both liners. The liners are drilled with holes to permit the coolant to flow over the inner surfaces. The holes are drilled at an angle in order to permit the cool air to spread over the full surface before it joins the main gas stream. The coolant is obtained from air bled from one of the higher pressure compressor stages.

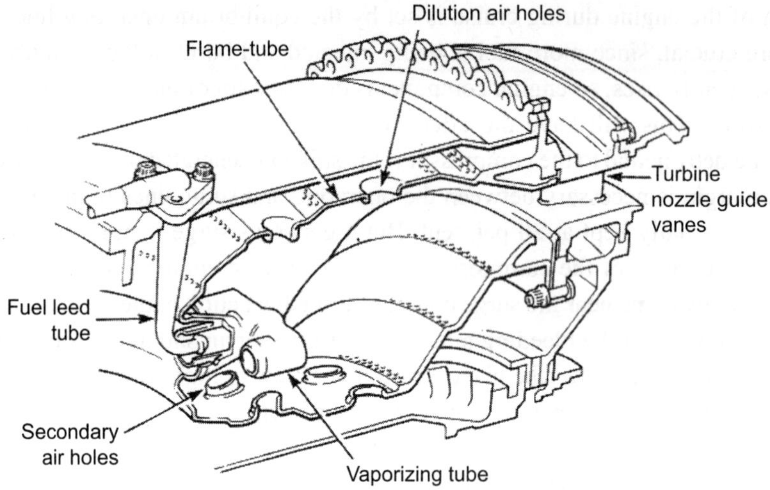

Fig. 6.9 Annular Combustor Design for Aviation Engine

Some other objectives to be achieved in the design of an aviation engine combustor are minimum pressure loss, stability of flame and equalized temperature of gases as they enter the turbine inlet nozzle vanes. Loss of flame during flight can pose significant problems, especially if the pilot is not successful in restarting the engine. Presence of non-uniform temperatures and hot spots in the gas stream as it approaches the turbine vanes can lead to thermal fatigue and premature failures in the liners and in the vanes. Bowing at the trailing edge of the vanes may also occur and cause loss in turbine work out put and efficiency. Fracture of thermal liners and subsequent splitting into pieces will allow the hot gases and flames to spread outside the main engine casing, thereby posing a major fire hazard for the engine and the airplane.

6.13 TURBINE MODULE

Axial turbines are the norm in aircraft power plants because of the higher mass flow capability. Besides meeting mechanical loads of driving the high and low pressure compressors and the fan of a turbofan engine (or the propellers of a turboprop engine), the turbine components must operate in the hostile environment of combustion gases at elevated temperatures. Turbine design will be controlled by the compressor's pressure ratio, quantity of energy to be extracted from the hot gases, rotor speed and maximum permissible diameter. Turbine stage designs tend to be based on a mixed combination of constant pressure and impulse principles. Use of single-crystal and directionally solidified alloys together with creative cooling techniques are of fundamental importance in a successful turbine design.

The rotor assembly consists of disks, blades and shaft segments at both ends. At the compressor end the shaft is of a conical design to permit attachment with the shaft driving the fan rotor, as seen in Figure 6.10 for a GE79 engine. A splined joint with a lock nut is favored for transmission of power from the turbine to the fan. The disks are bolted together, with spacers placed in between the disks. Varying thickness in the disks permits safe distribution of stresses that are caused by the blade centrifugal forces at the periphery of the disk, with the hub axial length being two to three times the length at the rim.

Bore sizing is crucial in that hoop direction stresses peak in the region, and must also permit placement of a vent tube through the center hole. Blades are inserted in machined slots and held in place either by clamps or retaining plates, with passage for cooling air directed towards holes machined in the base of the blade's dovetail.

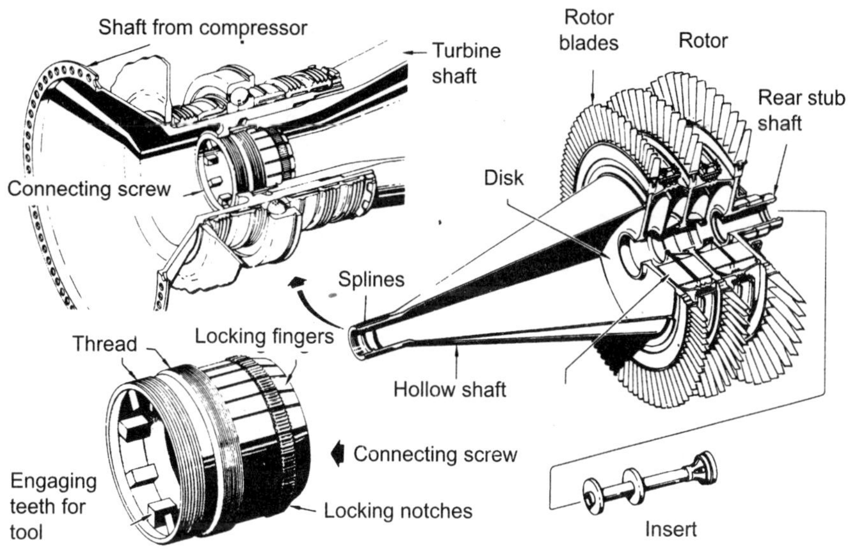

Fig. 6.10 GE79 Turbine Rotor (Hunecke, 1997)

Thermal barrier coating helps increase resistance of the base metal to elevated gas temperatures, while also providing protection from corrosion. A number of cooling methods are made use of to maintain metal temperatures well below those of the gas. Provision of cooling air to the rotating blades is at best a difficult proposition. Cooler air bled from the compressor flows through the inside of blades and vanes, and exit through holes drilled in the leading and trailing edges. Figure 6.11 illustrates the case for a high-pressure turbine blade. Contamination of cooling air from dust particles poses a serious problem, in that the exit holes may get clogged, thus creating hot spots and premature metal fatigue. In this respect, aircraft engines face a totally different situation when compared with industrial land based gas turbines provided with intake manifold filters to keep out the offending contaminants. One solution calls for cooling air to go through sharp turns before entering a swirl separator in which dust particles are whirled in the direction of shaft rotation, thus causing the particles to fly out by centrifugal action. Cooling air then proceeds through traps to separate the particles before entering the passages inside the blade.

Specially designed seals are required to prevent leakage of hot gases in the confined space where a shaft meets a stationary component. Relative thermal growth between the components aggravates the problem. Labyrinth seals have been expressly developed to resolve the problem (Figure 6.12). A number of grooves are machined into the rotor, separated from one another by a ridge with a knife-

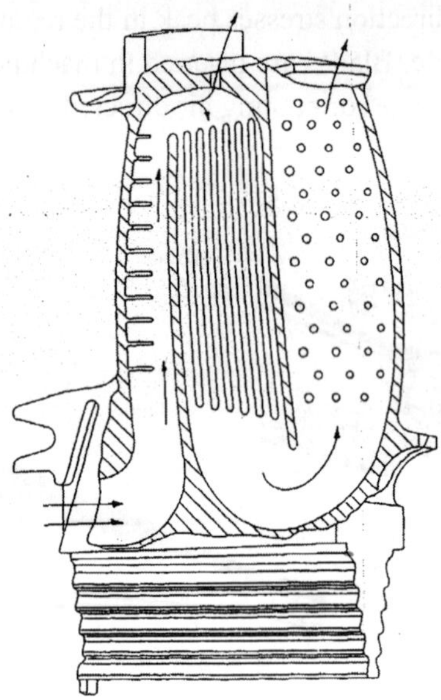

Fig. 6.11 Cooling Provision in High Pressure Turbine Blade

edged crest. The mating location on the stationary part has a flat cylindrical surface. Knife-edges cut into the rub surface of the stator to provide the sealing. When hot gases leak past the gap, throttling reduces the pressure from one knife-edge to the next. At the last location the throttled pressure equals the pressure on the aft side, thus effectively reducing the leakage rate to zero. The method may be applied at the blade tips as well as at the inner inter stage locations. Blade shrouds are designed with the cutting tip to control leakage flow. Softer honeycomb materials are more suitable in the turbine casing.

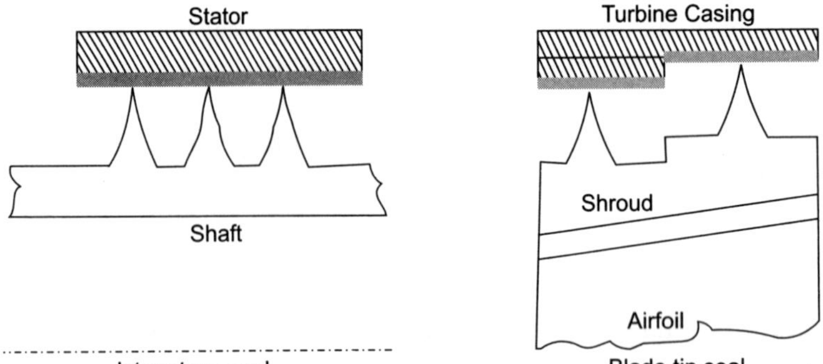

Fig. 6.12 Labyrinth Seal

Horizontally split casing design facilitates servicing of turbine components. On the other hand, General Electric's CF6-80C engine uses a 360° casing, thereby eliminating a number of parts required for joining the upper and lower halves at the horizontal flange. The turbine module is assembled by stacking the nozzle vane and rotor disk assemblies in sequential order. Removal of the complete turbine module and replacing it with another takes care of field related problems, with the module than serviced at the factory. Bore scope holes are conveniently placed, usually for each stage, on the casing shell for inspecting the internal parts of the turbine. Thus, if an engine is experiencing excessive vibrations (*ref*: Rangwala, 2004) or out of range exhaust gas temperatures, the internal inspection with a light probe facilitates diagnosis of the problem.

Control of radial clearances between blade tips and the casing is achieved by blowing cooler air on the external surface of the case in order to shrink the casing. Air is directed from holes in tubes placed around the case in the plane of the rotor stages. In the active clearance control system, generally employed for the gas generator turbine, temperature sensors and valves are employed to control the flow of cooling air during periods of varying temperature operation. Passive cooling systems for the low-pressure turbines merely blow cooler air on the shell. The expense of installing sensors and valves is not required in the latter approach, but the cool air is being continually robbed from the system. In either case performance gains are achieved, because leakage of hot gases past the blade tips is minimized.

6.14 THRUST AUGMENTATION FOR FIGHTER ENGINE

Fighter jets have the unique requirement of oftentimes taking off from a shorter than normal length runway, such as the deck of an aircraft carrier or in a battlefield area with its improper takeoff strip. The engines than have to develop greater thrust to permit the aircraft to become airborne and get into the climb mode.

Additional thrust is obtained by actuating the thrust augmenter, or afterburner, located at the aft end of the engine. Advantage is essentially taken of the exhaust gases exiting from the low-pressure turbine still having oxygen available to burn additional fuel. Special fuel injectors are arranged in the augmenter module to burn the fuel and create additional thrust. The added thrust allows the fighter aircraft to takeoff.

However, the procedure is an inefficient way of producing thrust. After exiting from the augmenter, the gases are at extremely high temperatures just prior to escaping and mixing with the surrounding air. It is not possible to put this additional energy input for some other tasks. The associated thermal energy of the gases is lost to the atmosphere, leading to considerable deterioration of thermodynamic efficiency. Military application of the thrust augmenter may be justified since there is no other way of getting the aircraft in the air where takeoff length is restricted. Commercial application of this method of augmenting the thrust can in no way be acceptable since fuel costs would increase dramatically.

6.15 THERMAL PROTECTION FOR HIGH PRESSURE TURBINE AIRFOILS

The stream of products of combustion discharging from the combustor section of an aviation engine are close to the melting point of the substrate of the turbine inlet vanes and the succeeding rotating stage blades. Thermal protection is required for the airfoils to protect them from thermal distress and to permit them to carry out their stated task. The stationary vanes are required to direct the flow of gases and feed them into the rotating blades at the precise angle of incidence.

Protection comes from two different sources. The airfoils are provided with thermal barrier coating. Made from materials such as oxides of aluminum and zirconium, multiple layers of coating are applied using specialized spraying methods. Care has to be taken to ensure the layers have uniform thickness and thoroughly bond to the metal substrate. The coatings physically isolate the metal substrate from the stream of harsh environment of hot and corrosive gases.

The airfoils also benefit from cool air extracted from the compressor which circulates in cooling passages built within the airfoils. The coolant is admitted into the stationary vanes at the outer diameter and exits through a number of holes machined in the airfoils. Delivery of the coolant to the rotating blades is more difficult, requiring some ingenious methods. Just below the dovetail and the airfoil portion of the blades, specially designed pockets are machined to admit the cool air. A cooling plate covers the dovetails of the blades, which also serves to prevent the blades sliding out due to axial thrust produced as the gases flow over the surfaces of the airfoil. In the vicinity of the admitting pockets of the blades the cooling plate is given a formed shape. Coolant from the compressor is delivered by a stationary tube and nozzle arrangement. As each blade pocket traverses the coolant nozzle, air enters and gets into the airfoil. A complex system of cooling passages is provided within the airfoil (Figure 6.11). A portion of the coolant gets into the passages that serve the leading edge of the blade, with the cool air impinging on the inside of the leading edge through a shower head form of spray. Some coolant circulates through serpentine passages that cools the suction and pressure surfaces of the airfoil before exiting through a number of holes. At the trailing edge, holes are drilled in a direction such that the coolant flow practically covers up the blade portion with a film of coolant.

Note that the trailing edge close to the region where the airfoil meets the platform is difficult to cool due to lack of available material for the coolant to reach. Also, sharp changes in the geometry in this area leads to high stress concentration due to centrifugal effects. A combination of mechanical and thermal stresses make this region susceptible to failures.

The disk to which the airfoils are attached at the rim also requires cooling. A second nozzle directs the coolant along the surface of the disk. As a consequence, reduced disk temperatures help in lowering hoop direction stresses arising from the centrifugal forces of the attached disks and also due to the disk's own mass.

6.16 ACCESSORIES AND GEAR BOX

A number of devices are required for an aviation engine to operate. These include air starter motor, fuel pump, lube oil suction and delivery pumps, oil filter and magnetic chip detector. All of the components operate at their speed, so speed reducers are required. The accessories and components are attached and housed in a gear box. The complete module is slung below the engine. Certain engine types have the accessories gear box attached below the high-pressure compressor section, but some other engine types have it mounted on the fan casing.

Specially designed links attach the accessories gear box to the engine casing. Individual structural links react to forces in the axial, radial and circumferential directions, and are provided with spherical seating at both ends to prevent transfer of overturning moments. This reduces stress concentration in the links and also in the engine casing mount points.

The drive for the module is obtained from the engine's high-pressure rotor through a vertically oriented tower shaft and a set of bevel gears. A similar set of bevel gears is provided at the gear box end of the tower shaft, which engage the horizontal shaft in the module. The drive train within the gear box than transmits mechanical power to all of the components.

6.17 NACELLE HOUSING FOR ENGINE

Aviation engines are placed inside a specially designed nacelle housing. The nacelle is made up of a complex system of housing doors. A number of devices are attached to the engine's casing. Together with tubes, valves and other appurtenances, the nacelle housing protects the components from the onrushing air while providing a path for the outside air to smoothen its flow.

The housing doors are split into two sections and hinged at the top of the engine. Maintenance work requires access to the engine's casing, and the doors may be open when required. The doors for the fan casing of a turbofan engine are in the form of half-cylindrical shells. The doors for the high-pressure section and for the low-pressure turbine are annular in configuration. The intent is to provide flow paths for the air exiting the fan casing and for the outside air flowing past the engine. Immediately aft of the fan casing is placed the thrust reverser module, which forms a part of the housing doors. After completion of the maintenance task the doors are deployed and locked with latches near the 6 o'clock location of the engine.

6.18 ATTACHMENT WITH AIRCRAFT

Two forms of attachment of turbofan engines are commonly used for both military and commercial transport planes. In some cases two engines are mounted at the rear of the aircraft fuselage. If a third engine is involved, it is attached at the top of the fuselage where the tail assembly meets. In other applications such as the Boeing 737, 777 and 787 aircrafts two engines are slung underneath the wings of the engine. Four engines are required for Airbus A380 and Boeing 747, with two engines on either side of the fuselage.

A disadvantage of mounting on the wings of the aircraft is reduced clearance between the ground and the base portion of the engine. Thus, if the airplane is not adequately levelled in the horizontal plane on the port and starboard sides while coming in for a landing, the engine runs the risk of rubbing against the ground. On the plus side, the engines are located in the region where the center of gravity of the entire aircraft lies along its length. This helps in the maneuvering of the aircraft. As the aircraft pitches and yaws while making turns to the right or to the left or with the nose pointing up or down during takeoff and landing, the maneuvering operation is easier.

A pylon is placed between the engine and the wing to facilitate attachment to mount points on both the engine casing and in the wing. Thus, the pylon is specially configured for different combinations of engine and aircraft types. A box like structure, the pylon houses tubes for fuel delivery form the aircraft's storage to the engine's combustor, for air bled from the compressor to pressurize the passenger cabin and for other pneumatic and hydraulic lines to control engine operation. A large number electrical wire harnesses for sensors, control valves, regulators and actuators also run through the pylon.

Connection between the pylon and the engine casing takes place in two planes, at the forward end in the plane where the compressor case front flange mates with the fan frame and at the rear end in the plane of the rear frame. The primary support point is at the forward end mount, which is in the form of a yoke of about 120° sector. Brackets are attached to the outer casing on the turbine rear frame. The rear mount attachment is approximately in the form of a coat hanger with structural links, with individual link reacting to loads such as axial thrust, bending in the engine casing and counter-rotating torque produced in the engine. Mount bolts attach the assembly with the brackets on the turbine frame's casing.

The structure of fighter jet aircrafts is designed such that the full engine is inserted into its body. Mount attachments at multiple points around the circumference of the engine casing ensure fixity with the aircraft structure.

6.19 SHORTCOMINGS IN COMPONENT LIFE PREDICTION

An interesting aspect of aviation engines is the need for accurate determination of useful life of rotating components. The option to replace parts after the failure mechanism sets in in aviation applications is not there, the part has to be retired before the part fails. Premature retirement of parts is also financially unacceptable due to the high cost factor. Hence, the need for precise determination of useful life of the parts is of great significance to turbine engine users.

Low cycle fatigue (LCF) has been a primary design consideration for bladed disk assemblies, and this has succeeded in controlling LCF related failures to a considerable extent. Problems due to high cycle fatigue (HCF) have also occurred, mostly due to blade passing frequencies. Prediction of useful operating life of rotating components in both the compressor and turbine, however, are based on accumulation of damage due to low-cycle fatigue.

A number of cumulative damage theories have been proposed in making estimates of the life of a blade subjected to variable stress amplitude (*ref*: Rao et al, 1999). A commonly used linear method in predicting fatigue life of a blade is based on work done by Palmgren (1924) and Miner (1945). Dynamic stress around a critical speed may be split into a convenient number of steps on either side of peak stress. Operation at a stress level S_i gives a life of N_i cycles. If the blade is subjected to n_i cycles, it suffers a damage fraction of $D_i = n_i/N_i$. Failure is then predicted to occur when $\Sigma(n_i/N_i) \geq 1$. The assertion is that the damage fraction at any stress level is linearly proportional to the ratio of number of cycles of operation to the total number of cycles that would produce failure at that stress level.

But a drawback to this theory is that it does not recognize the order of application of various stress levels, and damage is assumed to accumulate at the same rate at a given stress level without consideration of past history. Experimental evidence indicates non-linear accumulation of the fatigue damage occurs, and depends not only on the alternating stress level but the order in which the stresses are produced. Thus, a flight from New York to London with a full payload will produce a certain stress level in the rotating parts. But a flight from New York to Philadelphia with a partial payload is likely to develop lesser stresses.

A number of other theories have been proposed. For instance, the damage for each level of reversed sinusoidal stress amplitude is expressed by $D = (n/N)^m$, where the exponent m depends on the amplitude of the alternating stress. A specimen is deemed to have failed when D reaches a value c irrespective of the sequence in which the stresses are applied. The primary difficulty is the necessity of generating a lot of experimental data for the exponent m_i.

When stress levels are increased gradually, Palmgren-Miner's rule usually provides satisfactory results. When the stress levels are decreasing from a peak value, application of a factor simplifies the situation. Some other theories rely on the fracture mechanics theory for postulating initiation of permanent damage. Once initiated, fracture propagation is assumed to occur at stress levels that are well below the minimum stress required to initiate a crack.

6.20 REVIEW QUESTIONS

6.20.1 Why are roller bearings suitable for aviation engines and not for industrial power generation turbines?

6.20.2 How does weight requirements affect the design of an aviation engine?

6.20.3 Why is it necessary to reduce the speed of the propeller shaft in a turbo-prop engine?

6.20.4 Describe how variable stator vanes are actuated? What role do they play?

6.20.5 What are the advantages and disadvantages of an annular combustor?

6.20.6 When is it needed to increase thrust in a fighter jet engine and how is it accomplished?

6.20.7 Why do stator components in the hot path require thermal protection? How are they protected?

6.20.8 Mention some ways the high-pressure turbine components are thermally protected.

6.20.9 How is an aviation engine started?

6.20.10 Why is it essential to provide a nacelle housing for an aircraft engine?

6.20.11 What is the primary shortcoming in many of the methods used in predicting life of an aviation engine?

6.21 EXAMPLE PROBLEMS

Example Problem # 6.21.1: The fuel to air ratio f of a turbo-jet engine is 0.01969. If the engine produces specific thrust F_s 580 N-s/kg, calculate the specific fuel consumption in terms of kg/N-hour.

Solution: Specific fuel consumption $= f/F_s = 0.01969 \times 3600/580 = 0.1222$ kg/N-hour.

Example Problem # 6.21.2: An aviation engine has flow velocity of 575 m/s and density at turbine exhaust = 0.325 kg/m^3. Calculate the required nozzle area per unit flow rate.

Solution: Mass flow rate = $1/(0.325 \times 575) = 0.00535$ m^2/s-kg.

Example Problem # 6.21.3: Air enters a compressor at 0.485 bar at 9500 m in an aviation engine. If the compression ratio is 19.5, calculate the pressure at outlet.

Solution: Pressure at outlet = $19.5 \times 0.485 = 9.4575$ bar.

Example Problem # 6.21.4: Air flows into a compressor at the rate of 28.25 lb./s. If the fuel to air ratio is 0.025, calculate the fuel consumed in one hour if combustion efficiency is 0.9875.

Solution: Fuel consumed = $28.25 \times 0.025 \times 3600/0.9875 = 2574.7$ lb./hr.

Example Problem # 6.21.5: A hollow uniform disk has outer radius = 4.125 in., inner radius =1.25 in. and 1.75 in. width. Disk material density is 0.285 lb./in^3. Calculate the weight of the disk.

Solution: Weight of the disk:

$$W = \text{Volume} \times \text{density} = \pi R^2 l\rho$$
$$= \pi \times (4.125^2 - 1.25^2) \times 1.75 \times 0.285 = 24.2158 \text{ lb.}$$

6.22 EXERCISE PROBLEMS

Exercise Problem # 6.22.1: If air discharges from a compressor at 157.5 lb./in^2 in an aviation engine, calculate the pressure at the entrance if the compression ratio is 22.75.

Exercise Problem # 6.22.2: A 128 lb. disk is overhung from a weightless shaft supported on two bearings spaced 8.25 in. from each other. The disk is placed 5.75 in. from the near bearing. Calculate the bending moment in the shaft at the near bearing.

Exercise Problem # 6.22.3: If a turbine engine burns 3575.5 lb./hr of fuel while using air in the compressor at the rate of 34.95 lb./s, calculate the fuel to air ratio if the combustion efficiency is 0.975.

Exercise Problem # 6.22.4: Air is available for a compressor at 0.793 bar in an aviation engine. If the compression ratio is 28.5 and compressor efficiency is 89.5 per cent, calculate the pressure at outlet.

Exercise Problem # 6.22.5: A turbine nozzle area at its discharge end per unit flow rate of 0.00725 m^2/s-kg. Calculate the flow velocity if density at turbine exhaust = 0.4125 kg/m^3.

REFERENCES & BIBLIOGRAPHY

Boeing Commercial Airplane Development, "High Speed Civil Transport Study", NASA Contractor Report # 4233, Seattle, WA, 1989.

Douglas Aircraft Company, "Study of High Speed Civil Transport", NASA Contractor Report # 4235, Long Beach, CA, 1989.

Federal Register 47, number 251: 58462 – 58474, 1982.

Federal Register 55, number 155: 32856 – 32866, 1990.

Freeman, C., Cumpsty, N. A., "A Method for Prediction of Supersonic Compressor Blade Performance", ASME Gas Turbine Conference and Exposition, Toronto, 1989.

Garnier, V. H., Epstein A. H., Greitzer E. M., "Rotating Stall Anticipation and Initiation in Axial Compressors", ASME paper # 90-GT-156, New York, 1990.

Gysling, D. L., Dugundji, J., Greitzer, E. M., Epstein, A. H., "Dynamic Control of Centrifugal Compressor Surge Using Tailored Structures", ASME paper 90-GT-122, New York, 1990.

Harley, K. G., Burdsall, E. A., "High Loading Low Speed Fan Study II: Data and Performance – Unslotted Blades and Vanes", NASA CR 72667 (PWA-3653).

Hathaway, M., Gertz, J., Epstein, A. H., Strazisar, A., "Rotor Wake Characteristics of a Transonic Flow Fan", AIAA Journal 24, number 11, 1986.

Hobbs, D. E., Weingold, H. D., "Development of Controlled Diffusion Airfoils for Multi-Stage Compressors Applications", ASME Journal of Engineering for Power (106), page 271 – 278, New York, 1984.

Hunecke, K., "Jet Engines – Fundamentals of Theory, Design and Operation", Motorbooks International Publishers, Osceola, WI, 1997.

International Civil Aviation Organization (ICAO), Annex 16, Volume II, 1981.

Kerrebrock, J. L., "Aircraft Engines and Gas Turbines", The MIT Press, Cambridge, MA, 2nd edition, 1992.

Lefebvre, A. H., "Gas Turbine Combustion", Taylor and Francis, Philadelphia, 1999.

National Transportation Safety Board, "In Flight Loss of Propeller Blade – Aircraft Accident Report", NTSB/AAR-96/06, Washington, DC, 1996.

Nikkanen, J. P., Brooky, J. D., "Single Stage Evaluation of Highly Loaded High Mach Number Compressor Stages", NASA CR 120887 (PWA-4312).

Paduano, J. Epstein, A. H., Valavani, L., Longley, J. P., Greitzer, E. M., Gunette, G. R., "Active Control of Rotating Stall in a Low Speed Axial Compressor", ASME paper # 91-GT-88, New York, 1991.

Pratt and Whitney Aircraft, "The Aircraft Gas Turbine Engine and its Operation (PWA Operating Instruction 200)", revised May, 1974.

Rangwala, A. S., "Dynamic Analysis of Turbo-machinery", McGraw-Hill, New York, 2004.

Saravananamuttoo, H. I. H., Rogers, G. F. C., Cohen, H., "Gas Turbine Theory", Prentice Hall, Harlow, England, 2001.

Schlichting, H., "Boundary Layer Theory", McGraw-Hill, 1960.

Smith, C. J., "Affordable Nacelle Technologies for Future Turbofans", ASME paper # 95-GT-402, New York, 1995.

Suo, M., "Turbine Cooling", The Aerodynamics of Aircraft Gas Turbine Engines, editor G. C. Oates, AFAPL TR-78-52, Wright Patterson Air Force Base, Ohio.

Wadia, A. R., James, F. D., "F110-GE-132: Enhanced Power Through Low-Risk Derivative Technology", ASME paper # 2000-GT-578, New York, 2000.

Wagner, J. H., Johnson, B. V., Graziani, R. A., Yeh, F. C., "Heat Transfer in Rotating Serpentine Passages with Trips Normal to Flow", ASME paper # 91-GT-265, New York, 1991.

7

CHAPTER

HEAVY-DUTY STEAM TURBINE

7.1 INTRODUCTION

Approximately 50 per cent of the electric power produced in United States is obtained from coal-burning thermal plants. The coal is burned in a boiler to provide superheated steam at pressures sometimes exceeding 3500 lb./in^2. The steam drives large capacity steam turbines, which in turn drives an electric generator to develop electrical power. It is not possible to make use of the full potential of such high pressures in one section, so the turbine stages are split into high, intermediate and low-pressure sections. Coming from the super-heater of the boiler, the steam first is admitted into the high-pressure section. After developing mechanical power the steam may be redirected to the boiler for reheat, before admitting into the intermediate-pressure section. The steam may then crossover into the low-pressure modules. Since the flow volume of the steam at this point is considerably large, it may be necessary to use more than one low-pressure modules. The rotors of the high, intermediate and low-pressure sections are mechanically coupled, driving a single electrical generator to produce 750 MW or more of power. Small steam turbines ranging from 0.75 kW units are used as mechanical drives for pumps, compressors and other shaft driven equipment. Larger units of up to 1,500,000 kW (2,000,000 hp) turbines are used to generate electricity.

The arrangement of casings and shafts include single casing, tandem compound and cross compound turbines. Single casing units are the most basic style where a single casing and shaft are coupled to a generator. Tandem compound are used where two or more casings are directly coupled together to drive a single generator. A cross compound turbine arrangement features two or more shafts not in line driving two or more generators that often operate at different speeds. A cross compound turbine is typically used for many large applications.

The modern manifestation was invented by Sir Charles Parsons in 1884. Because the turbine generates rotary motion, it is particularly suited for use in an electrical generator – about 90% of all electricity generation in the United States (1996) is by use of steam turbines. The steam turbine is a form of heat engine that derives much of its improvement in thermodynamic efficiency through the use of multiple stages in the expansion of the steam, which results in a closer approach to the ideal reversible process.

While there are substantial differences in the design, complexity, application, steam conditions, and size of steam turbines, they all are fundamentally the same. They perform the same function, utilize similar major components and supporting systems, and are subjected to the same failure mechanisms. To support reliable turbine operation, there needs to be an effective infrastructure in place for monitoring the operating conditions, water/steam quality, and health of the steam turbine. Having and using written operating/maintenance procedures are essential with such large numbers of steam turbines in operation, and for conducting training for personnel on an ongoing basis.

7.2 STEAM PATH TECHNOLOGY

Steam turbines convert saturated, superheated, or supercritical steam from boilers or heat recovery steam generators (HRSG) into rotational torque and power. Consequently, steam turbines are utilized to drive a variety of equipment types and of numerous sizes and speeds. The turbines are found in just about every industry segment, including power generation, pulp and paper, iron and steel, combined heat and power, and chemical, oil and gas industries.

The turbine casing houses and supports the turbine rotor, labyrinth seals, and bearings. The casing is cast in two halves and bolted together with a metal to metal fit. This metal to metal fit is initially sealed using a sealing compound to form a gasket. In the event of a steam leak around the edge of the casing, a groove is machined around the peripheral of the turbine casing. This groove, known as the gunning groove, provides ability to inject a sealing compound for emergency sealing of the turbine casing.

Fig. 7.1 Steam Turbine Rotor for Power Plant

The turbine rotor is constructed of chromium molybdenum (chromoly) steel. The rotor wheels, which are mounted to the rotor shaft, carry the moving blades (Figure 7.1). The non-moving blades are attached to the turbine casing.

At the point where the turbine rotor penetrates the turbine casing, labyrinth seals are installed. In conjunction with the steam from the gland seal system, these seals prevent any steam from leaving the turbine casing and also prevent any air from entering the turbine casing and, subsequently, the main condenser.

7.3 THERMODYNAMICS OF STEAM TURBINES

The steam turbine operates on basic principles of thermodynamics using the Rankine cycle with superheat. The cycle may be split into following parts, as shown in Figure 7.2. In step 1-2 the working fluid is pumped from a low to a high pressure. In step 2-3 the high-pressure liquid enters a boiler where it is heated at constant pressure by an external heat source to become a dry saturated vapor. Step 3-3' requires the vapor to be superheated. In steps 3-4 and 3'-4' the dry saturated steam expands through a turbine to generate power. This decreases the temperature and pressure of the vapor, and some condensation may occur. In step 4-1 the wet vapor enters a condenser where it is condensed at a constant pressure to become a saturated liquid.

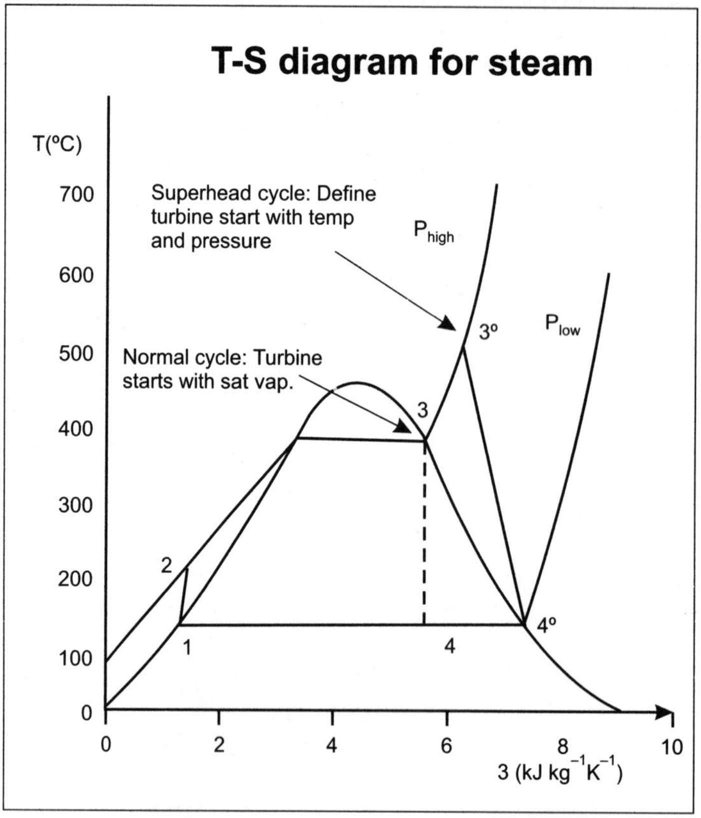

Fig. 7.2 Rankine Cycle with Superheat

Superheated vapor, point 3' (or dry saturated vapor, point 3, depending on application) enters the turbine after leaving the boiler at a high temperature and high pressure. The elevated temperature and highly pressurized steam is converted into kinetic energy using a nozzle, which may be a fixed nozzle in an impulse type turbine or fixed blades in a reaction type turbine. Once the steam exits the nozzle it moves at a high velocity, and is directed into the blades of the turbine. Force is imparted on the blades due to the pressure of the vapor on the blades, causing them to spin around the axis of the rotor. A generator, or other driven device, is coupled to the shaft, and the mechanical energy is converted into electrical power. The gas exits the turbine as a saturated vapor, or as a mixture of liquid and vapor depending on application, at a lower temperature and pressure. The vapors are then sent to the condenser to be cooled.

From the first law of thermodynamics, an equation can be developed to compare the rate at which work is developed per unit mass. Assuming there is no heat transfer to the surrounding environment and that the change in kinetic and potential energy is negligible when compared to the change in specific enthalpy, then:

$$(dW/dt)/(dm/dt) = h_1 - h_2 \tag{7.1}$$

where dW/dt is the rate at which work is developed per unit time, dm/dt is the rate of mass flow through the turbine and h_1 and h_2 are steam enthalpy at entry and discharge.

7.4 VELOCITY TRIANGLE

A velocity triangle paves the way for a better understanding of the relationship between the various velocities. In Figure 7.3, V_1 and V_2 are the absolute velocities at the inlet and outlet respectively, V_{f1} and V_{f2} are the flow velocities at the inlet and outlet, V_{w1} and V_{w2} are the swirl velocities at the inlet and outlet, and V_{r1} and V_{r2} are the relative velocities at the inlet and outlet. U_1 and U_2 are the velocities of the blade at the inlet and outlet respectively.

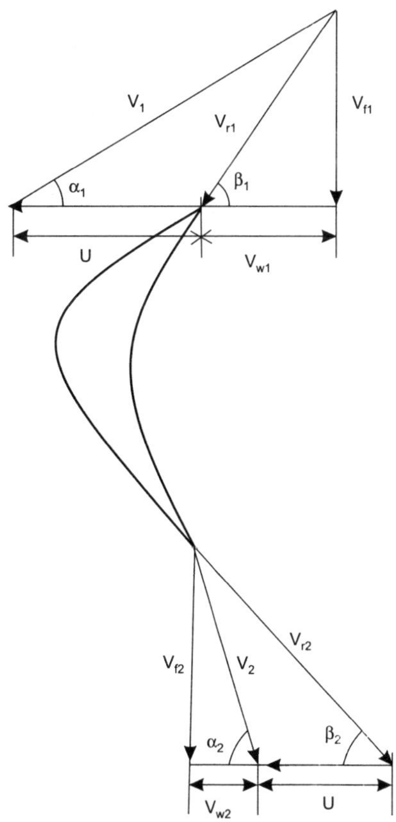

Fig. 7.3 Turbine Blade Velocity Triangles

According to the law of moment of momentum, the sum of the moments of external forces acting on a fluid which is temporarily occupying the control volume is equal to the net time change of angular momentum flux through the control volume. To understand, consider the swirling fluid entering the control volume at radius r_1 with tangential velocity V_{w1} and leaves at radius r_2 with tangential velocity V_{w2}. α is the guide vane angle and β is the blade angle. Then by the law of moment of momentum, the torque on the fluid is given by:

$$T = (dm/dt) \times (r_2 V_{w2} - r_1 V_{w1}) \quad (7.2)$$

Now for an impulse steam turbine, $r_2 = r_1 = r$. Therefore, the tangential force on the blades is: $F_u = (dm/dt) \times (V_{w1} - V_{w2})$. Work done per unit time or power developed is expressed by: $W = T \times \omega$, where ω is the angular velocity of the turbine. Blade peripheral speed is: $U = \omega \times r$, and hence work done per unit time or power developed is given by the expression:

$$W = (dm/dt) \times U \times (\Delta V_w). \quad (7.3)$$

7.5 NOZZLE AND BLADE EFFICIENCY

Blade efficiency η_b can be defined as the ratio of the work done on the blades to kinetic energy supplied to the fluid, or:

$$\eta_b = [\text{Work done/Kinetic energy supplied}] = 2UV_w/V_1^2 \quad (7.4)$$

Nozzle efficiency is: $\eta_{nozzle} = V_2^2/[2(h_1 - h_2)$, where the enthalpy (J/Kg) of steam at the entrance of the nozzle is h_1 and the enthalpy of steam at the exit of the nozzle is h_2.

A stage of an impulse turbine consists of a nozzle set and a moving wheel. The stage efficiency defines a relationship between enthalpy drop in the nozzle and work done in the stage. Then stage efficiency of the nozzle is given by the expression:

$$\eta_{stage} = [\text{Work done on blade/Energy supplied per stage}] = U\Delta V_w/\Delta h \quad (7.5)$$

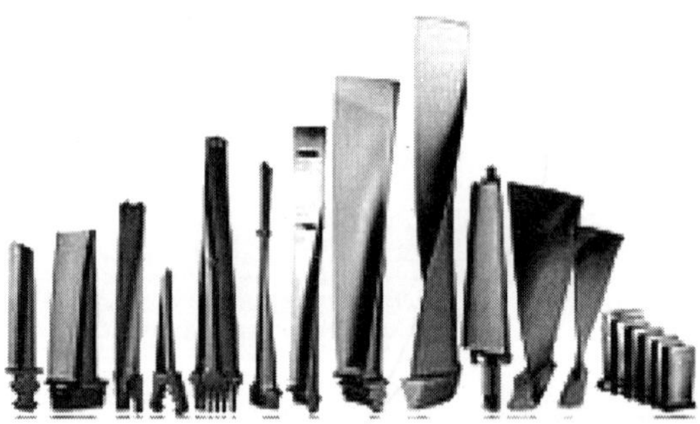

Fig. 7.4 A Selection of Turbine Blades

where the specific enthalpy drop of steam in the nozzle $\Delta h = h_2 - h_1$.

By the first law of thermodynamics, $h_1 + V_1^2/2 = h_2 + V_2^2/2$. Assuming V_1 is appreciably less than V_2, then: $\Delta h \sim V_2^2/2$. Furthermore, stage efficiency is the product of blade efficiency and nozzle efficiency, or: $\eta_{stage} = \eta_{blade} \times \eta_{nozzle}$.

To determine blade efficiency, energy input to the blades in a stage $E = \Delta h$, which is equal to the kinetic energy supplied to the fixed (*f*) and moving (*m*) blades. Or, E is enthalpy drop over the fixed blades, Δh_f, plus enthalpy drop over the moving blades, Δh_m. The effect of expansion of steam over the moving blades is to increase the relative velocity at the exit. Therefore, the relative velocity at the exit, V_{r2}, is always greater than the relative velocity at the inlet, V_{r1}.

In terms of velocities, the enthalpy drop over the moving blades is given by: $\Delta h_m = (V_{r2}^2 - V_{r2}^2)/2$, and contributes to a change in static pressure. The enthalpy drop in the fixed blades, with the assumption that the velocity of steam entering the fixed blades is equal to the velocity of steam leaving the previously moving blades, is given by the expression: $\Delta h_f = (V_1^2 - V_0^2)/2$, where V_0 is the inlet velocity of steam in the nozzle.

Now, V_0 is very small and hence can be neglected. Therefore, $\Delta h_f = V_1^2/2$ and energy input $E = \Delta h_f + \Delta h_m$, and $E = V_1^2/2 + (V_{r2}^2 - V_{r1}^2)/2$.

A widely used design has half degree of reaction or 50% reaction, and is known as Parson's turbine. This consists of symmetrical rotor and stator blades. For this turbine the velocity triangle is similar and so: $\alpha_1 = \beta_2$ and $\beta_1 = \alpha_2$. Also, $V_1 = V_{r2}$ and $V_{r1} = V_2$. Thus, for a Parson's turbine, $E = V_1^2 - V_{r1}^2/2$. From the velocity triangle, and expressing E in terms of V_1, U and α_1, blade efficiency takes the form:

$$\eta_b = [2U(2V_1\text{Cos}\ \alpha_1 - U)]/[V_1^2 - U^2 + 2V_1U\text{Cos}\ \alpha_1] \tag{7.6}$$

Figure 7.4 shows a selection of commonly used turbine blades and Figure 7.5 shows the composite velocity diagram for the inlet and outlet of a Parson's steam turbine design.

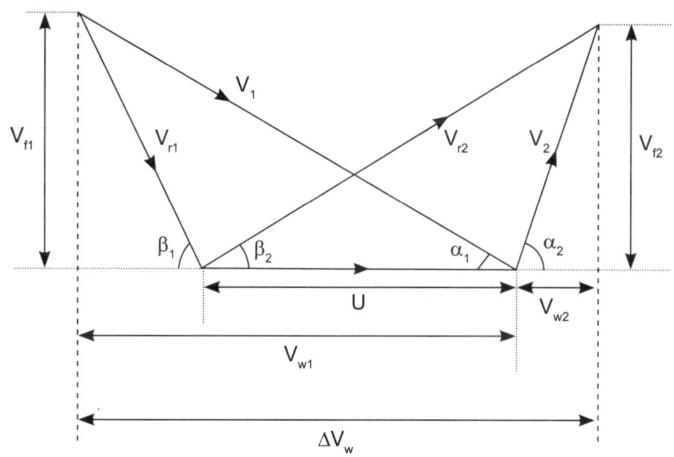

Fig. 7.5 Velocity Diagram

7.6 TURBINE EFFICIENCY

To measure how well a turbine is performing can best be gauged by its isentropic efficiency. The parameter compares the actual performance of the turbine with the performance that would be achieved by an ideal isentropic turbine, with a few underlying assumptions. Heat lost to the surroundings is assumed to be zero. The starting pressure and temperature is the same for both the actual and the ideal turbines, but at turbine exit the energy content, or the specific enthalpy, for the actual turbine is greater than that for the ideal turbine because of irreversibility in the actual turbine. The specific enthalpy is evaluated at the same pressure for the actual and ideal turbines in order to give a good comparison between the two. The isentropic efficiency is found by dividing the actual work by the ideal work:

$$\eta_t = \frac{h_1 - h_2}{h_1 - h_{2s}} \tag{7.7}$$

where h_1 is the specific enthalpy at state one, h_2 is the specific enthalpy at state two for the actual turbine and h_{2s} is the specific enthalpy at state two for the isentropic turbine.

No steam turbine is truly isentropic, however, with typical isentropic efficiencies ranging from 20 to 90 per cent, depending on the application of the turbine. The interior of a turbine comprises several sets of blades, or 'buckets' as they are more commonly referred to. One set of stationary blades is connected to the casing and one set of rotating blades is connected to the shaft. The sets intermesh with certain minimum clearances, with the size and configuration of sets varying to efficiently exploit the expansion of steam at each stage.

To maximize turbine efficiency the steam is expanded while doing work in a number of stages. These stages are characterized by how the energy is extracted from them and are known as either impulse or reaction turbines. Most steam turbines use a mixture of the reaction and impulse designs: each stage behaves as either one or the other, but the overall turbine uses both. Typically, higher pressure sections are impulse type and lower pressure stages are reaction type.

7.7 TURBINE CLASSIFICATION

There are several classifications for modern steam turbines. These types include condensing, non-condensing, reheat, extraction and induction.

Condensing turbines are most commonly found in electrical power plants. These turbines exhaust steam in a partially condensed state, typically of a quality near 90 per cent, at a pressure well below atmospheric, to a condenser. Non-condensing or back pressure turbines are most widely used for process steam applications. The exhaust pressure is controlled by a regulating valve to suit the needs of the process steam pressure. These are commonly found at refineries, district heating units, pulp and paper plants, and desalination facilities where large amounts of low pressure process steam are available.

Reheat turbines are also used almost exclusively in electrical power plants. In a reheat turbine, steam flow exits from a high pressure section of the turbine and is returned to the boiler where additional superheat is added. The steam then goes back into an intermediate pressure section of the turbine and continues its expansion. Extracting type turbines are common in all applications. In an extracting type turbine, steam is released from various stages of the turbine, and used for industrial process needs, or sent to boiler feed-water heaters to improve overall cycle efficiency. Extraction flows may be controlled with a valve, or left uncontrolled. Induction turbines introduce low pressure steam at an intermediate stage to produce additional power.

The relationship between pressure and velocity describes the type of compounding. This refers to the use of blading which causes a series of pressure drops and a series of velocity drops. Single axial flow calls for the steam to flow in only one direction, parallel to the axis of the turbine rotor. Steam enters the forward end of the turbine and exhausts through the aft end of the turbine. On a dual flow turbine the steam enters in the center of the turbine rotor and flows both forward and aft simultaneously.

7.8 STEAM TURBINE TYPES

The Curtis stage (see Figure 7.6) is designed to be a power rotor, extracting a large amount of energy out of the steam. As main steam enters the HP turbine, it first passes through the nozzle block containing the nozzles. The velocity of steam is increased and the pressure is decreased as the steam passes through these nozzles.

The remaining stages of the HP turbine are a series of Rateau stages. A single Rateau stage consists of a nozzle diaphragm followed by a row of moving blades. The nozzle diaphragm separates the stages of an impulse turbine and provides support for the nozzles. The nozzles within the nozzle diaphragm serve the same purpose as the nozzles within the nozzle block. As steam passes through the nozzle, velocity is increased and pressure is decreased. After leaving the nozzle, steam then enters the moving blades where once again work is extracted from the steam. As work is extracted from the steam, its velocity will once again decrease although its pressure is not affected. Even though there is a velocity increase and a velocity decrease in each Rateau stage, the overall velocity from the inlet of the first Rateau stage to the exhaust of the final Rateau stage is not changed. In contrast, there is a pressure drop in each Rateau stage, resulting in an overall pressure drop from the inlet of the first Rateau stage to the exhaust of the final Rateau stage. This overall pressure drop causes the Rateau staging to be considered as pressure compounded.

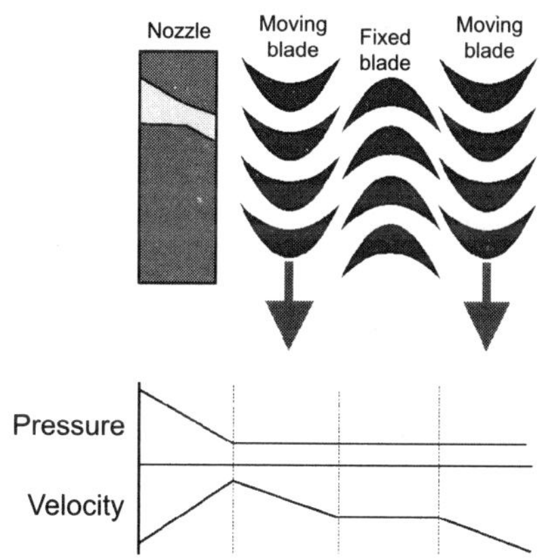

Fig. 7.6 Pressure, Velocity Profiles in Curtis Stage

7.9 IMPULSE AND REACTION TURBINES

An impulse turbine has fixed nozzles that orient the steam flow into high speed jets. These jets contain significant kinetic energy, which is converted into shaft rotation by the bucket-like shaped rotor blades, as the steam jet changes direction. A pressure drop occurs across only the stationary blades, with a net increase in steam velocity across the stage. As the steam flows through the nozzle, its pressure falls from inlet pressure to the exit pressure (atmospheric pressure, or more usually, the condenser vacuum). Due to this high ratio of expansion of steam, the steam leaves the nozzle with a very high velocity. The steam leaving the moving blades has a large portion of the maximum velocity of the steam when leaving the nozzle. The loss of energy due to this higher exit velocity is commonly called the carry over velocity or leaving loss.

In an impulse turbine the only time a pressure drop occurs is when steam passes through a nozzle. After steam passes through the nozzles, it passes through the first set of moving blades, where work is extracted from the steam and causing the velocity to drop. After passing through the moving blades, the steam then passes through the non-moving blades. The only purpose the non-moving blades serve is to redirect steam from the first set of moving blades to the second set of moving blades.

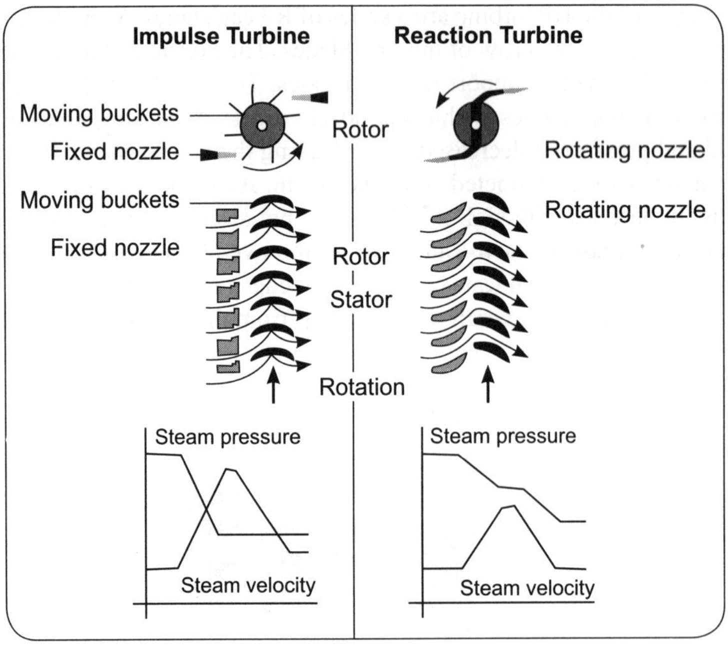

Fig. 7.7 Impulse and Reaction Steam Turbine Stages

The non-moving blades do not have any effect on the pressure or the velocity of the steam passing through them in an impulse turbine. After leaving the non-moving blades the steam passes through another set of moving blades. This setup of a nozzle followed by a set of moving blades, non-moving blades, and moving blades makes up a single Curtis stage. After steam exits the nozzle, there are no further pressure drops. However, across both sets of moving blades there is a velocity drop. This causes the Curtis stage to be classified as velocity compounded blading.

In the reaction turbine, the rotor blades themselves are arranged to form convergent nozzles. This type of turbine makes use of the reaction force produced as the steam accelerates through the nozzles formed by the rotor. Steam is directed onto the rotor by the fixed vanes of the stator. It leaves the stator as a jet that fills the entire circumference of the rotor. The steam then changes direction and increases its speed relative to the speed of the blades. A pressure drop occurs across both the stator and the rotor, with steam accelerating through the stator and decelerating through the rotor, with no net change in steam velocity across the stage but with a decrease in both pressure and temperature, reflecting the work performed in the driving of the rotor. Figure 7.7 provides a comparison between the impulse and reaction turbine stages. Figure 7.8 provides comparison of efficiencies between impulse and reaction turbines as a function of cosine of angle of α_1 absolute velocity at the blade inlet, shown in Figure 7.3.

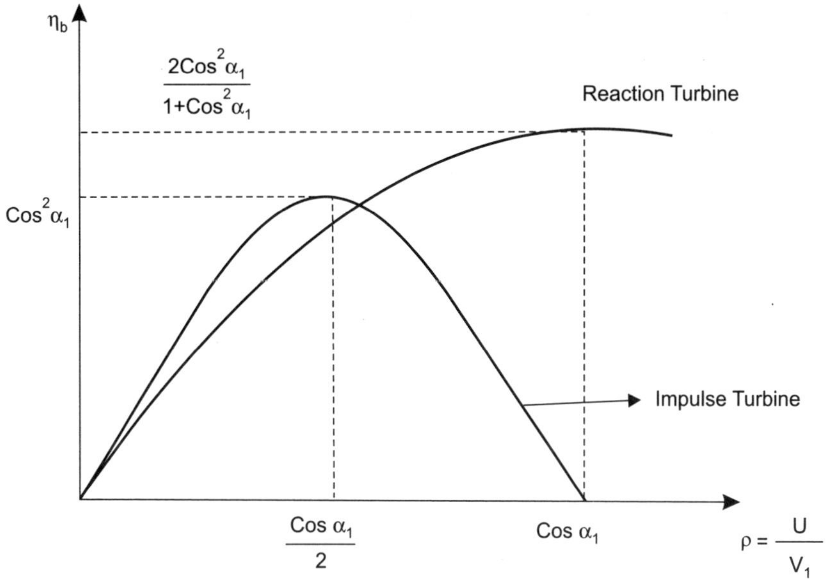

Fig. 7.8 Comparing Efficiencies of Impulse and Reaction turbines

7.10 HIGH-PRESSURE TURBINE SECTION

Heavy duty steam turbines may be categorized into high-pressure (HP), intermediate-pressure (IP) and low-pressure (LP) turbine sections. Though each section is designed differently, there are several components which all three have in common.

The HP turbine is the first main engine section to receive steam from the main steam system. It is designed to efficiently extract work out of high pressure steam. The HP turbine is a pressure-velocity compounded, single axial flow, non-condensing impulse turbine.

During normal operations, main steam first enters the steam chest. The amount of steam then allowed to flow from the steam chest into the turbine is controlled by the throttle valve located in the bottom of the steam chest. After passing through the throttle valve, steam then passes through the nozzle block. The nozzle block causes the pressure of steam to drop while increasing steam velocity. The Curtis stage extracts work from the steam and sends the steam to the Rateau stages. After exhausting from the final Rateau stage, steam flows to the IP section before going through the crossover pipe into the LP turbine. The reaction blading of the IP and LP sections extract more work from the steam. After the steam exhausts from the LP turbine blading, it flows through the exhaust trunk into the main condenser.

The type of blading used on the HP turbine is impulse blading because it extracts more work from the high pressure steam than reaction blading. Impulse blading is in the shape of a half moon. As steam impacts the moving blade, it pushes the blade forward. This impact causes the steam to lose velocity without losing pressure. In order to efficiently extract the maximum amount of work out of the steam, two different types of impulse stages are used. The Curtis stage is the first stage of the HP turbine. The Curtis stage is designed to initially extract a large amount of work out of the steam as it enters the turbine. The remaining stages of the HP turbine are Rateau stages.

7.11 LOW-PRESSURE TURBINE SECTION

In the absence of an intermediate turbine section, the LP turbine is located next to the HP turbine. The LP turbine is a pressure compounded, either single or dual axial flow, condensing reaction turbine.

On board ships where space is a consideration, the LP turbine is designed to be a dual flow turbine (Figure 7.9). Steam enters the center of the turbine from the crossover pipe and flows across the reaction blading in two opposite directions. This configuration reduces axial thrust on the turbine and allows for a smaller turbine installation. On ships where space is not a concern, a single flow turbine is used.

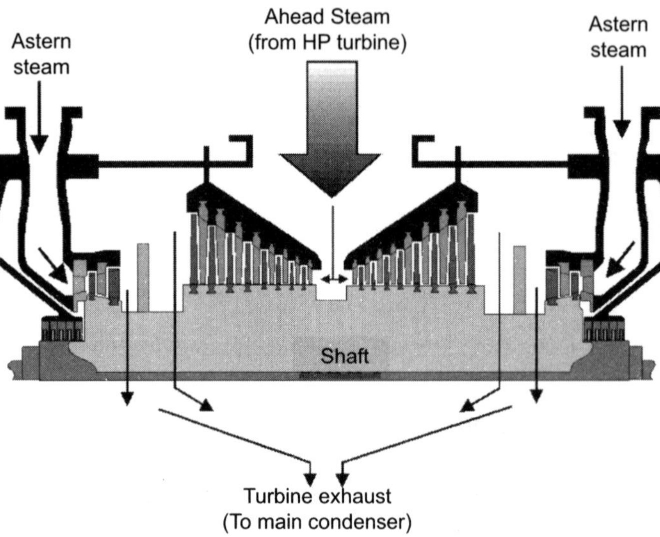

Fig. 7.9 Low-Pressure and Astern Turbine Section

As in the HP turbine, the direction of steam flow of steam is parallel to the turbine rotor. Unlike the HP turbine, the LP turbine exhausts into the main condenser. Because the LP turbine exhausts into its own dedicated condenser, it is considered a condensing turbine.

7.12 ASTERN TURBINE FOR SHIPS

The astern turbine is designed to propel the ship in the astern direction. The concept is the same as in a car designed to go both forward and reverse. The automobile could be designed with two totally separate transmissions, one for forward and one for reverse. Instead, one transmission capable of going both forward and reverse is used. The same concept exists on the ship's main engines. The ship is not designed with two engines per shaft. Rather, it is designed with one engine per shaft.

In an automobile, the ability to go in reverse is contained within the transmission. On the ship's main engine, the ability to go astern is contained within the LP turbine. The astern turbine is designed as an integral part of the LP turbine rotor. On a double flow LP turbine, the ahead elements of the LP turbine are located towards the center of the LP turbine. The astern elements are located on the forward and aft end of the LP turbine rotor. On a single flow LP turbine, the astern elements are located on the forward end of the LP turbine. The astern turbine is a single axial flow, velocity compounded, condensing impulse turbine consisting of one or two Curtis stages located on the forward and/or after end of the LP turbine.

During astern operations the steam will want to flow into the ahead elements of the LP turbine. Similarly, during ahead operations the steam will want to flow into the astern elements of the LP turbine. If this were permitted, the rotation created by the ahead elements would be hindered by the steam acting on the astern elements. To prevent this, a deflector plate is installed to provide a physical barrier and prevent steam from the ahead elements from impinging on the astern elements, and vice versa.

Fig. 7.10 Steam Turbine Under Installation

7.13 POWER GENERATION STEAM TURBINE

All three turbine sections, HP, IP and LP sections of power generation steam turbines utilize moving blades that are attached to the turbine rotor. Non-moving blades are either attached directly to the turbine casing or else they are located inside a nozzle diaphragm.

Short strips of metal called shrouding are attached to the outer edges of the blading. This shrouding is used to assist in maintaining rigidity of the blades. In addition, there is minimal clearance between the shrouding and the turbine casing, preventing the steam from leaking around the outer edges of the turbine blading. Figure 7.10 shows a power generation steam turbine under erection.

7.14 MARINE PROPULSION

Large steam turbines are used to propel ocean going bulk carriers. The ship's main propulsion turbines are designed to efficiently convert the thermal energy of steam into useful mechanical energy to cruise through the water. As work is extracted from steam, its pressure decreases. The high-pressure turbine is designed to efficiently extract work out of the pressurized steam as it initially enters the main propulsion turbines. The low-pressure turbine is designed to efficiently extract work out of steam which is exhausting out of the high pressure turbine at a lower pressure.

In ships, the compelling advantages of steam turbines over reciprocating engines are smaller size, lower maintenance, lighter weight, and lower vibration. A steam turbine is only efficient when operating in the thousands of RPM, while the most effective propeller designs are for speeds less than 100 RPM; consequently, precise and thus expensive reduction gears are usually required, although several ships, such as Turbinia (see Figure 7.11), had direct drive from the steam turbine to the propeller shafts. Another alternative is turbo-electric drive, where an electrical generator run by the high-speed turbine is used to run one or more slow-speed electric motors connected to the propeller shafts. The purchase cost of reduction gears is offset by much lower fuel and maintenance requirements and the small size of a turbine when compared to a reciprocating engine having an equivalent power. However, diesel engines are capable of higher efficiencies: propulsion steam turbine cycle efficiencies have yet to break 50 per cent, yet diesel engines routinely exceed this percentage, especially in marine applications.

In marine applications there are various other components of the HP turbine which must be considered. The aft end of the HP turbine is rigidly mounted to the frame of the ship. The forward end is mounted using either sliding feet (similar to what is used on the boiler) or using a flexible I-beam. The mounting is designed to support the weight of the forward end of the HP turbine as well as compensate for the expansion and contraction encountered during start up and securing.

Nuclear-powered ships and submarines use a nuclear reactor to create steam. Nuclear power is often chosen where diesel power would be impractical (as in submarine applications) or the logistics of re-fuelling pose significant problems (for example, icebreakers). It has been estimated that the reactor fuel for the Royal Navy's Vanguard class submarine is sufficient to last 40 circumnavigations of the globe-potentially sufficient for the vessel's entire service life. Nuclear propulsion has only been applied to a very few commercial vessels due to the expense of maintenance and the regulatory controls required on nuclear fuel cycles.

Fig. 7.11 The *Turbinia*, 1894, First Steam Turbine-Powered Ship

7.15 STEAM TURBINE FOR RAILROAD LOCOMOTIVE

Nearly one hundred years back, railroad locomotive engines were sometimes driven by a steam turbine. The main advantages of a steam turbine locomotive are better rotational balance and reduced hammer blow on the track. However, a disadvantage is less flexible power output, so that turbine locomotives were best suited for long-haul operations at a constant output power.

The first steam turbine rail locomotive was built in 1908 by a firm in Milan, Italy. In 1924 Krupp built the steam turbine locomotive T18 001, operational in 1929, for Deutsche Reichsbahn.

7.16 LONG SLENDER BLADES IN LAST STAGE OF LOW-PRESSURE SECTION

As the highly pressurized steam coming from the super-heater gradually works its way through the rotating and stationary cascades in the HP, IP and LP sections, performing work to drive the generator rotor, the steam loses pressure. This is accompanied by increase in the flow volume. Reaching the last stage of the LP turbine, the flow volume is so large it necessitates use of long and slender blades. The large blades provide the required cross-sectional flow area to handle the flow volume. Blades made of nickel-chrome alloys are 48 in. long in large power generator installations.

The large span of these blades causes them to lose torsional rigidity. As the blades pass the wake shed by the stationary vanes, excessive twisting of the blades occurs along the span. If the blades are not constrained to prevent the twisting, torsional fatigue will dramatically reduce the useful life of the blades. If a last stage LP turbine blade were to fail during regular operation, the centrifugal force from the separated segment spinning at 3600 rpm would be large enough to destroy the whole turbine, and indeed the whole power generation station. Fragments of the rotor would be flying out through the casing with tremendous kinetic energy and cause widespread destruction.

Shrouds are used to increase the rigidity and decrease twisting along the length of the blades. Adjacent blades are connected with shrouding elements, in effect permitting the blades to lean against other and reduce the deformation individually and collectively in the blades. The shrouds are located at the tip of the blades, at one-third span from the tip and at two-third span from the tip.

The configuration of the shrouds takes different forms. At the tip it may be in the form semi-circular strips of steel. At the tip of the blade, tenons are machined integral to the blade, which engage with similarly shaped cutouts in the shroud strip. At the one-and two-third span, one method calls for tie wires, which may be in the form of a circular rod. A hole with a boss is provided in the blade profile. The circular rod, split at one point, passes through the holes to form a nearly continuous ring. Another method calls for protrusions on both blade surfaces. The ends of the protrusion on adjacent blades contact each other to reduce elastic twisting and bending motion in the blades.

7.17 TURBINE STEAM CHEST

The steam chest, located on the forward upper half of the HP turbine casing, houses the throttle valve assembly. This is the area of the turbine where main steam from the boiler's super-heater section first enters the main engine. The throttle valve assembly regulates the amount of steam entering the turbine. After passing through the throttle valve, steam enters the nozzle block.

The optimal approach to condition assessment, regardless of the component involved, is to use a programmatic approach, and steam chest condition assessment is no exception. Steam chests typically vary significantly from one to the next in shape and complexity; consequently, stress distributions vary and damage occurs first, and is most advanced at the high stress regions. An ongoing implementation of Non-Destructive Evaluation (NDE) has little technical justification if the intent is to identify high stress locations through flaw detection. Flaws manifest themselves at both macro and micro levels. Therefore, inspection typically includes surface inspection using liquid dye penetrant and/or magnetic particle inspection for macro damage. Metallographic replication for micro damage, plus ultrasonic inspection for volumetric inspection of subsurface flaws and flaws at otherwise inaccessible surfaces are part of the evaluation.

In a systematic approach, the first step is to accurately understand the stresses of the steam chest to determine the appropriate areas requiring examination and monitoring. Only after identifying the critical areas on the steam chest must attention turn to defining the optimal techniques and procedures to examine the areas identified. By implementing a focused inspection that concentrates on the critical areas, the scope, cost, and the frequency of the inspection is significantly reduced.

The best method is most often dictated by access constraints and limitations at the region of interest. Significant strides have been made in the use of advanced ultrasonic testing techniques such as linear phased array (LPA) and annular phased array (APA) ultrasonic inspection for sizing cracks in some of the least accessible areas. In many cases, once identified, the damage can subsequently be monitored periodically with only local removal of insulation. Disassembly of the valve is not required on an ongoing basis, nor is full insulation removal in most cases. Finally, once damage has been identified and characterized, be it early form cavitation through to defined cracks, the model used initially to identify the inspection locations is then used to assess the damage in terms of growth rates and failure potential. This information is utilized for a complete 'fitness for service assessment' of the unit. This would include definition of re-inspection intervals, monitoring requirements, and possibly to assess repair/replace options and schedules. A robust life assessment program may include upfront analysis of the steam chest to identify problem areas including modeling of the valve, focusing baseline inspection of identified potential problem areas, analyze for fitness for service, and continued monitoring of critical areas of the valve.

7.18 OTHER COMPONENTS

Turbine casing drains remove the condensate from the turbine casing during warm-up, securing, maneuvering and other low flow conditions. Two sentinel valves are installed on the LP turbine. One sentinel valve is located on the crossover pipe leading to the LP turbine and the second is located on the forward end of the LP turbine casing. Both of these sentinel valves warn the operator of over-pressurization of the LP turbine. A sentinel valve does not relieve system pressure. It only acts to provide an audible alarm in the event of over-pressurization of the LP turbine. Some crossover pipes also have relief valves installed.

The moving steam imparts both a tangential and axial thrust on the turbine shaft, but the axial thrust in a simple turbine is unopposed. To maintain the correct rotor position and balancing, this force must be counteracted by an opposing force. Either thrust bearings can be used for the shaft bearings, or the rotor can be designed so that the steam enters in the middle of the shaft and exits at both ends. The blades in each half face opposite ways, so that the axial forces negate each other but the tangential forces act together. This design of rotor is called two-flow or double-exhaust. This arrangement is common in low-pressure casings of a compound turbine to accommodate the larger flow volume.

In order to support the weight of the turbine and to maintain radial and axial alignment, two different types of bearings are used. Turbine journal bearings maintain the radial alignment of the turbine and supports the weight of the rotor. Bearings are spherically seated allowing for slight radial misalignment during installation only. They are located on the forward and aft end of the turbine rotors.

Turbine thrust bearings absorb any axial thrust created in the turbine and also maintain the axial position of the rotor in the casing. The thrust bearings are double acting, segmented shoe, Kingsbury type thrust bearings. They are usually located on the forward end of each turbine rotor.

A flexible coupling transmits the torque from the turbines to the reduction gears, if required. The flexible couplings are designed so that any thrust created in the turbines will not be transmitted to the reduction gears. They also allow for slight radial misalignment and provide a means of disconnecting the turbines from the main reduction gears.

7.19 PARAMETER MONITORING OF STEAM TURBINES

In order to operate the main engines safely, various parameters must be constantly monitored. These main engine indicators will give the operator a good overview of the operating condition of the main engine.

The rotor position indicator (RPI), located on the forward end of the turbine rotor, is a device that indicates a safe distance between fixed and moving blades inside the turbine and gives indication of thrust bearing wear. As the thrust bearing naturally wears, this reading will gradually increase.

The HP turbine steam chest pressure indicator shows available steam pressure to the main engine from the boiler. This pressure gauge senses the pressure of the steam entering the steam chest. The gauge is located on the throttle board.

First stage pressure is measured at the point where steam is exhausting the first stage nozzle block. As the throttle valve is opened, admitting more steam to the turbine, this pressure will increase, and will decrease as the throttle valve is shut. The gauge is located on the throttle board. During speed changes, the throttle-man controls the amount of steam admitted to the turbine. This gauge allows the throttle-man to monitor the amount of steam admitted to the engine.

First stage temperature indicator shows the temperature of the first stage of the HP turbine. During astern operations on board a ship, there is no flow of steam through the HP turbine to cool the turbine blading. However, due to the HP turbine being connected to the reduction gear, it will still rotate in the reverse direction. As the HP turbine blading passes through the air inside the turbine casing, friction is created. This is known as windage. During extended astern operations, windage will create large amounts of heat. If the turbine rotor overheats, damage will occur to the turbine blading and rotors. The watch-standers monitor this temperature in order to be aware of any overheating which may be occurring. Windage also occurs on multi-shaft ships when a shaft is trailing.

HP turbine exhaust pressure indicator shows the pressure of steam exhausting from the HP turbine before it enters the LP turbine. At low speeds, the vacuum of the main condenser may be pulling the steam through the turbines, and this gauge may indicate a vacuum.

Main condenser vacuum indicator shows vacuum (pressure below atmospheric pressure) in the main condenser. The main condenser is designed to operate under a vacuum. If a decrease in vacuum occurs, the main engine will no longer operate efficiently. If vacuum is totally lost, this could result in damage to the main engine. This low pressure area is the most efficient place for the turbine steam to exhaust.

Main engine lube oil pressure indicator shows oil pressure at the most remote bearing, farthest away from the lube oil pump. In the event system pressure is lost, this bearing will normally be the first one to lose lube oil pressure. Bearing oil outlet thermometers give an indication of the temperature of the oil leaving the bearings. If a bearing overheats, the Babbitt will begin to break down and cause a bearing failure. By monitoring bearing temperature, a watch-stander will notice any bearings which are abnormally hot and will be able to take corrective action in order to prevent any further bearing damage from occurring.

Sight flow indicators (SFI) are located on the outlet of the bearing and allow the watch-stander to monitor oil flow through a bearing. The SFI is constructed with small glass windows which permit the watch-stander to look into the indicator. Normally a flow of oil can be seen through the SFI. In the event lubrication is lost to the bearing, the watch-stander will not see a flow of oil through the SFI. All bearings have both a sight flow indicator and a temperature gage attached to them. A watch-stander monitoring the RPI's, SFI's and bearing oil temperature on main engines is conducting what is called a 'three point check' and can quickly report the satisfactory or unsatisfactory condition of the turbines.

7.20 STARTING OPERATION

When warming up a steam turbine for use, the main steam stop valves (after the boiler) have a bypass line to allow superheated steam to slowly bypass the valve and proceed to heat up the lines in the system along with the steam turbine. Also, a turning gear is engaged when there is no steam to the turbine to slowly rotate the turbine to ensure even heating and prevent uneven expansion. After first rotating the turbine by the turning gear, allowing time for the rotor to assume a straight plane to prevent bowing, the turning gear is disengaged and steam is admitted to the turbine, while slowly rotating the turbine at 10 to 15 RPM to warm the turbine.

Any imbalance of the rotor can lead to vibration, which in extreme cases can lead to a blade breaking away from the rotor at high velocity and being ejected directly through the casing. To minimize the risk, it is essential that the turbine be very well balanced and turned with dry steam that is, superheated steam with a minimal liquid water content. If water gets into the steam and is impacted on the blades (moisture carry over), rapid impingement and erosion of the blades can occur and lead to imbalance and catastrophic failure. Also, water entering the blades will result in the destruction of the thrust bearing for the turbine shaft. To prevent this, along with controls and baffles in the boilers to ensure high quality steam, condensate drains are installed in the steam piping leading to the turbine. Modern designs are sufficiently refined to a level that problems with turbines are rare and maintenance requirements are relatively small.

7.21 SPEED REGULATION

Alternating electrical power must be produced at a fixed frequency of 60 cycles/s (or Hz), although in many parts of the world 50 cycles/s is favored. To achieve this, the turbine rotor is made to run at a nearly fixed speed of 3600 rpm for 60 Hz electrical signal. For 50 Hz the corresponding rotor speed is 3000 rpm.

The control of a turbine with a governor is essential, as turbines need to be run up slowly, to prevent damage. Certain applications such as the generation of alternating current electricity require precise speed control. Uncontrolled acceleration of the turbine rotor can lead to an over-speed trip, which causes the nozzle valves that control the flow of steam to the turbine to close. If the valves fail to control for any reason, then the turbine may continue accelerating because of the huge mass and associated inertia of the rotor until it breaks apart, often spectacularly. Turbines are expensive to make, requiring precision manufacture and special quality materials.

During normal operation in synchronization with the electricity network, power plants are governed with a five per cent droop speed control. This means the full load speed is 100 per cent and the no-load speed is 105 per cent. This is required for stable operation of the network without causing the generator to hunt between the upper and lower speed points, leading to drop-out of the generator from the power grid. Normally, the changes in speed are minor. Adjustments in power output are made by slowly raising the droop curve through increased spring pressure on a centrifugal governor. Generally, this is a basic system requirement for all power plants because the older and newer plants have to be compatible in response to the instantaneous changes in frequency without depending on outside communication.

7.22 MAINTENANCE CONSIDERATIONS

No regulatory maintenance practices are available in North America and Europe, or intervals specified for non-nuclear steam turbines regardless of the industry or application. As such, the frequencies and tasks are defined by the turbine manufacturers, consultants, industry organizations, plant personnel, plant process requirements, or insurers based on past experience. In Japan, there are regulatory requirements for periodic maintenance.

However, the recommended scheduled maintenance requirements for steam turbines are quite similar. For establishing longer time intervals between major overhaul outages, there are a number of different approaches which are utilized worldwide. Regardless of the approach, it is important that the methodologies effectively establish the overhaul intervals based on the highest risk portions of the steam turbine.

Maintenance tasks and frequencies should be prioritized towards those portions of the steam turbine that have the highest risk. This usually means protecting the steam turbine from over-speeds, water induction, loss of lube oil, corrosive steam, and sticking valves that could cause major damage to the turbine. Related tasks include conducting internal inspections of the turbine flow-path, shells and rotors for failure mechanism damage [creep, erosion, corrosion, fatigue, thermal fatigue, and stress corrosion cracking (SCC)] to detect damage early enough to prevent a subsequent major failure.

Never place any part of the body near rotating machinery. Although it is unlikely anyone will ever attempt to grab the main shaft while it is rotating at 200 rpm, there are other things to be considered. While the main engine jacking gear is engaged, the shaft is rotating at a very slow rate. Despite this slow rotation, watch-standers still should not be permitted to do any type of work to the main shaft, such as painting or cleaning the main shaft.

Oil leaks must be corrected at their source. Spills of any kind must be wiped up immediately and wiping rags disposed of immediately or stored in fire safe containers. Failure to observe safety with any petroleum product can result in a major Class B fire. Promptly reinstall shaft guards, coupling guards, deck plates, handrails, flange shields and other protective devices removed as interferences immediately after completion of maintenance on machinery, piping, valves or other system components.

An open main engine presents special safety precautions. While the main engine is open, an E-5 or above is required to stand guard. A security area is established around the main engine using ropes and signs. No tools are permitted within the security area without first being inventoried. Before personnel are permitted to enter the security area, they must remove all jewellery, securely fasten eyeglasses and tools to their body using lanyards, and all clothing fasteners must be covered with tape. As an added safety precaution, ensure all warfare and rank devices are removed before entering the security area. This precaution prevents inadvertent introduction of anything that could cause damage to the turbines.

7.23 FAILURE MODES IN STEAM TURBINE

There have been numerous cases of steam turbine failures worldwide. The highest frequency events have been loss of lube oil incidents, while the highest severity events have been over-speed events. Typically, higher frequency and higher severity events have been blade/bucket failures, particularly in the low pressure (LP) section of the turbine where the blading experienced a number of failure mechanisms (SCC, erosion, foreign object damage (FOD)) which ultimately led to failure.

The technologies being incorporated into new steam turbines are more sophisticated, require operation at higher pressures and temperatures, and generally have smaller clearances between stationary and rotating parts to improve efficiency. While these technologies have not caused any large losses, the inherent risk exposures are increasing and the in-service experience with these technologies needs to be monitored.

Safety is of overriding importance while operating the main engine. Failure to observe safety precautions can result in damage to the main engine, personnel injuries, or even death. Even though many safety precautions seem to be common sense, many times personnel fail to consider the results of their actions.

7.24 TEST CODES

British, German, other national and international test codes are used to standardize the procedures and definitions used to test steam turbines. Selection of the test code to be used is an agreement between the purchaser and the manufacturer, and has some significance to the design of the turbine and associated systems.

In the United States, American Society of Mechanical Engineers (ASME) has developed several performance test codes on steam turbines. These include the ASME PTC 6-2004, Steam Turbines, ASME PTC 6.2-2011, and Steam Turbines in Combined Cycles, PTC 6S-1988, Procedures for Routine Performance Test of Steam Turbines. These ASME performance test codes have gained international recognition and acceptance for testing steam turbines.

7.25 REVIEW QUESTIONS

7.25.1 What are the underlying assumptions made in the derivation of isentropic efficiency in a steam turbine?

7.25.2 How does efficiency of a turbine stage relate with enthalpy drop and work done in the stage?

7.25.3 How does a Rateau stage differ from a Curtiss stage, and where do they find usage in a steam turbine?

7.25.4 What are the primary differences between reaction and impulse turbine stages?

7.25.5 How can long turbine blades be stiffened to prevent them from twisting?

7.25.6 What causes steam chests to be prone to failure, and what steps may be taken to ensure their integrity?

7.25.7 How does thrust develop in the axial direction of a steam turbine rotor, and what is required to counteract it?

7.25.8 What causes a rotor to bow, and how is it avoided?

7.25.9 What are the priorities in establishing maintenance tasks for a steam turbine?

7.25.10 Describe methods and devices used to prevent disruption of lubrication oil to the main bearings.

7.25.11 List some major failure mechanisms that can lead to catastrophic breakdowns of turbine systems.

7.26 EXAMPLE PROBLEMS

Example Problem # 7.26.1: A blade section is located 27.5 in. from the shaft centerline. If the rotor speed is 3600 rpm, determine the tangential speed of the section.

Solution: Tangential speed $U = 2\pi \times r \times N/60 = 2\pi \times (27.5/12) \times 3600/60$
$= 863.5$ ft/s.

Example Problem # 7.26.2: A blade section located 21.5 in. from the shaft centerline of a Parsons steam turbine has steam entering at absolute velocity 752.27 ft/s at an angle 41.25°. If the rotor speed is 3600 rpm, calculate the blade efficiency.

Solution: From equation (7.6), $\eta_b = [2U(2V_1\text{Cos }\alpha_1 - U)]/[V_1^2 - U^2 + 2V_1U\text{Cos }\alpha_1]$

$U = 2\pi \times r \times N/60 = 2\pi \times (21.5/12) \times 3600/60 = 675.1$ ft/s.

$V_1 = 752.27$ ft/s. and $\alpha_1 = 41.25°$, Cos 41.25° =0.7521

Hence blade efficiency:

$$\eta_b = [2 \times 675.1 \times \{2 \times 752.27 \times 0.7521 - 675.1\}]/[752.27^2 - 675.1^2 + 2 \times 752.27 \times 675.1 \times 0.7521]$$
$$= 0.7051$$

Example Problem # 7.26.3: Calculate the tangential force on a blade if the steam flow rate in a steam turbine stage is 113.3 lb/hr, and swirl velocities V_{w1} and V_{w2} are 103.5 ft./s and 87.9 ft./s.

Solution: Tangential force on the blades is:

$$F_u = [(dm/dt)/g] \times (V_{w1} - V_{w2})$$

Then: $F_u = (113.3) \times (103.5 - 87.9)/32.2 = 54.89$ lb.

Example Problem # 7.26.4: A steam turbine rotor's speed is 5400 rpm and blade section height from shaft centerline is 18.75 in. Calculate the power developed if the difference in swirl velocities (ΔV_w) = 54.95 ft./s and if the steam flow rate is 245.25 lb./s.

Solution: $U = 2\pi \times r \times N/60 = 2\pi \times (18.75/12) \times 5400/60 = 883.13$ ft/s.

From eq. 7.3, power developed is: $W = [(dm/dt)/g] \times U \times (\Delta V_w)$.

Then: $W = [(245.25)/32.2] \times 883.13 \times (54.95) = 369609$ ft-lb./s.

Example Problem # 7.26.5: Determine the torque developed in an impulse turbine stage where radius is 18.25 in., steam flow rate is 379.55 lb./s, swirl velocity at inlet is 148.45 ft./s and at outlet is 178.49 ft./s.

Solution: From eq. 7.2, power developed is:

$$T = (dm/dt) \times (r_2 V_{w2} - r_1 V_{w1})$$

Radius $r_2 = r_1 = r$ for an impulse turbine.

Hence:

$$\begin{aligned} T &= (dm/dt) \times (r_2 V_{w2} - r_1 V_{w1}) \\ &= (379.55/32.2) \times (18.25/12) \times (178.49 - 148.45) \\ &= 538.51 \text{ ft-lb.} \end{aligned}$$

7.27 EXERCISE PROBLEMS

Exercise Problem # 7.27.1: Calculate the steam flow rate if a steam turbine rotor's speed is 4500 rpm and blade section height from shaft centerline is 16.78 in, power developed is 278500 ft-lb/s and difference in swirl velocities is 102.87 ft/s.

Exercise Problem # 7.27.2: The torque developed in an impulse turbine stage is 5125.0 ft-lb. Find the steam flow rate if tangential speed is 47.25 in., swirl velocity at inlet is 239.65 ft/s and at outlet is 98.45 ft/s.

Exercise Problem # 7.27.3: A blade section is located 17.5 in. from the shaft centerline. If the tangential speed of the section is 912.25 ft/s, calculate the rotor speed.

Exercise Problem # 7.27.4: The tangential force on a turbine blade is 118.75 lb, and the steam flow rate is 123.87 lb./hr. Determine the difference in the swirl velocities V_{w1} and V_{w2}.

Exercise Problem # 7.27.5: The torque developed in an impulse turbine stage is 42115.0 ft-lb. The steam flow rate is 452.65 lb/s. Calculate the blade radius if swirl velocity at inlet is 372.65 ft/s and at outlet is 112.45 ft/s.

REFERENCES AND BIBLIOGRAPHY

American Society of Mechanical Engineers, ASME PTC 6S-1988, Procedures for Routine Performance Test of Steam Turbines, New York, 1988.

American Society of Mechanical Engineers, ASME PTC 6-2004, Steam Turbines, New York, 2004.

American Society of Mechanical Engineers, ASME PTC 6.2-2011, Steam Turbines in Combined Cycles, New York, 2011.

Boyce, M., "Handbook for Cogeneration and Combined Cycle Power Plants", ASME Press, New York, 2002.

Brown, C. J., ASME Paper # 2000-GT-595, New York, 2000.

Chmielewski, R., Jacobucci, S., Harkins, W., Kuten, P., Wu, S., Berruti, A., McArthur, J., "Unique Combined Cycle Design Caters to Plants with Cyclical Demand Profiles", Power Engineering, January, 2002.

Cofer, J. I., "Advances in Steam Path Technology", ASME Paper # 95-CTP-2, New York, 1995.

Introduction to Naval Engineering ISBN 0-87021-320-2.

Leyzerovich, A., "Large Power Steam Turbines: Design and Operation", Volume 1, Pennwell Books, Tulsa, OK, 1997.

Leyzerovich, A., "Large Power Steam Turbines: Operation", Volume 2, Pennwell Books, Tulsa, OK, 1997.

McCloskey, T. H., "Turbine Steam Path Damage: Theory and Practice, Turbine Fundamentals", Volume I, EPRI, 1999.

McCloskey, T. H., "Turbine Steam Path Damage: Theory and Practice, Damage Mechanism", Volume II, EPRI, 1999.

NAVOSH Manual OPNAVINST 5100.19B VOL II Principles of Naval Engineering NAVPERS 10788 Series.

8

CHAPTER

INDUSTRIAL AND POWER GENERATION GAS TURBINES

8.1 INTRODUCTION

Gas turbines are a practical and economic way for utility and industrial service. Advanced design units offer high firing temperatures, low NO_x and improved efficiency. Newly developed units have a heavy-duty single shaft combustion turbine that combines the proven efficiency and reliability concepts with the latest advances in aerodynamic technology. Designed for both simple and combined cycle applications, the turbines can operate on all conventional turbine fuels and on coal derived low BTU gas produced in an integrated gasification plant.

The exhaust from the gas turbine is typically in the 900 – 1100° F (482 – 593° C) range. The gases have enough energy to produce steam in a waste heat boiler (WHB), or a heat recovery steam generator (HRSG). This aspect is made use of in the 'combined steam and gas turbine cycles'. The units find considerable favor in the electric power generation industry, primarily because of its ability to operate at thermal efficiencies nearing 60 per cent.

The steam turbine may be provided with dual pressure steam cycle since exhaust gas temperature entering the HRSG is high enough. If the exhaust temperature from the turbine is low, the level of available oxygen is sufficient enough to permit auxiliary firing of additional fuel in the HRSG to increase the amount of steam produced. Besides producing electricity, 'cogeneration or combined heat and power (CHP) systems' make use of steam in process and petro-chemical plants for diverse tasks such as drying of materials and providing heating in buildings.

Maintenance costs and availability are two of the most important concerns to the equipment owner. A maintenance program that optimizes the owner's costs and maximizes equipment availability must be instituted. For a maintenance program to be effective, a general understanding of the relationship between operating plans and priorities for the plant is necessary. Manufacturer's recommendations regarding the number and types of inspections must also be taken into account.

Changes in environmental factors play a major role in the operation of an engine. Gas turbines destined for desert environment or high altitudes are more susceptible to variable operating parameters in the engine or in the air. The thermodynamic characteristics of air such as specific heat, density, pressure, temperature, and humidity play a considerable role in the performance of the engine.

8.2 SIMPLE CYCLE GAS TURBINE

In evaluating the performance of an ideal cycle for a gas turbine, two specific parameters that control the work output and efficiency stand out: compression ratio and maximum cycle temperature. This may be observed from Figure 8.1. Commonly referred to as the Joule or Brayton cycle, points 1-2-3-4 represent the ideal cycle. The isentropic cycle efficiency then takes the form:

$$\eta = 1-\left(\frac{1}{r}\right)^{\left(\frac{\gamma-1}{\gamma}\right)} \tag{8.1}$$

where r = pressure ratio $p_2/p_1 = p_3/p_4$, and γ denotes the ratio of specific heats c_p/c_v. Note that for isentropic processes, $T_2/T_1 = r^{(\gamma-1)/\gamma} = T_3/T_4$.

Thus, the efficiency of the ideal cycle depends primarily on the compression ratio and the ratio of specific heats of the working fluid. Efficiency η is shown as a function of compression ratio r for air ($\gamma = 1.4$) in Figure 8.2.

The useful output that a given plant is capable of delivering, W, is of interest in establishing its size. The difference between the output from the turbine and work required to drive the compressor is:

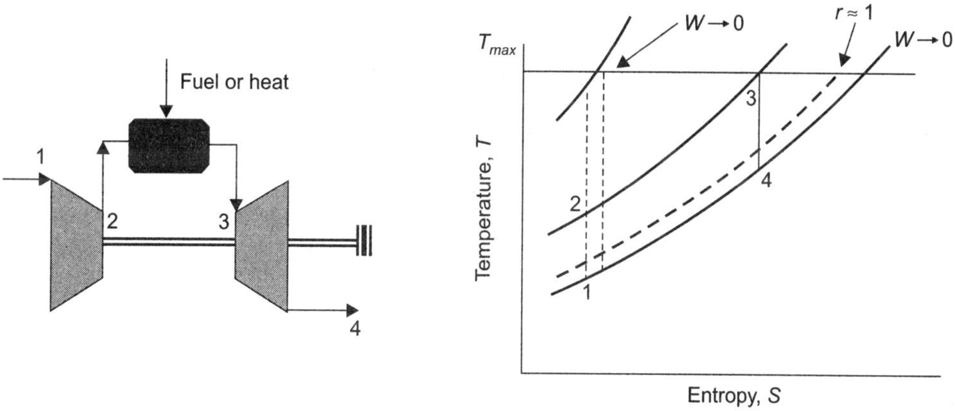

Fig. 8.1 Simple Gas Turbine Cycle (*left*), Temperature-Entropy Diagram (*right*)

$$\frac{W}{c_p T_1} = t\left(1-\frac{1}{r^{(\gamma-1)/\gamma}}\right)-\left(r^{(\gamma-1)/\gamma}-1\right) \tag{8.2}$$

where $t = T_3/T_1$. T_1 is the normal atmospheric temperature that does not change much. The work output, a function of r and t, is expressed in non-dimensional form, W/c_pT_1, and is plotted in Figure 8.3. Maximum temperature T_3, or the ratio t, will be set by the material characteristics of the components in the hot path to withstand the imposed stresses, such as combustor liner, rotating and stationary airfoils and turbine disks. In the absence of protection from elevated temperatures, the ratio t is limited to about 3.5, but with the use of thermal barrier coating and advanced cooling techniques, the value of t may be raised to 5.5.

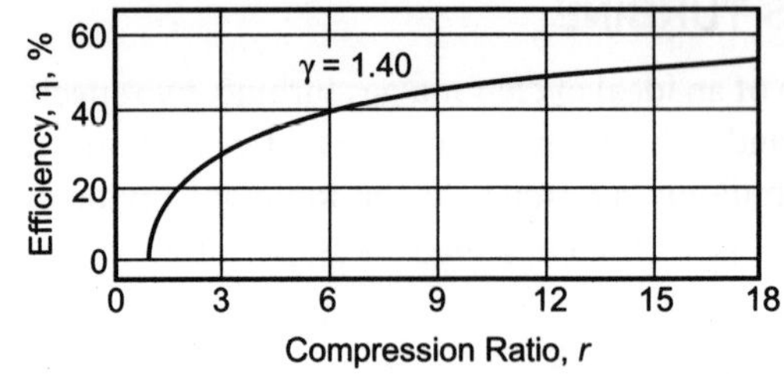

Fig. 8.2 Simple Gas Turbine Cycle Efficiency

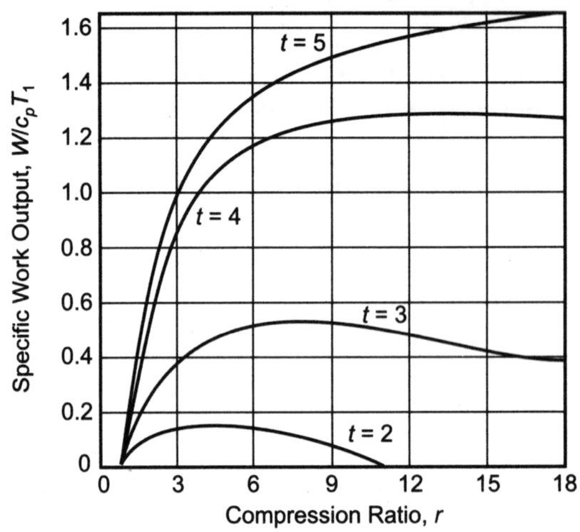

Fig. 8.3 Simple Gas Turbine Cycle Work Output

8.3 EFFECTS OF PRESSURE AND TEMPERATURE

Performance of the engine is generally rated at a standard temperature of 59° F (15° C), although the engine still has to meet the minimum thrust at higher temperatures. At lower ambient temperatures the power developed is higher, and thus requires more careful adjustment of the power lever actuator. The reason behind the deteriorated thrust at higher air temperature is the increased activity among the molecules as they move farther apart, thus reducing the amount of air entering the engine's inlet.

Pressure is dependent on the number of molecules present in a volume. At higher pressures the air has greater density, which causes more molecules to enter and to increase the amount of air in the engine system. The relation between pressure, temperature and density is expressed by: density ratio = KP/T, where K is a constant, P is pressure, and T is absolute temperature.

Water vapor weighs five-eighth as much as dry air, so increased humidity of ambient air decreases the mass of air entering the engine. Gas turbines generally operate with an excess of air needed for combustion. If the dry air temperature is high, for instance on a hot day in a desert region, addition of water at the inlet is justified to lower the temperature and thus increase the mass flow. But when the dry air temperature is low during the nighttime, addition of water will not help to increase the airflow into the compressor.

8.4 COMBINED CYCLE OPERATION WITH STEAM AND GAS TURBINES

The Brayton-Rankine cycle combines the gas turbine with a steam turbine, and has many benefits for electric utilities and for process industries requiring steam. The hot gases from the exhaust of the gas turbine are employed in a fired boiler to generate superheated steam to drive a steam turbine. Figure 8.4 shows the schematic diagram for a turbine installation in the combined cycle mode of operation.

A heat recovery steam generator is used to recover heat from a hot gas stream. Essentially a heat exchanger, it is used in a combined-cycle power station where hot exhaust from a gas turbine is fed into the heat exchanger to generate steam. This combination produces electricity more efficiently than either the gas turbine or steam turbine alone. Another application is in diesel engine combined cycle power plants, where hot exhaust from a diesel engine generates steam.

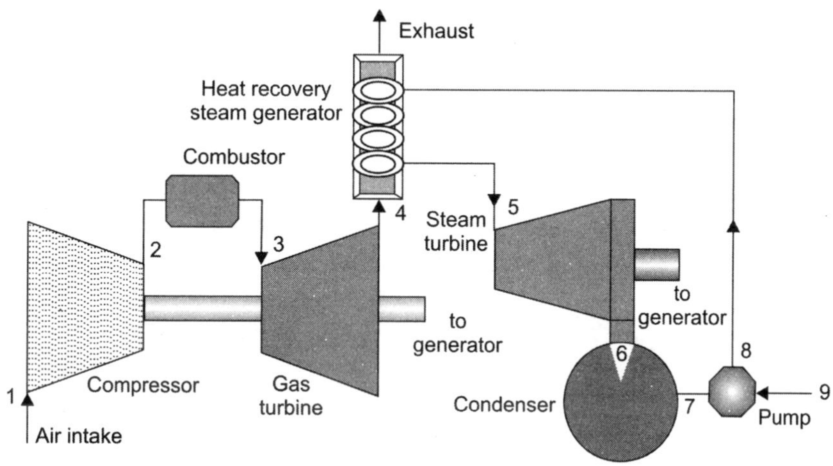

Fig. 8.4 Combined Gas and Steam Cycle Plant

The combined cycle power generation plant permits, as one of its features, quick startup as well as quick response to load change. Many heat recovery steam generators are manufactured based on the proven dynamic characteristic simulation technology, and can follow the quick plant startup and load change with a sufficient margin. Since gas turbines commonly use natural gas or light oil as fuel, their exhaust gas is cleaner than that from conventional thermal power generation facility.

Breakthroughs in energy efficiency continue to be made by the power generation industry. General Electric Power System's H turbine is designed to operate at 60 per cent thermal efficiency in a combined cycle mode. Performance gains derive from advance materials and a new cooling system. This enables the H turbine to operate at 2,600° F firing temperature. Since fuel expense is easily the main component of cost of the electric power, the increase in efficiency has significant economic implications.

8.5 HEAT RECOVERY STEAM GENERATOR

The Heat Recovery Steam Generator (HRSG) comes in numerous shapes, designs, configurations, and arrangements. In a water tube (as opposed to a fire tube) type heat recovery unit the process fluid, steam or water, is on the inside of the tube, with the products of combustion on the outside. The products of combustion are normally at or close to atmospheric pressure.

Modular HRSGs can be categorized by number of ways, such as direction of exhaust gases flow or number of pressure levels. Based on the flow of exhaust gases, HRSGs are categorized into vertical and horizontal types. In the horizontal type, exhaust gas flows horizontally over vertical tubes, whereas in vertical type HRSGs, the gases flow vertically over horizontal tubes. Based on pressure levels, HRSGs may be categorized into single pressure and multi pressure. Single pressure HRSGs have only one steam drum, and steam is generated at single pressure. Multiple pressure HRSGs employ double or triple pressure steam drums. Triple pressure HRSGs consist of three sections: an LP (low-pressure) section, a reheat/IP (intermediate-pressure) section, and an HP (high-pressure) section. Each section has a steam drum and an evaporator section where water is converted to steam. This steam then passes through superheaters to raise the temperature and pressure past the saturation point. Figure 8.5 shows an HRSG unit for an electrical power generation system.

Fig. 8.5 Heat Recovery Steam Generator for Electric Power Generation

In the design of an HRSG, a theoretical heat balance provides the relationship between the tube side and shell side processes. Heat balance calculations permit mathematical prediction of flow velocity, pressure, temperature and density of the water and steam, and also of the flue gases emitting from the gas turbine at selected locations in the overall heat exchanger. The heat calorific value of the fuel, amount of flue gases exiting the gas turbine and flow losses on account of friction are to be included in the heat balance calculations. Before computing the heat balance, the arrangement of tube side components making up the HRSG unit must be decided. The three primary coil types of an

HRSG are the evaporator, superheater and the economizer. Other components may perform related heat exchange services.

8.6 COMPONENT ARRANGEMENT IN HEAT RECOVERY STEAM GENERATOR

A single-pressure HRSG is schematically shown in Figure 8.6. The generalized sketch permits viewing of the water and steam flow in the feed-water circuit and the turbine exhaust in the flue gas circuit as the two flow in opposed directions. The blow-down shown in the diagram from the steam drum refers to extraction of water contaminated from built-up solids such as salts, iron oxide, silt and other extraneous material that finds its way and collects in the bottom of the drum.

Modern HRSG units are more complex then this simple arrangement, with the components placed in many configurations to achieve the desired results. The range of arrangements that the coils may be placed is only limited by the users imagination and the constraints of the temperature approaches.

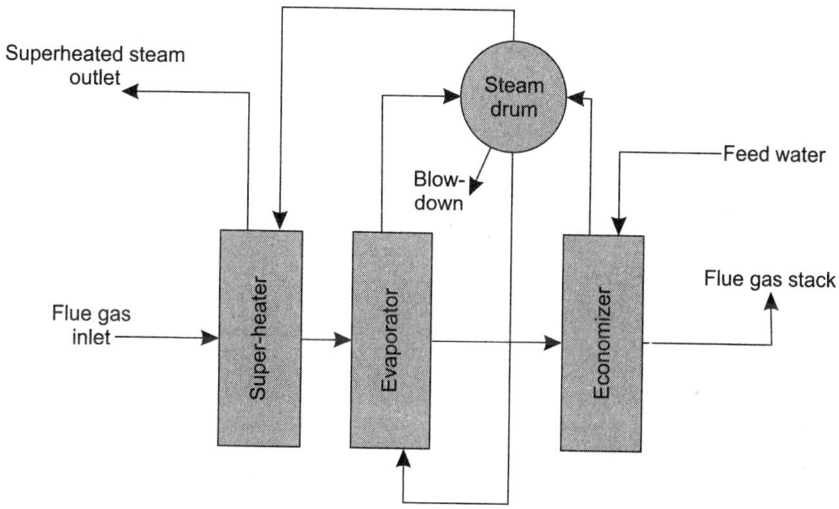

Fig. 8.6 Schematic of Single-Pressure HRSG Unit

The most important component is the evaporator section, since without this coil (or coils), the unit would not be an HRSG. In the evaporator coils, the effluent (water), passing through the tubes is heated to the saturation point at the pressure it is flowing. The evaporator section type is important since it generally defines the overall configuration of the HRSG unit. Although there are many types of HRSGs, five general evaporator types are commonly used: D-Frame, O-Frame, A-Frame, I-Frame and horizontal tube evaporator, as shown schematically in Figure 8.7.

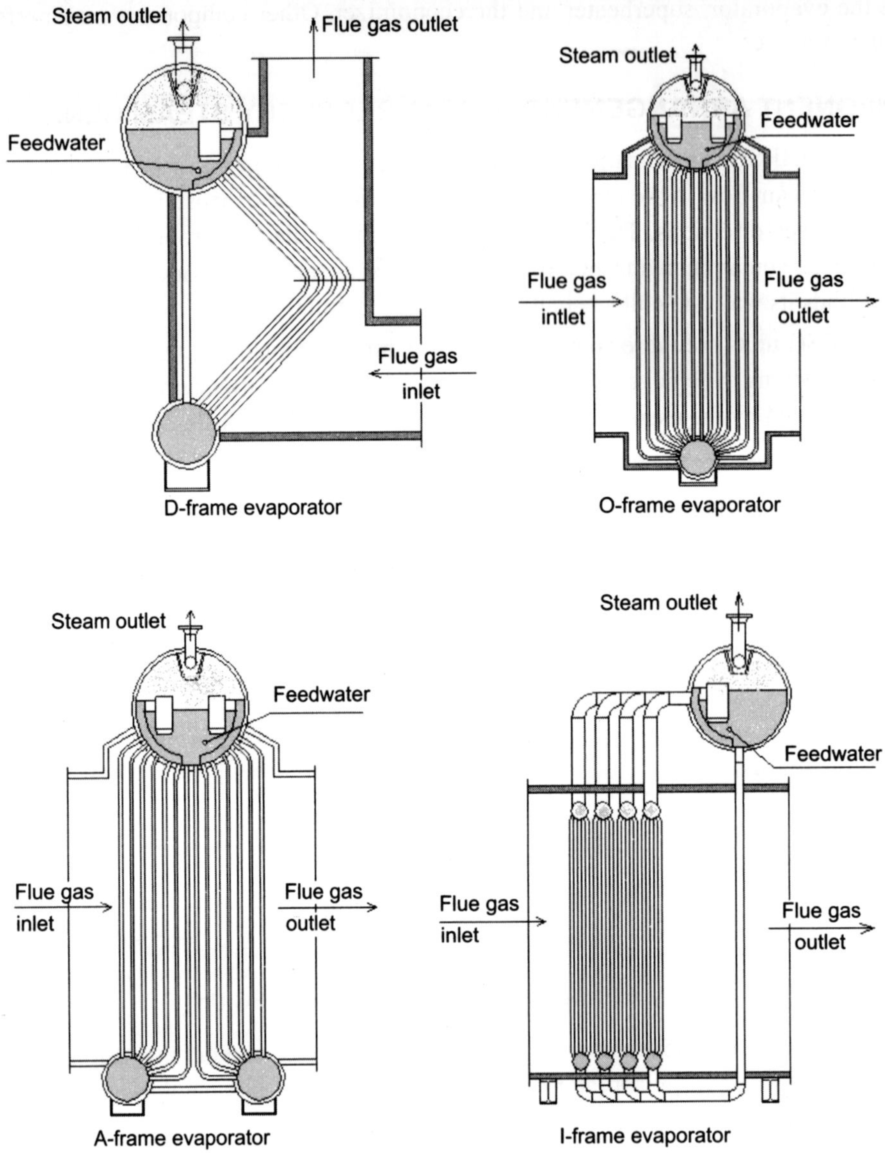

Fig. 8.7 Evaporator Types for HRSG

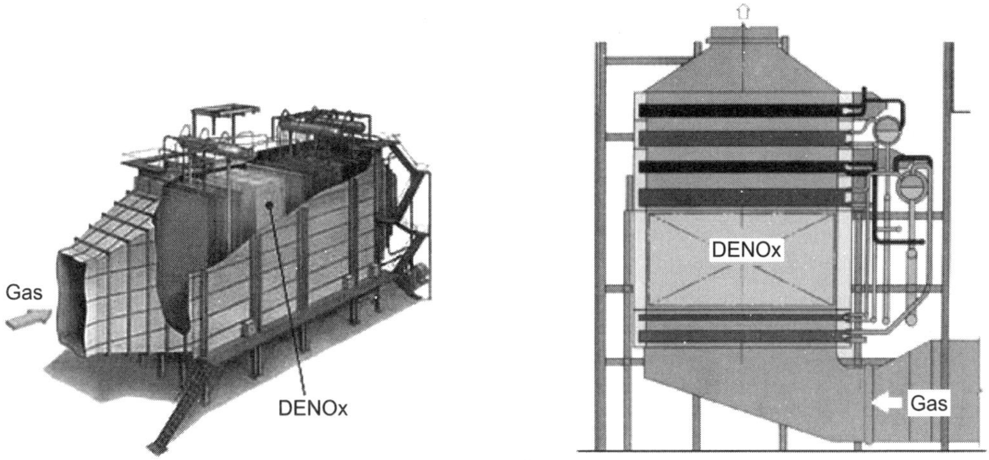

Fig. 8.8 Heat Recovery Steam Generators with DENOx Units

The economizer section, also called pre-heater or preheat coil, pre-heats the feed-water prior to admittance into the system. Fresh feed-water is needed to replace the steam (vapor) being removed from the system through the superheater or steam outlet and the water loss through blow-down.

The superheater section of the HRSG is used to dry the saturated vapor being separated in the steam drum. In some units, it may only be heated a little above the saturation point, while in other units it may be superheated to a significant temperature for additional energy storage. The superheater is normally located in the hotter gas stream, in front of the evaporator.

Both the economizer and superheater designs normally follow along with the evaporator type that is being used. The horizontal tube designs would normally be used for the D-Frame evaporator if the gas flow is vertically up at the outlet.

HRSGs can also have diverter valves to regulate incoming flow. This allows the gas turbine to continue to operate when there is no steam demand or if the HRSG needs to be taken offline.

Emissions controls may also be located in the HRSG. Some may contain a Selective Catalytic Reduction (SCR) system to reduce nitrogen oxides or a catalyst to remove carbon monoxide. The inclusion of an SCR affects the layout of the HRSG. NO_x catalyst performs best in temperatures between 650°F and 750°F. This usually means that the evaporator section of the HRSG will have to be split and the SCR unit placed in between the two sections (Figure 8.8).

8.7 AUXILIARY FIRING IN HEAT RECOVERY STEAM GENERATOR

Supplemental, or duct, firing in HRSGs permits installation of more burners to provide additional energy to produce extra steam, hence increasing the output of the steam turbine. Generally, duct firing provides electrical output at lower capital cost. It is therefore often utilized for peaking.

Packaged HRSGs are shipped as fully assembled units from the factory. They can be used in waste heat or turbine (usually under 20MW) applications. The packaged HRSG can have a water-cooled furnace, which allows for higher supplemental firing and better overall efficiency.

8.8 STEAM AND ELECTRIC POWER COGENERATION

Since a greater part of a fuel's thermal energy is lost in the exhaust gases of a prime mover, converting waste energy to electric power or steam improves the efficiency of the system, which provides the underlying concept for co-generation. Waste heat can be recovered economically in many different processes. Petro-chemical plants, iron and steel foundries and paper and pulp mills require modified power generation units to directly convert wasted thermal energy into useful mechanical and electrical power. Use of heavy crude, pulverized coal, sawdust and many other by-products of industries support the use of externally fired gas turbines, permitting protection of hot gas path components from chemical and particulate corrosion.

Gas turbine co-generation is far more efficient than a standard steam utility plant, with 75 per cent of the thermal energy utilized to generate power and heat. In comparison, fossil steam plants average 35 per cent thermal efficiency, with 45 to 48 per cent consumed as losses in the condenser and 13 to 15 per cent in the boiler. The exhaust gas from the turbine produces steam in a heat recovery steam generator (HRSG) or a waste heat boiler (WHB). Dampers may be provided between the gas turbine and HRSG in the event operation is desired in the simple cycle mode. Use of regenerators with gas turbines is common, helping reduce heat input in the combustor when operating at low pressure ratio and low firing temperature.

Classification of co-generation systems is not facilitated by the many different possibilities. Lack of fresh water, for example in the deserts of North Africa and the Mojave, calls for desalination plants. Co-generation and hybrid processes may be combined to generate electric power and desalt seawater. A gas turbine's exhaust can be used for steam generation; the steam may then be gainfully employed to drive steam turbine, or used in the distillation's brine heater.

In an interesting twist, co-generation may be used to cool inlet air to the gas turbine for improving its power output when ambient temperature during the daytime is high. At night when the air temperature falls, the additional turbine output drives a chilling and refrigeration plant, and the resultant ice is used to cool intake air during days. The process is well adaptable to desert areas where air temperature at midday approaches 120° F, then plunges to 40° F at night. Cooling of intake air may be of evaporative, refrigeration chilling or inlet fogging types. The effectiveness of the systems varies with relative humidity of the inlet air. Absorption chillers have a low coefficient of performance, but can extract heat at temperatures as low as 250° F. Merchant power plants generally have mechanical chillers with a higher performance coefficient. A centrifugal screw or reciprocating compressor pressurizes the refrigerant for complete condensation in a condenser. After passing through a pressure reduction valve, the liquid refrigerant is partially vaporized. Latent heat of vaporization then cools water to form ice flakes, which in turn cools the turbine inlet air; under appropriate conditions, ice formation may be bypassed to directly cool the incoming air. Storage facility for the ice is required during the night for use during the day.

8.9 INDUSTRIAL COMBUSTION TURBINE

A common configuration of combustion turbine calls for a bolted construction rotor that is supported in two bearings. The bearings may have compounded form, with tilting pads in the lower half and a fixed arc element on the upper side. The compressor segment of the rotor is assembled from spigotted disks bolted together with through bolts. Alignment and torque transmission are assured by employing radial pins between the disks. The turbine rotor section is made up of disks provided with curvic clutches, and are bolted together by 12 through bolts. The curvic clutch is machined in the form of uniformly spaced teeth protruding axially from the flat face of the disk. The teeth engage and interlock with a similar pattern machined on the face of the adjacent disk, providing a slippage-free joint under the action of the clamping load from the through bolts.

The combustion system consists of cans arranged in a circular pattern about the shaft centerline. Stability of the flame and uniformity in the distribution of fuel flow between the combustors are monitored by thermocouples located downstream of the last turbine stage. Ultra-violet sensors may aid in detecting malfunctions in the combustor when at load and sensing of ignition during the startup mode.

The casings are split horizontally to permit performance of maintenance with the rotor in place. The inlet and compressor casings are made of nodular cast iron and cast steel, while the combustor, turbine and exhaust casings are of alloy steel. The inlet end bearing housing is supported from eight radial struts. At the exhaust end the bearing is supported by tangential struts that respond gradually during transient conditions, and maintain alignment between the rotor and the bearing through rotation of the housing about its axis to accommodate thermal growth. The arrangement provides an additional benefit of reducing thermal stresses in the struts. The fairings around the struts are configured in the form of an airfoil to enhance aerodynamic performance.

Since the turbine operates at extremely high inlet temperatures, cooling air for the rotor is extracted from the compressor discharge, and is externally cooled and filtered before returning to the torque tube casing to cool the disks and the first, second and third stage blades and vanes. Filtration is deemed essential for eliminating blockage by suspended particles of the intricate cooling passages inside the blades. Bleed air from the compressor is also used to cool the blade ring cavities and to cool and purge the inter-stage disk cavities to prevent ingestion of hot blade path gases.

Compressor diaphragms are coated to improve aerodynamic performance and to obtain protection from corrosion. The stationary vanes and rotating blades for the first two stages of the turbine may be provided with a thermal barrier coating. The operating firing temperature level of 1425° C is selected to be commensurate with the capabilities of the super alloys for the components in the hot path and with the cooling schemes. The cycle pressure ratio is chosen to maximize power output during simple cycle operation and efficiency in the combined cycle mode. With a compression ratio of 19.2:1, the potential combined cycle efficiency is 58 per cent. Cycle airflow rate is controlled through the annular flow area at the turbine exit, since the last stage blade stress level is directly proportional to it. A flow rate of 1200 lb./second results in a conservatively stressed blade and higher power output.

The compressor is equipped with variable inlet guide vanes for improving the low-speed surge characteristics and to enhance part-load performance in combined cycle applications. The rotating blades are controlled diffusion airfoils, made of multiple circular arc forms. Stationary blades are fabricated in two 180° segments for easy removal, and provided with sealing at the inner shroud.

Moderate aerodynamic loads are used for the four turbine stages operating at a higher peripheral speed. Aerodynamic airfoil shapes are obtained from a fully three-dimensional viscous analysis code. The third and fourth row blades are shrouded. The first row stationary vanes are individual precision cast of IN939 alloy, and can be removed from access man-ways without lifting the cylinder cover. Inner shrouds are supported from the torque tube casing to limit flexural stresses and distortion. Vane segments for the other rows are supported in a separate inner ring.

The dry low NO_x combustor operates at 25 ppm NO_x level at 1260° C turbine inlet temperature. Steam cooling is used for reduced emissions at higher firing temperatures. By eliminating the transition cooling air, virtually all the combustion air is introduced into the primary zone of the combustor to maintain the flame temperature.

8.10 FACTORS INFLUENCING MAINTENANCE AND EQUIPMENT LIFE

Several factors influence equipment life, and should be accounted for in maintenance planning. Starting cycle, power setting, fuel and level of steam or water injection are key factors in determining the maintenance interval requirements, since they play a key role in the life of critical gas turbine parts.

Consider a natural gas fired turbine operating on continuous duty and without water or steam injection as the baseline for setting the maximum recommended maintenance intervals. Maintenance factors may be obtained for operation differing from the baseline for increased levels of maintenance. Thus, a factor of two implies half of the baseline interval of maintenance.

Most heavy-duty gas turbines are capable of withstanding severe duty. Selected combustion components, hot-gas-path parts and rotor assemblies may require qualified off-site shop service. Most other components can be refurbished onsite. Onsite maintenance can be readily attended for the following components:

- Casings, shells and frames split on machine's horizontal plane. Internal parts may be accessed after lifting the upper half.
- Compressor stator vanes and variable inlet guide vanes slide tangentially out of the casing once the upper casing is removed, with the rotor staying in place after lifting the upper half of inlet casing.
- The turbine first stage nozzle assembly can be similarly inspected, repaired, replaced or removed when the upper-half of the turbine shell is lifted, with the rotor unaffected. Later stage nozzle assemblies are in some designs fitted with the upper half of the turbine shell. Turbine blades may also be inspected and removed.
- Prior to assembly in the spool, rotating turbine buckets are moment weighed and programmed to be located in the proper circumferential position of the turbine wheel to obtain dynamic balance. The blades may then be replaced without the need to remove or rebalance the rotor assembly.
- Journal bearing housings and liners are split in the horizontal plane for inspection and replacement if necessary. The lower half of the bearing liner can be removed with the rotor in place.
- Seals and shaft packings are not attached to the main bearing housings and casing structures, and may similarly be removed and replaced.
- Many combustor designs have fuel nozzles, combustion liners and flow sleeves that can be removed for inspection and service without lifting any casings.
- Filters and coolers and other accessories are accessible subassemblies to permit inspection and maintenance of individual parts. Visual and bore-scope inspection and clearance measurement of hot gas-path components may also be made without removal of the gas turbine outer casings and shells.

8.11 BORE-SCOPE INSPECTION

Radially aligned holes are machined in the compressor casings and turbine shells for visual inspection of intermediate compressor rotor stages, turbine buckets and turbine nozzle partitions with an optical bore-scope. The radial holes permit insertion of the optical bore-scope into the compressor or turbine flow path area.

Bore-scope ports are generally located a few inches above the horizontal split-line flange at three to four locations, both on the compressor and turbine shells. Leading and trailing edges of stator and rotor airfoils down to the root may then be conveniently inspected for damage arising from fatigue, thermal or foreign object ingestion.

Only if the bore-scope inspection reveals damage to the internal parts of the rotor and casing can the lifting of the casings and shells from a turbine unit be necessitated. A recommended interval for a planned bore-scope inspection program may be prepared following initial base line inspections and review of operating parameters from sensors. Bore-scope inspections can only be carried out when the machine is off-line, so adjustment of intervals may be based on operating experience of a turbine unit's mode of operation, fuel used and indications from previous bore-scope inspections. Scheduled outages may then be pre-planned, while also permitting acquisition of parts needed for replacement.

8.12 FUEL

Gas turbines are capable of operating on natural gas and a variety of liquid fuels. Residual oils will impact maintenance requirements for the units. Heavier hydrocarbon fuels have a maintenance factor ranging from three to four for residual fuel and two to three for crude oil fuels. The liquid fuels generally release a higher amount of radiant thermal energy, which results in a subsequent reduction in combustion hardware life. If untreated, liquid fuels frequently contain corrosive elements such as sodium, potassium, vanadium and lead that can accelerate hot corrosion of turbine nozzles and blades. Certain elements in the fuels do not burn completely, and form deposits either directly or through compounds formed with inhibitors used to prevent corrosion. These deposits lead to deteriorated performance, and can require more frequent maintenance.

Refined distillates of liquid hydrocarbons do not generally contain high levels of corrosive elements, but harmful contaminants can creep in during transportation. Distillate fuel oils can get contaminated in a couple of different ways. Saltwater ballast often mixes with no. 2 distillate fuel during sea transport. Another source of contamination of the distillate fuel is when transported to site in tankers or pipelines that previously transported contaminated fuel, chemicals or leaded gasoline. Experience with distillate fuels indicates that maintenance factor for hot gas path can range from as low as one (equivalent to natural gas) to as high as three. In the absence of operating experience, a maintenance factor of 1.5 may be advisable for hot gas path components for operation on distillate oil. Contaminants in liquid fuels also tend to affect the life of auxiliary components such as fuel pumps and flow dividers.

Gas fuels may be considered the optimum fuel and have the least negative impact on turbine maintenance. Dry Low NO_x (DLN) combustion systems have brought to the forefront the significance of proper fuel quality. If the requirements for fuel specifications are properly adhered to, the combustion system will operate reliably while maintaining manufacturer's warranties. Carryover from liquid hydrocarbon can expose hot-gas-path hardware to severe over-temperature conditions and consequent thermal distress. Gas scrubber systems and superheating the gaseous fuel prior to firing can control this potential issue.

Besides fuel, suspended contaminants such as sand and rust particles can enter the turbine through the air intake when the air filters are not working at optimum efficiency. Steam or water injected for

NO_x emission control or for power augmentation and carryover from evaporative coolers are another source. Occasionally, sources of contaminants other than the fuel can lead to hot-gas-path degradation equal to that seen with fuel-related contaminants. Turbine manufacturers define limits for maximum concentrations of contaminants for fuel, air and steam/water.

8.13 WATER OR STEAM INJECTION

Water or steam injection for emissions control or power augmentation can impact parts lives and maintenance intervals even when the quality of water or steam meets specifications. The effect of the added water on the hot-gas transport properties is largely responsible for this. The higher gas conductivity increases the heat transfer to the buckets and nozzles, leading to higher metal temperature and reduced parts lives. The impact on the useful life of parts from steam or water injection is related to the turbine's control system. If the control system reduces the firing temperature when water or steam is injected, this counters the effect of the higher heat transfer on the gas side, and does not affect bucket life. On the other hand, when the control system leaves firing temperature unaffected with water injection level, the unit's output increases. At the same time, however, it decreases parts life. Turbines used in peaking applications are generally left uncontrolled during injection of water or steam. Yearly accumulated operating hours are low, so reduced parts lives are justified by the power advantage. The two modes may be described as operation in dry and wet control curves. For wet control curve operation, the maintenance factor may be increased to two at 5% water injection.

Still another factor associated with water or steam injection arises from higher aerodynamic loading on the turbine components. The injected water increases the cycle pressure ratio. The extra burden will increase the downstream deflection rate of the second and third stage nozzles, which will reduce the repair interval for the two components. Improved alloy material with higher creep strength for the two stages can minimize this factor.

8.14 INLET FOGGING

One of the ways some users increase turbine output is through the use of inlet foggers. Foggers inject a large amount of moisture in the inlet ducting, exposing the forward stages of the compressor to a continuously moist environment. Operation of a compressor in such an environment may lead to long-term degradation of the compressor due to fouling, material property degradation, corrosion and erosion. Experience has shown that fouling of the compressor can be severe with inlet foggers, depending on the quality of water used, ducting material, and condition of inlet silencer. Evaporative cooler carryover and excessive water washing can produce similar effects.

Long-term material property degradation usually results from operating the compressor in a wet environment. For certain compressor blades the presence of moisture will reduce blade fatigue strength by as much as 30 per cent, while subjecting the blades to corrosion. Further reductions in fatigue strength will result if the environment is acidic and if pitting is present on the blade. Pitting is corrosion-induced, and blades with pitting can see material strength reduced to 40 per cent of its virgin value. The presence of moisture also increases the crack propagation rate in a blade if a flaw is present.

Water droplets in excess of 25 microns in diameter will cause leading edge erosion on the first few stages of the compressor. This erosion, if sufficiently developed, may lead to blade failure. Additionally, the roughened leading edge surface lowers the compressor efficiency and unit performance. It is recommended to check for erosion and pitting of the compressor blades after every 100 hours of water wash. Utilization of inlet fogging or evaporative cooling may also introduce water carryover or water ingestion into the compressor. Although the design intent of evaporative coolers and inlet foggers should be to fully vaporize all cooling water prior to its ingestion into the compressor, evidence suggests that on some systems the water is not being fully vaporized (for instance, streaking discoloration on the inlet duct or bell mouth). If this is the case, then the unit should be inspected every 100 hours of combined water wash, inlet fogger and evaporative cooler operations.

8.15 OPERATION AT EXTREME LOW TEMPERATURE

Compressor rotors are susceptible to failure arising from the material becoming brittle when exposed to extremely low temperatures. The problem becomes acute if a gas turbine goes through a start early in the morning in the winter in places close to the Arctic Circle, when temperatures come close to or dip below 40° F/40° C. During shutdown, the substrate temperature will be close to outside ambient conditions. As the engine goes through the start cycle, the incoming rush of air from the intake plenum will result in further chilling effect of the rotor. The heating effects of fuel burned in the combustor and raising the temperature of the rotor cannot be felt for some period since it takes time for the igniter to initiate the flame and the operation to stabilize.

Land-based gas turbine rotors are heavy in weight, so manufacturers prefer to use low cost materials for the compressor spool. Alloys that can withstand the low temperatures without experiencing brittleness and subsequent failures are difficult to justify since manufacturing cost will be high. At the same time, the turbine user would not be willing to wait to start the engine till the air temperature rises to more conducive levels.

One available option in such cases calls for heating the incoming air at the intake plenum. Gas-fired or electric air heaters may be placed to avoid the cold draft directly impacting the compressor spool.

Still other problems associated with extreme low ambient temperatures will come from liquid fuels and bearing lubricant oil. The increased density of the liquid fuel may not permit the fuel to go through the fuel manifold and injectors. Similarly, lubricant oil becomes too viscous to allow it to be pushed through the pipes. It then becomes necessary to heat the liquids prior to pumping.

8.16 CYCLIC EFFECTS

Thermal fatigue in combination with mechanical fatigue is the dominant limiter of life for peaking machines, while creep, oxidation and corrosion are the dominant limiters of life for continuous duty machines. Interactions of these mechanisms may be considered, but to a great extent are second order effects. So gas turbine maintenance requirements may be based on independent counts of starts and hours. Whichever criterion limit is first reached determines the maintenance interval. The recommendations for inspection fall within the design life expectations. Thus, components verified to be acceptable for continued use at the inspection point will have low risk of failure during the subsequent operating interval.

An alternative approach converts each start cycle to an equivalent number of operating hours (EOH) with inspection intervals based on the equivalent hours count. But the method usually comes up with longer intervals when more frequent maintenance inspections may be necessary. Midrange duty applications, with hours per start ratios of 30–50, are particularly penalized by this approach. As an illustration, consider General Electric's MS7001EA gas turbine operating on gas fuel, at base load conditions with no steam or water injection or trips from load. The unit will be assumed to operate 4000 hours and 300 starts per year. The operator needs to perform hot gas path inspection after four years of operation, with starts being the limiting condition. Performing maintenance on this same unit based on an equivalent number of hours criterion would require a hot gas path inspection after 2.4 years. Next, consider a continuous duty application of 8000 operating hours and 160 starts per year. Based on count of starts and hours, the hot gas path inspection needs to be performed after three years of operation with the operating hours being the limiting condition for this case. However, the equivalent number of hour criterion would set the hot gas path inspection after 2.1 years of operation.

Operating factors impact the hours-based maintenance criteria. For the starts-based maintenance criteria, operating factors associated with the cyclic effects produced during startup, operation and shutdown of the turbine are considered. Operating conditions other than the standard startup and shutdown sequence can potentially reduce the cyclic life of the hot gas path components and rotors, and may require more frequent maintenance and parts refurbishment or replacement.

8.17 FIRING TEMPERATURE

Significant operation at peak load, because of the higher operating temperatures, will require more frequent maintenance and replacement of hot-gas-path components. For an MS7001EA turbine, from a bucket parts life standpoint, each hour of operation at peak load firing temperature (+100°F/56°C) is the same as six hours of operation at base load. This type of operation will result in a maintenance factor of six. Note that a linear relationship does not prevail, as a +200°F/111°C increase in firing temperature would have an equivalency of six times six, or 36:1.

Higher firing temperature reduces hot-gas-path parts lives while lower firing temperature increases parts lives. This provides an opportunity to balance the negative effects of peak load operation by periods of operation at part load. However, it is important to recognize that the non-linear behavior described above will not result in a one for one balance for equal magnitudes of over and under firing operation. Rather, it would take six hours of operation at –100°F/56°C under base conditions to compensate for one hour operation at +100°F/56°C over base load conditions.

It is also important to recognize that a reduction in load does not necessarily mean decreased firing temperature. In heat recovery applications, where steam generation drives overall plant efficiency, the load is first reduced with closed variable inlet guide vanes to reduce inlet airflow while maintaining maximum exhaust temperature. In this case, firing temperature does not decrease until load is reduced below approximately 80% of rated output. By the same token, a turbine running in simple cycle mode maintains fully open inlet guide vanes during a load reduction to 80 per cent, and will experience over a 200°F/111°C reduction in firing temperature. The hot gas path parts life effects for these different modes of operation will thus differ. Similarly, turbines with DLN combustion systems will have their inlet guide vanes turned down and inlet bleed heat reduced to extend operation to part load conditions.

8.18 INSPECTING HOT GAS PATH COMPONENTS

The hot gas path inspection includes the full scope of the combustion chamber and a detailed inspection of the turbine nozzles, stationary stator shrouds and turbine buckets. To perform this inspection, the top half of the turbine shell must be removed. Prior to shell removal, proper machine centerline support using mechanical jacks is necessary to assure proper alignment of rotor to stator, obtain accurate half-shell clearances and prevent twisting of the stator casings.

In addition, all combustion transition pieces and the first stage turbine nozzle assemblies must be removed. Removal of the second and third stage turbine nozzle segment assemblies is optional, depending upon the results of visual observations and clearance measurement. The buckets can usually be inspected in place. Also, it is usually worthwhile to fluorescent penetrant inspect (FPI) the bucket vane sections to detect any cracks. Internal turbine radial and axial clearances (opening and closing) must be measured during any hot gas path inspection. Re-assembly must meet clearance diagram requirements to ensure against rubs and to maintain unit performance. Typical hot gas-path inspection requirements for all machines are:

- Inspect and record condition of first, second and third-stage buckets. If required, turbine buckets must be removed according to instructions, and note condition. Buckets with protective coating should be evaluated for remaining coating life.
- Inspect and record condition of first, second and third stage nozzles.
- Inspect and record condition of later stage nozzle diaphragm packings.
- Check seals for rubs and clearance deterioration. Record bucket tip clearances.
- Inspect bucket shank seals for clearance, rubs and deterioration.
- Check turbine stationary shrouds for clearance, cracking, erosion, oxidation, rubbing and build-up.
- Check and replace any faulty wheel-space thermocouples.
- Enter compressor inlet plenum and observe condition of the forward section of compressor and IGVs, looking for corrosion, bushing wear evidenced by excessive clearance and vane cracking.
- Enter the combustion wrapper, and observe with bore-scope condition of blading in aft end of the axial flow compressor.
- Visually inspect the turbine exhaust area for any signs of cracking or deterioration.

Since the first stage turbine nozzle assembly is exposed to the direct hot-gas discharge from the combustion process, it is subjected to the highest gas temperatures in the turbine section. Hence, cracks in the nozzle and oxidation occur frequently. Second and third-stage nozzles are exposed to high gas bending loads, and in combination with the operating temperatures, can lead to downstream deflection and closure of critical axial clearances. As a general rule, first stage nozzles will require repair at the hot-gas path inspection. The second- and third-stage nozzles may require refurbishment to re-establish proper axial clearances.

Normally, turbine nozzles can be repaired several times to extend life and it is generally repair cost versus replacement cost that dictates the replacement decision. Coatings play a critical role in protecting the buckets operating at high metal temperatures, ensuring the full capability of the high strength superalloy. This will ensure bucket rupture life meets design expectations. This is particularly true of cooled bucket designs that operate above 1985°F (1085°C) firing temperature. Significant exposure of the base metal to the environment will accelerate the creep rate and can lead to premature replacement through a combination of increased temperature and stress and a reduction in material strength. This degradation process is driven by oxidation of the unprotected base alloy. Surface degradation due to corrosion or oxidation is not only a performance issue but also a factor in bucket life.

The best coatings available have a finite life, and the condition of the coating will play a major role in determining bucket replacement life. Refurbishment through stripping and recoating is an option for extending bucket life, but if recoating is selected, it should be done before the coating has breached to expose base metal. If recoating is not performed at the hot gas path inspection, the run-out life of the buckets would generally extend to the major inspection, at which point the buckets would be replaced. Some gas turbine manufacturers recommend recoating of the first stage buckets at each hot gas path inspection. Visual and bore-scope examination of the hot gas path parts during the combustion inspections as well as nozzle deflection measurements will allow monitoring distress patterns and progression. This makes part-life predictions more accurate.

8.19 REVIEW QUESTIONS

8.19.1 Name some reasons that cause a gas turbine to trip and cease operation. How does a trip affect the turbine unit?

8.19.2 Discuss the role played by the compression and temperature ratios in the performance of a gas turbine.

8.19.3 How does ambient relative humidity and temperature affect the operation of a gas turbine?

8.19.4 Name the advantages and disadvantages of gas turbines operating in the simple and combined cycle modes of operation.

8.19.5 What are the major components and their role in an HRSG?

8.19.6 Under what circumstances can auxiliary firing in a HRSG be considered advantageous?

8.19.7 Name the limiting factors for life of turbine components operating on continuous basis and when run for peaking conditions. Why are they different?

8.19.8 Is it reasonable to expect higher firing temperature of certain magnitude and duration to compensate for lower firing temperature of the same magnitude and duration when service intervals are being considered?

8.19.9 What form of distortion is likely to occur in the second and third stage nozzle vanes? How does it take place?

8.19.10 Explain why the trailing edge of turbine blades in the region where the airfoil meets the dovetail platform is susceptible to cracks.

8.19.11 How does chipping of coatings on turbine blades affect the operating life of the blades?

8.20 EXAMPLE PROBLEMS

Example Problem # 8.20.1: Calculate the cycle efficiency for simple-cycle gas turbine with the compressor developing compression ratio of 11.5.

Solution: From equation (8.1), $\eta = 1 - \left(\frac{1}{r}\right)^{\left(\frac{\gamma-1}{\gamma}\right)}$. Assume $\gamma = 1.4$,

then: $$\eta = 1 - \left(\frac{1}{11.5}\right)^{\left(\frac{1.4-1}{1.4}\right)} = 0.502$$

Example Problem # 8.20.2: Determine the compression ratio required for a simple-cycle gas turbine if the ratio of specific heats is 1.392 and isentropic efficiency is 0.5875.

Solution: From equation (8.1), $\eta = 1 - \left(\frac{1}{r}\right)^{\left(\frac{\gamma-1}{\gamma}\right)}$.

then: $$0.5875 = 1 - \left(\frac{1}{r}\right)^{\left(\frac{0.392}{1.392}\right)} = 1 - \left(\frac{1}{r}\right)^{0.2816}$$

Hence, $(1 - 0.5875)^{(1/.2816)} = 1/r = 0.0431$, so $r = 23.208$

Example Problem # 8.20.3: Calculate the specific work output of a simple-cycle gas turbine if the ratio of specific heats is 1.392, compression ratio is 17.5, inlet air temperature is 290 K and turbine inlet temperature is 1225 K.

Solution: From equation (8.2), $$\frac{W}{c_p T_1} = t\left(1 - \frac{1}{r^{(\gamma-1)/\gamma}}\right) - \left(r^{(\gamma-1)/\gamma} - 1\right)$$

Temperature ratio $t = 1225/290 = 4.224$

Then: $$\frac{W}{c_p T_1} = 4.224\left(1 - \frac{1}{17.5^{(1.392-1)/1.392}}\right) - \left(17.5^{(1.392-1)/1.392} - 1\right)$$

$$= 1.098522$$

Example Problem # 8.20.4: Determine the turbine inlet temperature of a simple-cycle gas turbine if specific work output is 0.955, the ratio of specific heats is 1.392, compression ratio is 22.0 and inlet air temperature is 290 K.

Solution: From equation (8.2), $$\frac{W}{c_p T_1} = t\left(1 - \frac{1}{r^{(\gamma-1)/\gamma}}\right) - \left(r^{(\gamma-1)/\gamma} - 1\right)$$

Then: $$0.955 = t\left(1 - \frac{1}{22.0^{(1.392-1)/1.392}}\right) - \left(22.0^{(1.392-1)/1.392} - 1\right)$$

$$= 0.5812t - 1.3880$$

Hence: $t = 4.031$

So, turbine inlet temperature = 290 × 4.031 = 1169 K

8.21 EXERCISE PROBLEMS

Exercise Problem # 8.21.1: Calculate the specific work output of a simple-cycle gas turbine if compression ratio is 37.25, the ratio of specific heats is 1.385, inlet air temperature is 290 K and turbine inlet temperature is 1189 K.

Exercise Problem # 8.21.2: Determine the compression ratio required if the isentropic efficiency is 0.6441 and ratio of specific heats is 1.388 for a simple-cycle gas turbine.

Exercise Problem # 8.21.3: Calculate the cycle efficiency for simple-cycle gas turbine with the compressor developing compression ratio of 29.75.

Exercise Problem # 8.21.4: Determine the turbine inlet temperature of a simple-cycle gas turbine if specific work output is 1.1285, the ratio of specific heats is 1.388, compression ratio is 27.3 and inlet air temperature is 293 K.

REFERENCES & BIBLIOGRAPHY

Boyce, M., "Handbook for Cogeneration and Combined Cycle Power Plants", ASME Press, New York, 2002.

Brown, C. J., ASME Paper # 2000-GT-595, New York, 2000.

Chmielewski, R., Jacobucci, S., Harkins, W., Kuten, P., Wu, S., Berruti, A., McArthur, J., "Unique Combined Cycle Design Caters to Plants with Cyclical Demand Profiles", Power Engineering, January, 2002.

Ganapathy, V., "HRSG's for Gas Turbine Application", Hydrocarbon Processing, August, 1987.

Leyzerovich, A., "Large Power Steam Turbines: Design and Operation", Volume 1, Pennwell Books, Tulsa, OK, 1997.

Leyzerovich, A., "Large Power Steam Turbines: Operation", Volume 2, Pennwell Books, Tulsa, OK, 1997.

Maunsbach, K., Issakson, A., Yan, J., Svedberg, G., Eidensten, L., "Integration of Advanced Gas Turbines in Pulp and Paper Mills for Increased Power Generation", ASME Paper # 00-GT-020, New York, 2001.

Ragland, T. L., "Industrial Gas Turbine Performance Up-rates: Tips, Tricks and Traps", ASME Paper # 97-GT-409, New York, 1997.

Rangwala, A. S., "Dynamic Analysis of Turbo-machinery", McGraw-Hill, New York, 2004.

Saravananamuttoo, H. I. H., Rogers, G. F. C., Cohen, H., "Gas Turbine Theory", Prentice Hall, Harlow, England, 2001.

Sawyer, T., "Gas Turbines – Volumes I, II and III", International Gas Turbine Institute, Atlanta, ASME, 1982.

Southall, L., McQuiggan, G., "New 200 MW Class 501G Combustion Turbine", ASME Paper # 95-GT-215, New York, 1995.

Valenti, M., "Reaching for 60 Per cent", Mechanical Engineering, pages 35-39, ASME, New York, April, 2002.

Van Leuwen, V., "Solar Turbines Inc. Taurus 60 Gas Turbine Development", ASME Paper # 94-GT-115, New York, 1994.

Wood, M. I., "Developments in Blade Coatings: Extending the Life of Blades? Reducing Lifetime Costs?" CCGT Generation, IIR Ltd., March 1999.

9
CHAPTER

AERO-DERIVATIVE ENGINE FOR SHIP AND OFFSHORE OIL PLATFORM

9.1 INTRODUCTION

Turbines of all types perform yeoman service in a variety of situations and under widely differing operating conditions. Nowhere is this more abundantly clear than onboard ships and on offshore oil platforms. Under stormy conditions, a marine vessel may be buffeted and tossed around by strong winds and waves taller than the ship itself. The vessel, powered either by steam or gas turbine, is required to produce maximum power so that the ship can stay on course as it struggles with the weather. Failure of the ship's propulsion system under such extreme conditions can cause the vessel to be thrown off course by possibly hundreds of miles.

Reciprocating steam or diesel engines have been used on board ships for a long time, but turbines provide significant advantages. Multiple cylinders of large bore and stroke invariably result in large weight and require substantial amount of floor space while extending over a number of decks. The reciprocating piston and crank rod masses result in significant inertia loads, which have to be borne by the ship's super structure. Steam and gas turbines, on the other hand, are lighter in weight and provide superior power to weight ratio. Also, there are no inertia forces and moments to contend with, resulting in reduced vibrations. However, the turbine's rotor speed has to be reduced, calling for precision-machined gears to bring it down to that of the ship's propeller. Steam or diesel driven reciprocating engine's running speed is within the propeller's speed range of between 85 and 110 rpm.

Gas turbines derived from aviation engine technology find wide acceptance for power production on offshore oil platforms. The platforms have onboard a large number of pumps, compressors, lifting equipment and other electrical machines. Since most of the platforms are operating far from the shoreline, electrical power sourced from around the coast is not an option. With large amounts of natural gas and liquid fuel coming from the seabed, the operators rely on gas turbines to produce the electrical power. Figure 9.1 shows an offshore oil platform exposed to stormy weather.

A number of changes are incorporated in turbine engines destined for aviation applications to produce electrical power through the power turbine section. The front fan and the fan exhaust are eliminated. The design is focused on producing mechanical power for the driven equipment, such as an electrical generator. Aero-derivative engines also are widely used for pipeline transmission.

Fig. 9.1 Offshore Oil Platform in Storm

9.2 APPLICATIONS OF LIGHTWEIGHT AVIATION ENGINE TECHNOLOGY

Offshore oil platforms typically are supported on four very large columns that rest on the sea floor, with a platform deck some 35 to 50 feet above the water line. A large array of several different types and size of pumps, compressors, motors, hoists and cranes is mounted on the platform deck. The equipment is intended to ensure all of the massive columns are balanced and the platform structure lies in the horizontal plane. The pumps and compressors continuously feed in or remove seawater in and out of large number of tanks built within the columns to maintain the structure dynamically balanced. Thus, in the event of storms, gale winds cannot readily push the rig from its moorings to upset the oil production efforts.

Aviation related gas turbine engines must necessarily be as light in weight as possible if the aircraft is to get off the ground and stay aloft in the air. Aircraft engine manufacturers resort to designing and manufacturing all components with thin walls to maintain overall engine weights within prescribed limits. A direct consequence is the components are flexible and have low natural frequencies of vibrations. The vibrations induce fatigue damage. So the components, especially the rotating blades, have limited life. Periodic replacement of the parts at a faster clip is essential.

9.3 ELECTRIC POWER FOR OFFSHORE OIL PLATFORM

Oftentimes offshore platforms are located several miles from the coastline, so the installation needs electrical power generation capacity to drive the equipment and machinery of the platform. Depending on the size of the installation, a set of turbine generators may be required to provide the electrical power. The turbines burn natural gas coming out of the sea floor.

A necessary requirement to ensure the rig is not top-heavy calls for the engines to be light in weight. Aviation engine manufacturers also produce aero-derivative engines that are low in weight. The fan of a turbo-fan engine is eliminated, and with other changes, make the aero-derivative engine suitable for applications such as power generation and pipeline pumping on offshore oil platforms and for propulsion of ships.

Since the gas coming from the seabed may be mixed with slugs of crude oil or other liquid hydrocarbons, an essential requirement calls for the combustors to burn different grades of fuel.

9.4 PROPULSION POWER FOR NAVAL AND MERCHANT MARINE SHIPS

Figure 9.2 shows an LM2500 marine gas turbine, a simple-cycle, high-performance engine. Derived from GE's CF6-6 aircraft engines, the engine has a gas generator, a power turbine, attached fuel and lube oil pumps, a fuel control and speed governing system, associated inlet and exhaust sections, and lube and scavenge systems, as well as controls and devices for starting and monitoring engine operation.

The gas generator consists of a 16-stage, 18:1 pressure ratio compressor with seven stages of variable stators and inlet guide vanes; a fully annular combustor with externally mounted fuel nozzles, and a two-stage, air-cooled high-pressure turbine which drives the compressor and the accessory gearbox. A six-stage, aerodynamically coupled low-pressure power turbine is driven by the gas generator's high-energy exhaust gas flow. This in turn powers the output shaft of the marine gas turbine.

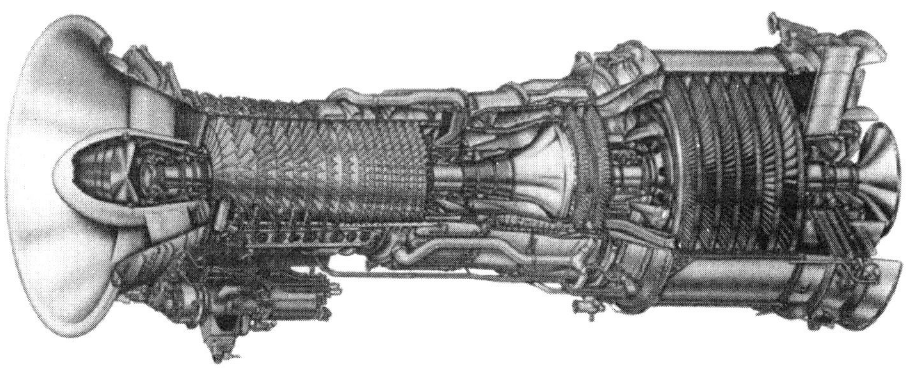

Fig. 9.2 General Electric LM2500 Aero-derivative Gas Turbine

The 10,300 lb turbine engine features a simple modular design, marine corrosion-resistant materials and minimum size, weight and space requirements per developed horsepower. The LM2500 delivers high thermal efficiency and low fuel and airflow per horsepower produced. The engine may be housed in a high shock-resistant, thermally and acoustically insulated enclosure on a mounting base of the type shown in Figure 9.3. The enclosure attenuates noise in the engine room and provides sensors for inlet icing and fire detection. It also houses fire extinguishing equipment. GE furnishes these modules to the US Navy and other navies.

The simple modular design of the LM2500 incorporates many features to maximize shipboard maintainability. The split compressor casing allows in-place blade and vane replacement. In-place hot section maintenance and accessible external fuel nozzles are also useful features. On an average, the engine performance may be considered for 60Hz power and operating at 59°F, sea level, 60% relative humidity, no inlet/exhaust losses, using liquid fuel of lower heat value of 18,400 BTU/lb. The engine delivers 33,600 horsepower at power turbine speed of 3600 rpm. Specific fuel consumption is 0.373 lb/shp-hr, and thermal efficiency is 37 per cent. Engine length is 6.52 m and 2.04 m height.

9.5 GAS COMPRESSION FOR PIPELINE PUMPING

Pipeline compressors are designed with a choice of axial or conventional opposed inlets. Aerospace assembly techniques provide ease of maintenance with full removal, causing no disturbance to the main piping. Pipeline compressors are preferable for the large flow rates required by natural gas transmission applications.

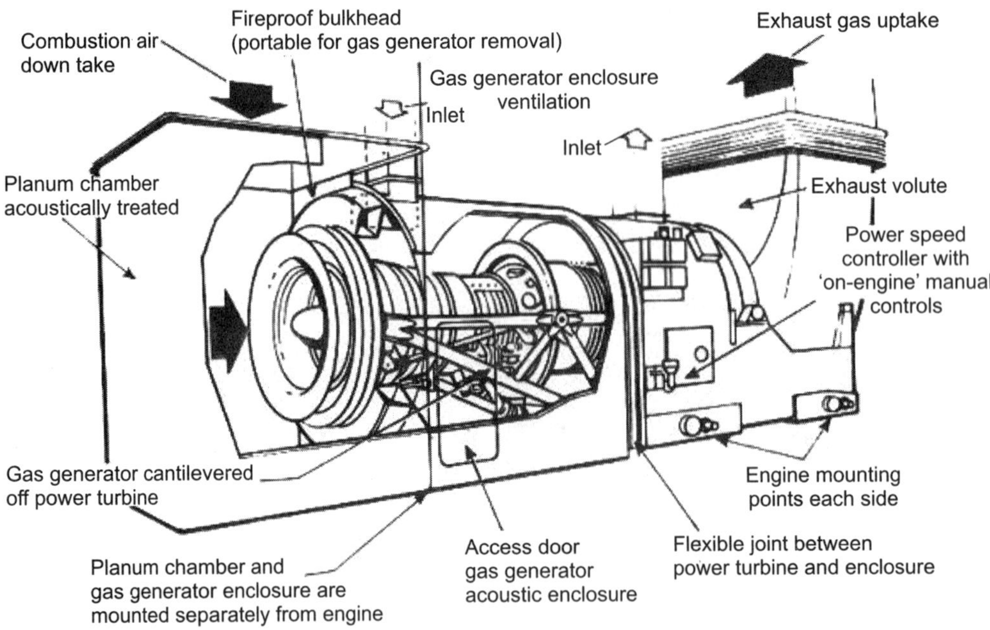

Fig. 9.3 Aero-derivative Engine within Enclosure

Compressor aerodynamic performance has been refined to the point where conventional side inlet compressors offer up to 89 per cent polytropic efficiency. Customized aerodynamic engineering of the gas path may be needed for specific compressor requirements. Conventional side inlet pipeline compressors can accommodate up to five stages of compression, although only one or two stages are usually required for pipeline transmission duty.

An axial inlet pipeline compressor features horizontal discharge and a single-stage, spherically shaped cast casing. The weight may be only two-thirds as much as a conventional pipeline compressor. Multi-stage barrel compressors provide the oil and gas industry with natural gas gathering, re-injection, lift, depletion, storage/withdrawal and re-pressurization solutions. Barrel compressors are preferred for high pressure ratios and lower flow rates experienced in these applications.

9.6 POWER FOR HEAVY MILITARY VEHICLE

Heavy military vehicles, such as the US Army's Abrams Main Battle Tank and the Crusader self-propelled howitzer, require an efficient hybrid propulsion system to meet large and continuous power needs. The LV100 recuperated turbine engine is well suited to provide electric power for advanced armored motor vehicles where volume and weight are a premium. Mission requirements may call for sustained high power to maintain speed even on a slope. To avoid easy detection low emissions and noise characteristics of the vehicle are mandatory. To top it all, since the fuel tank uses valuable under the armor volume, fuel consumption over a wide range of power settings has to be low.

Aircraft engines are designed for lowest possible weight and best performance in the 80 to 100 per cent power range for fixed wing and helicopter applications, where operation at low idle speed is of little relevance. Vehicular engines need to be optimized in the idle to 60 per cent of maximum power output range. A recuperated cycle is advantageous when the engine spends a considerable amount of time in the idle mode of operation. In terms of cyclic usage, vehicular operation subjects the engine to cycle counts many times higher than for aviation applications. Another interesting feature calls for military vehicles to be designed for continuous medium to high ground shocks during field exercises.

Besides a good volume to shaft horsepower ratio, the turbine must contain related components to address a number of constraints and needs:

- Means for power augmentation to cover short-term needs
- Components with wide operating mass flow range
- Parts designed for high thermal cycle count and shock loads
- Durability and ease of maintenance of engine system.

Goals of improvement in specific power and fuel consumption as outlined in the US military's Integrated High Performance Turbine Engine Technology program for simple cycle aircraft engines can be obtained by pushing to ever increasing levels of temperatures and pressure ratios. Cycle temperatures approaching stoichiometric limits and pressure ratios of 60 have been proposed. At maximum power the simple cycle can get close to reaching the performance levels of the recuperated cycle, but at much higher specific power output. Since vehicular recuperated engines optimize at much lower pressure ratios, the feature reduces engine complexity, reduces the number of stages and results in comparatively higher component efficiencies.

If the LV100 engine's combustion system can be modified to operate on natural gas and equipped with a catalytic device, NO_x emissions may be expected to fall into the ultra-clean regime, levels that cannot be reached with diesel engines. The system may then be attractive for commercial applications. Of particular interest are space and weight sensitive installations such as oil drilling platforms, emergency power supply, power generators in rural areas for supplemental power and for helicopter airlift.

9.7 INTEGRATED ELECTRICAL PROPULSION AND POWER SERVICE

An integrated electric propulsion and power service system provides for greater flexibility, efficiency and survivability of naval ships. Examples of this concept include the type-45 destroyer program for the Royal Navy and DD (X) program for the US Navy.

The US Navy system comprises of a General Electric LM2500 gas turbine engine directly coupled to a Brush Electric Company synchronous generator. The unit is capable of producing 21.6 MW at 0.8 power factor. To determine system response during start up a ramp load is conducted during system test to characterize power system interface, stability and control performance. Step loads are required in the form of propulsion motor acceleration at an average of 4 per cent power per second up to steady state condition. System operating conditions are 25, 50, 75 and 100 per cent of shaft output power of the propulsion motor. Figure 9.4 provides details of turbine performance during the 100 per cent ramp. Turbine performance and voltage regulation were exceptionally well controlled. The generator power turbine speed followed the US Navy's 3.33 per cent droop curve standard with very little deviation during each ramp up operation. Since motor load is proportional to the third power of speed, motor acceleration was shaped to provide a linear power ramp and a cubic speed profile.

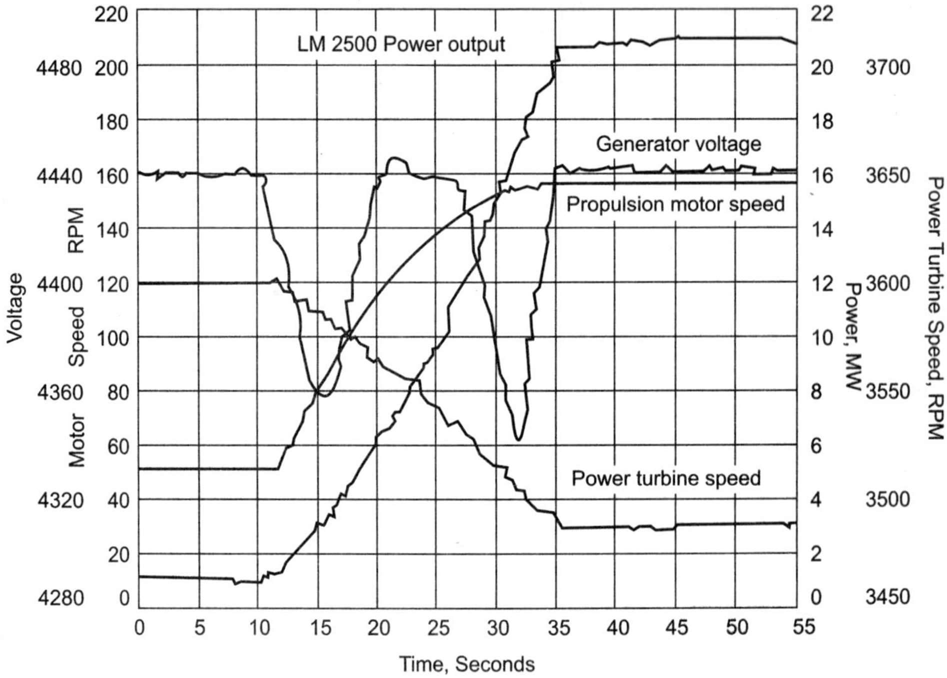

Fig. 9.4 Turbine Generator and Propulsion Motor Characteristics — Ramp Acceleration

Ramp unloads were also carried out during the system test to a similar set of characteristics as in the ramp up mode, but with the engine decelerating. Since the motor converter is non-regenerative, the motor's deceleration is restricted, and ramps down at a rate of 2 per cent power per second. Figure 9.5 illustrates this feature, with the deceleration time period more extended than the period during acceleration. The motor slows down in a linear speed profile as opposed to a cubic acceleration schedule.

Step unloading of the power generating system represents a considerably more severe situation than a gradually unloaded ramp operation. The engine is driving at a constant mechanical power and the generator exciter is providing a constant excitation when the electrical load is suddenly lost. Frequency and voltage of the machine overshoot momentarily, hence mechanical power input to the generator must be reduced rapidly to prevent over speed and return the power turbine to normal speed. This is accomplished by adjusting the fuel-metering valve. In a similar manner, the automatic voltage regulator must rapidly reduce excitation current to the generator to return the voltage to a normal level.

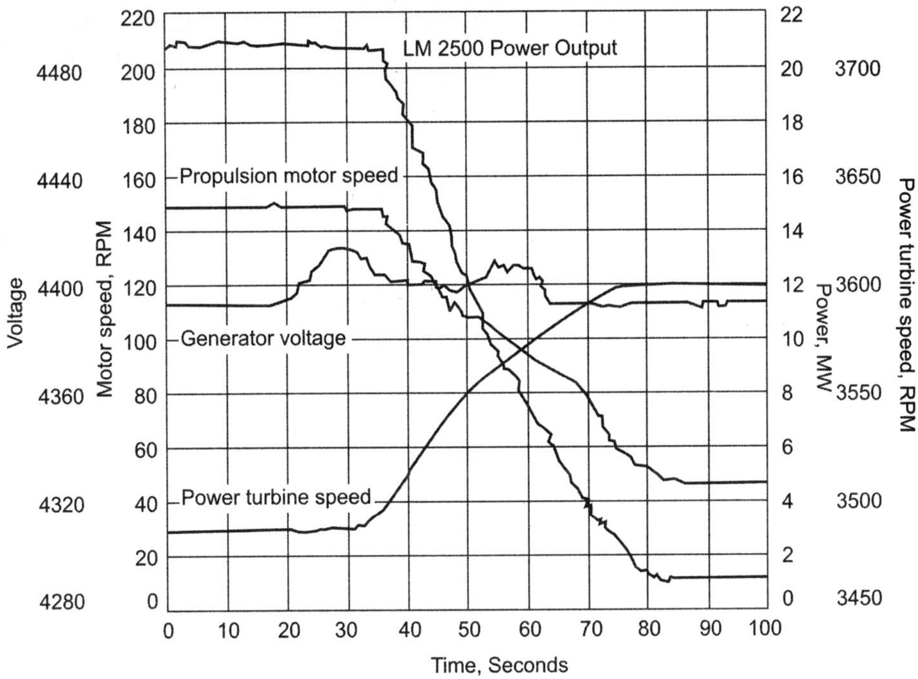

Fig. 9.5 Turbine Generator and Propulsion Motor Characteristics — Ramp Deceleration

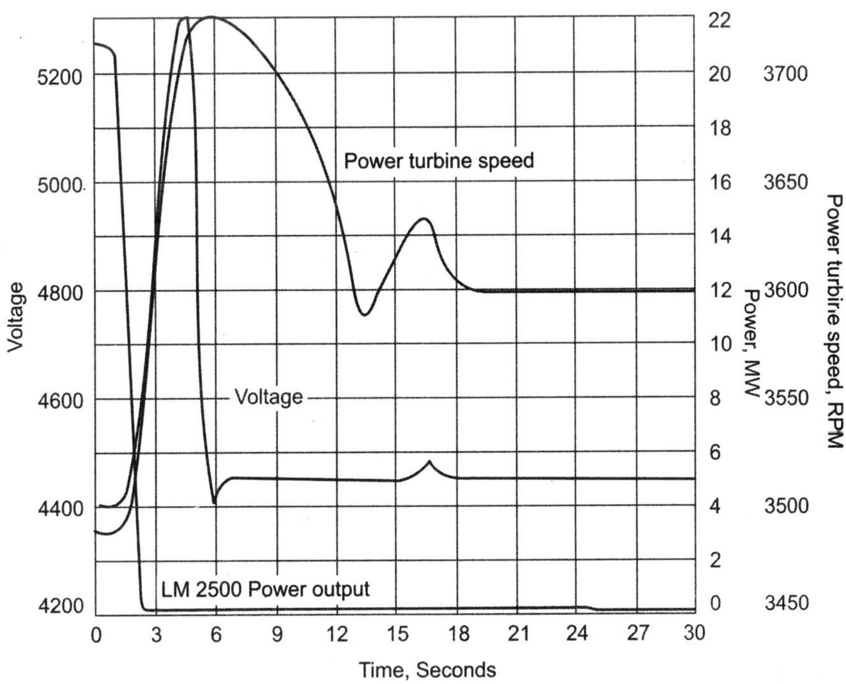

Fig. 9.6 Turbine Generator and Propulsion Motor Characteristics — 100% Step Unload

Upon loss of motor load at 100 per cent power the frequency of the LM2500 generator rises from its initial steady state droop setting of 58.0 Hz, or 3,480 rpm, shown in Figure 9.6. The upper steady state tolerance limit is 63 Hz (3,780 rpm) or 5%, which is met. The final frequency is 60 Hz (3,600 rpm), which represents the no load frequency.

9.8 TURBINE OPERATION IN MARINE ENVIRONMENT

Operation of turbo-machines at sea is hazardous from a number of considerations. When the sea is calm and the breeze gentle, benign conditions will permit the machine to run smoothly. Under storm conditions the vessel may be expected to be tossed around both in the horizontal and vertical planes.

The pitch and yaw motions of the ship will be transmitted to the turbine's rotating assembly through the bearings and their foundations. Aside from the normal torsional load and related shear stress, the rotor will be subjected to bending moments and create normal stresses. All of the components in aviation technology derived engines must necessarily be thin-walled to save weight. Consequently, the shafts, bearings and casings will experience substantial amount of bending and twisting. The clearance between the rotating blade tips and the distorted casing may be lost, causing the blade tips to rub and to chip.

Ingestion of salt water through the intake and exhaust manifolds, access doors, and instrumentation and maintenance access portals represent another source that can seriously compromise operation of the gas turbine and electrical generating equipment. In spite of the many extraordinary measures taken to protect the structural integrity of the equipment, the environment in which operation takes place takes an unusual toll on the on operating efficiency as also longevity.

9.9 ACCELERATED CORROSION FROM SEA AIR AND WATER

Corrosion problems arise due to the process metal going through complex changes. To best describe corrosion, recall that iron ore has a chemical composition of two iron atoms bonded with three oxygen atoms. Mined out of the ground, it is a brownish-red powder that is refined and smelted to create iron. Plain iron can be further processed and combined with other elements to get different types of steels and its alloys.

However, the iron wants to return to its normal state as iron ore, iron oxide, or rust, the state in which iron is most stable. Most metals used in manufactured products want to do the same and return to the natural state.

Galvanic corrosion is an electrochemical reaction between two or more different metals. The metals must be different because one must be more chemically active (or less stable) than the others for a reaction to take place. Galvanic corrosion occurs from electrical exchange. All metals have electrical potential because all atoms have electrons, which have an electrochemical charge.

Galvanic corrosion of the more chemically active metal can occur whenever two or more dissimilar metals that are connected by actually touching each other, or through a wire or metal part when immersed in a conductive solution. Saltwater, freshwater with high mineral content, and polluted freshwater are very conductive. Conductivity increases with water temperature, which explains why boats in Florida experience more corrosion than boats in Maine.

Pitting corrosion is a localized attack, and penetration into steel can be deep in relation to the area under attack (Figure 9.7). This is one of the most insidious forms of corrosion and can be most intense on mill-scaled or croated steel.

Fig. 9.7 Damage from Localized Corrosion

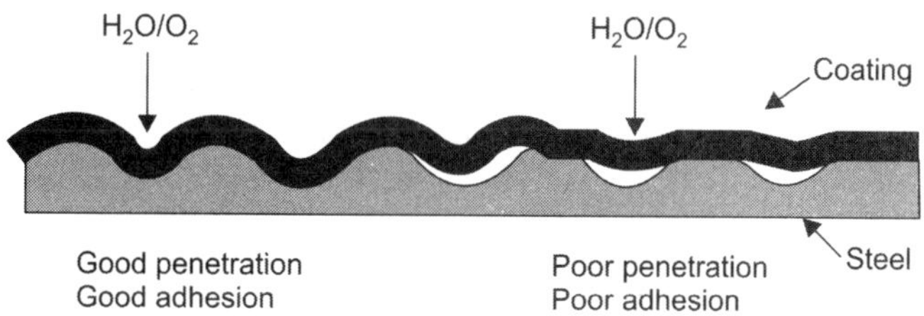

Fig. 9.8 Coating Penetration Related Problems

Bacterial corrosion, also referred to as micro-biological or anaerobic (without oxygen) corrosion, may occur if local operating conditions support bacterial activity. Bacterial corrosion is most likely to occur in stagnant (anaerobic) water, a hydrocarbon feed source such as oil or coatings, in presence of sulfates and in operating temperature between 20 to 40°C.

Seawater attacks iron mostly due to the presence of salt compounds of sodium, calcium, magnesium and other elements. Even when not coming into direct contact with seawater, exposure to salt water spray and air is liable to increase the incidence of corrosion in gas turbines. Air coming through the inlet plenum will carry abundant salt water particles and affect components in the bell-mouth and compressor sections.

Preventing corrosion requires the elimination or suppression of the chemical reactions described earlier. There are two principal methods, used in conjunction, cathodic protection and coatings.

The effectiveness of coatings in preventing corrosion depends upon many factors, for example, coating type, end use and operating environment. Many theories exist to explain the mechanics of corrosion protection by coatings.

Coatings do not simply acted as a barrier to oxygen and water. Research has proved that oxygen and water, at a level sufficient to initiate the corrosion reaction, can indeed permeate through intact coatings. Current theories indicate water permeating through a coating to the steel surface can cause displacement of the coating from the steel, allowing corrosion to occur (Figure 9.8). Low permeability and good 'wet adhesion', or adhesion under immersion, are widely believed to be the single most important aspects of corrosion control by coatings.

For maximum coating adhesion, steel surfaces must be clean, dry and free of oil, rust, salts and other contaminants. Whilst increasing surface roughness results in a greater surface area and potentially increased coatings adhesion, this is not always the case. If the coating does not 'wet out' and penetrate into all of the pores/crevices on the steel surface, there will be areas of coating not in direct contact with the steel. In such cases, more rapid oxygen and water passage through to the steel will occur, resulting in corrosion and subsequent coating detachment.

Penetration of the coating into the surface or roughness profile can be improved by reducing the relative surface tension of the coating, reducing viscosity of coating, modifying coating solvent blend, increasing coating curing time and increasing coating application temperature.

Highly cross linked, chemically curing systems are likely to have relatively low permeability characteristics. Pigmentation can also have a significant effect. Coating permeability decreases as a function of increasing pigment volume in the film up to a critical pigment volume concentration. Pigment levels above this level results in voids in the coating, and permeability will increase.

9.10 CONTROLLING ENGINE VIBRATIONS

Since wall thickness is restricted due to weight requirements, all of the parts in aero-derivative engines tend to be flexible. This results in excessive vibrations, and high fatigue inducing stresses under the influence of dynamic loads.

Engine vibrations may be controlled either by reducing the amplitude of the dynamic loads or by introducing damping to control the dynamic response. A major source of dynamic loads is rotating unbalance. When a rotating part lacking in adequate dynamic balancing is operating in close proximity of any of the natural frequencies of the rotor and bearing assembly, flexural activity increases. This induces dynamic stresses, and may reach a point where failure mechanism sets in. The dynamic motion may be controlled by fine tuning the balancing to acceptable levels. Another possibility is to avoid operating near a resonant frequency, but from practical considerations that may not be a viable option.

The lighter weight of the rotor permits use of ball or roller bearings to support the rotor. The small axial length of these bearings is also conducive to keeping the overall length, and hence the weight, of the engine system. But rolling element bearings are inherently free of friction, a major source of damping. The friction coefficient of rolling element bearings is so low that it is necessary to introduce external damping to reduce dynamic activity when the rotor has some residual unbalance. Squeeze-film dampers are commonly used in turbine engines derived from aviation technology. Acting in tandem with the rolling element bearing, the damper is capable of reducing the motion of the rotating system.

9.11 VIBRATIONS AND COMPONENT DYNAMIC STRESS

A rotating shaft transmitting torque at steady speed is an example of a part subjected to dynamic stresses. The weight of the shaft and attached disks will lead to bending stresses in the shaft substrate. Bending will develop compressive stresses in the upper fibers and tensile stresses in the lower surface of a horizontal rotating assembly. When the shaft turns through 180°, the position is reversed. The stress in the upper and lower surfaces of the shaft will change from compression to tension on one side and tension to compression on the opposite side. The reversal in the normal stresses takes place for each shaft rotation. Thus, a shaft spinning at 3600 rpm will experience an equal number of such stress reversals.

Not all dynamic loads result in complete reversal in the stress direction. The reversal may be partial, for instance from zero to maximum, for a load cycle. Fatigue in the metal is induced under the influence of the time, or cycle, varying stress pattern. The amount of damage incurred depends primarily on the magnitude of the stress variation and the number of cycles. When the fatigue limit of the material is reached, it gives way and causes the part to rupture.

The limit when the part fails also depends on many associated factors. Operating temperature of the metal, presence of corrosive elements, surface finish of the component and the manner in which the load is applied are some factors that can accelerate or retard the failure mechanism.

9.12 PERFORMANCE ENHANCEMENT BY PARAMETER MANIPULATION

Pressure losses are encountered in a gas turbine system when air filtration, silencing, evaporative coolers, chillers in the inlet, or exhaust heat recovery devices are inserted. Effects vary with the unit. A pressure drop of 4 in. (10 m bar) of water at the inlet may be accompanied by a decrease in the output power and efficiency of around 1.5 and 0.5 per cent, respectively, may be expected. The same pressure drop at the exhaust of a gas turbine will reduce the output power and efficiency by around 0.4 per cent.

The type of fuel used also has an impact on the performance of the gas turbine. In addition to being a clean fuel that leaves behind little or no residue after combustion, natural gas also offers the advantage of producing more power output then distillate oil. Since a higher concentration of water vapor is present in the products of combustion of natural gas due to a higher hydrogen-carbon ratio, the resulting higher specific heat is responsible for a difference of as much as 2 per cent more power production.

Gaseous fuels having lower heating values than natural gas can significantly alter the performance of a gas turbine. When the heating value falls, the fuel flow must increase to provide additional heat. However, the compressor does not pressurize the additional mass flow. Since the compressor power is not affected by this change, the turbine and the output power of the machine are increased. Some other side effects may also be expected, including:

- An increase in mass flow through the turbine increases the power developed by the turbine. The compressor absorbs some of this increased power, but the net result is an increase in the pressure ratio across the compressor. The compressor then operates closer to the surge limit.
- Because the increase in turbine power has the potential to take the turbine and the equipment in the power train above their 100 per cent rating, durability of the equipment rated at higher limits may be required in some cases.
- A higher volume of low or medium heating value (BTU) coal gases is usually required at elevated temperatures, which have a lower density. In tandem with the increased fuel flow, the consequent increase in the volumetric flow will require increments in the size and cost of the piping, fittings, and valves.
- Gases having low heating values (BTU) have a propensity to be saturated with moisture

before delivery to the turbine. This results in an increase in the heat transfer coefficients of the combustion products, leading to an increase in the substrate temperature in the turbine.

- The amount of air required for burning the fuel increases as the heating value decreases. Gas turbines having high firing temperatures may not be able to operate using low-heating-value fuel.

Each model of a gas turbine will require a set of guidelines for any application. The list includes specification of flows, temperatures, and output power to ensure performance, efficiency, and life of the machine. In most applications involving lower-heating-value fuel, it is assumed that the efficiency and power output will be equal to or higher than the ones obtained using natural gas. On the other hand, in applications where higher heating value fuels such as refinery gases are contemplated, the efficiency and output power will be less than or approach those obtained using natural gas.

Water and steam injection have been used during the last few decades to reduce NO_x emissions. This technique involves injecting water or steam in the cap area, or the head end, of the combustor liner. The output power and efficiency will increase due to the additional mass flow. However, each machine has limits on the amount of water or steam injected. These are imposed to protect the combustor and turbine section. Steam injection can increase the output power and efficiency by 20 and 10 per cent, respectively. Water injection can increase the output power by 10 per cent. However, it has very little effect on efficiency because more fuel is needed to raise the water to the combustor's temperature level.

Many gas turbines require air to be bled from the compressor. For example, aircrafts use the air from the auxiliary power unit for starting the main engines. Air from the main engines is used for items as diverse as cabin pressurization and cooling the combustor liner and the blades and vanes in the high-pressure turbine section. In general, up to 5 per cent of the flow can be extracted from the discharge casing of the compressor. This can be done without modification to the casings or on-base piping. Higher flow (from 16 to 20 per cent) can be extracted from the compressor. However, this requires modifications to the casings, piping, and controls. Extracting air from the higher-pressure stages of the compressor has a significant effect on the performance of the machine. The rule of thumb is that every 1 per cent of air extraction reduces the power output by 2 per cent. But when the intent is to lower the metal temperatures, the higher temperature at the turbine inlet aids in increasing the thermal efficiency of the unit. This provides ample justification for extracting the air from the compressor.

The performance values for a machine are normally given for base load ratings. The American National Standards Institute (ANSI) B 133.6 Ratings and Performance guideline defines operation at base load at the rate of 8000 hours in a year with 800 hours per start, and peak load operation at the rate of 1,250 hours per year with 5 hours per engine start. Because of their ease in starting, gas turbines are the preferred option for meeting peak load requirements.

Since peak-load operating hours are shorter, increasing the firing temperature can increase the power output. Shorter operating periods will necessitate more frequent inspections of the gas turbine, but this disadvantage is justified since operation at peak levels is generally more cost-effective. The additional power is then generated in periods of higher power cost for the consumer. Generators also have peak ratings. These are obtained by operating at a higher power factor or temperature increase. The firing temperature can be adjusted between the base and the peak conditions. They are chosen to maximize the power output while remaining within the limits of the repair interval of the turbine hot section.

Since the mass flow of air entering the gas turbine determines the amount of fuel for proper combustion, ambient air conditions will determine the amount of power produced. Figure 9.9 illustrates the case of ambient air temperature affecting the output power, heat rate (reciprocal of thermal efficiency), heat consumption, and the exhaust flow for a typical single-shaft heavy-duty gas turbine. Another example is the altitude of the location where the gas turbine is operating. Airflow and power output of a gas turbine will decline with increasing altitude due to the decrease in the barometric pressure. The reduction in these parameters is proportional to the decrease in the air density. Typically, the decrease in airflow and output power of a gas turbine is 1 per cent for every 100-meter increase in the altitude, although the heat rate and other cycle parameters will not experience any changes.

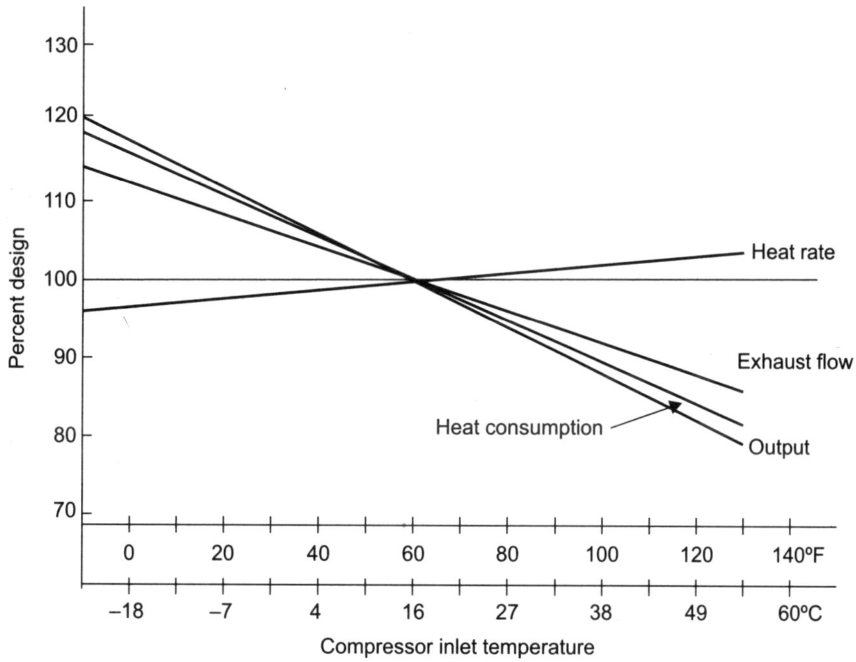

Fig. 9.9 Effect of Ambient Temperature (Courtesy: General Electric)

Because density of humid air is lower than that of dry air, an increase in ambient humidity will reduce the power output and efficiency of a gas turbine. An increase in specific humidity of 0.01 kg of water vapor/kg dry air will typically reduce the power output and efficiency by 0.0015 and 0.0035 per cent, respectively. Although sometimes considered negligible, water vapor content is of practical importance when water or steam injection is essential for controlling the level of nitric oxide (NO_x) in the exhaust gases. This humidity effect is mainly caused by the control system approximation of the firing temperature. Some gas turbine control systems reduce the power when ambient humidity increases. However, on some multi-spooled aero-derivative engines the power generated is increased because the discharge temperature from the gas generator is used to regulate the fuel flow. The increased level of fuel flow raises the temperature of the moist air to the set-point temperature. The increase in the fuel flow will increase the speed of the gas generator rotor. The operating speed of the gas generator is independent of the speed of the power turbine rotor. Thus, the increase in fuel flow rate compensates for the decrease in density of the moisture-laden air.

9.13 ROTOR-STATOR CLEARANCE MANAGEMENT

A number of methods have been devised for ensuring the optimum gap between the rotating airfoil tips and the stationary casing. The approaches, broadly divided into passive and active control methods, rely upon thermal, pneumatic and mechanical means.

Thermal systems rely on the stator's material characteristics, in particular the thermal coefficient of expansion, and operating condition of the engine to obtain the same level of growth in the rotor and the stator. Frequently used on smaller engines, thermal expansion of the shroud at lower temperatures is controlled to better match the centrifugal growth of the rotor during takeoff while the casing is still gaining temperature. During aircraft climb the casing is at a higher temperature and is required to grow at a slower pace, and the rotor is only growing from thermal effects. However, the passive thermal means lacks in accuracy and reliability in achieving optimum clearance for the minimum clearance condition necessary for takeoff and for maneuvers.

Passive pneumatic systems are actuated by gas pressures generated in the engine or by hydrodynamic effects. The designs are based on floating shroud segments and blade tip cooling air discharge. Development of force from a hydrodynamic system, however, suffers from the limited surface area of blade tips and high positioning and alignment tolerance requirements. Aggravating the situation are the problems of friction and secondary seal hang-up, and the concepts become unattractive. Pneumatic systems relying on flexible piping or bellows are subject to high cycle fatigue from the flexing motion, thin walls of the materials needed for operation, vibration, thermal, and pressure loadings in the high-pressure turbine's environment.

Active systems remain the mainstay for controlling clearances in modern gas turbines. Active clearance control methods also depend on thermal, pneumatic and mechanical means, either singularly or in combination with electro-mechanical, magnetic, and other devices for the functioning.

Thermal designs depend on compressor and fan air to respectively heat and expand or cool and shrink the supports for shroud seal segments to change the blade tip clearance. The drawback of the system is slow thermal response, which does not provide for further gain from clearance closure during cruising conditions. Improvements in the system have been made through the use of more thermally efficient materials and geometric changes in the components. One method calls for attaching the seal segments to a thermally efficient channel structure that is cooled on the inside and decoupled from the substantial casing flange for obtaining a faster and more responsive control system.

Active mechanical designs combine linkages and some form of actuating mechanism, which may be hydraulic, electro-mechanical, or electro-magnetic for shifting the tip clearance. One system calls for a segmented shroud with the segments attached to an actuating ring through linkages and radial spokes. When the ring is turned in any direction, the segments move radially inward or outward in unison.

Active pneumatic concepts for clearance control depend on the engine's own internal pressure, or pressure from other sources, and bellows and valves. Load is directly applied on the flexible and sealed shroud segments to change the radial tip clearance. Precise balancing of the pressure loads is essential for proper operation, and high cycle fatigue issues may also arise. Since pressure requirements are likely to be substantial, the method may detract from efficiency considerations.

9.14 COMBUSTION AND EMISSIONS CONTROL

All combustion processes generate exhaust gases composed of carbon monoxide CO), carbon dioxide (CO_2), water vapor (H_2O), oxides of nitrogen (NO_x), oxygen, nitrogen, unburned hydrocarbons (UHC) and particulate matter in the form of carbon. Because of their impact on people's health and the environment, government agencies have passed aggressive rules to control emission of pollutants from turbines. Carbon monoxide is poisonous even in small quantities, unburned hydrocarbons are objectionable because of toxicity, NO_x produces acid rain, and smoke and soot lead to visibility concerns. Of considerable interest is the formation of ozone in the troposphere above the earth by the pollutants. Recent research indicates a number of respiratory problems are a direct consequence of excessive pollutants in the air.

Carbon monoxide is formed when combustion of fuel occurs in the absence of sufficient oxygen. Combustion of a lean fuel mix may cause dissociation of CO_2 to form CO. Inadequate residence time in the primary zone, improper mixing of air and fuel and chilling of the products of combustion by cooling air near the liner walls are also responsible for incomplete combustion of the fuel and high concentrations of CO. Unburned fuel particles sometimes leave the combustor for the same reasons, when the parent fuel is degraded into other components of different molecular weight. Soot in the form of a fine powder is produced when re-circulating burned gases move back towards the flame, closer to the spray. Local regions of the fuel vapor then are surrounded by oxygen deficient gases at elevated temperature, and emerge from the combustor as smoke (*ref*: Lefebvre, 1999). Soot and smoke formation becomes more pronounced at higher pressures. Chemical reaction rate tends to rise with increase pressure, causing combustion to initiate earlier and a greater portion of the fuel to burn in the region near the spray. Fuel properties such as viscosity and volatility can also affect smoke and soot production because drop size, penetration and evaporation are affected. A decrease in the hydrogen content of the fuel is known to raise smoking tendencies.

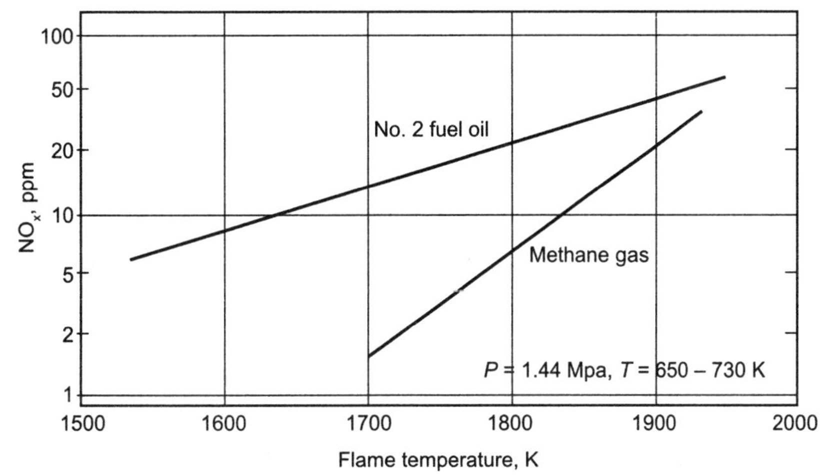

Fig. 9.10 NO_x Variation with Flame Temperature (Snyder, 1994)

Production of NO_x occurs mostly in flame zones where the temperature exceeds 1850 K. Oxygen molecules in the air split up to combine with nitrogen rather than with the fuel to form NO, especially when the charge is fuel lean. Production of NO reduces at an exponential pace as the flame temperature reduces. Liquid fuels tend to generate higher levels of NO_x than gaseous fuels for the same flame temperature, but at higher temperatures this difference is practically absent. Figure 9.10 illustrates the difference based on experimental tests. Air temperature at the inlet to the combustor plays a role, since it affects the flame temperature. Larger residence time within the combustion chamber tends to increase NO_x emissions, except when the rate of formation is low due to considerably lean mixtures. Figure 9.11 provides data from results determined by Anderson (1975) for premixed air-fuel combustion.

Reduction in the levels of CO and UHC can be achieved through more complete combustion of the fuel, which also has the advantage of raising overall efficiency of the unit. Re-circulation of the airflow to obtain better mixing of fuel and air with an equivalence ratio of 0.8 in the primary zone helps towards this objective. Restricting the amount of air for cooling of the liner surface in the primary zone also can be effective in reducing CO and UHC. Improved fuel atomization and increasing residence time in the primary zone are also factors for enhancing combustion efficiency. Soot and its accompanying smoke can be controlled by inserting more air in the primary zone to obtain better mixing. Exhaust smoke will then diminish by preventing formation of fuel-rich regions in the flame.

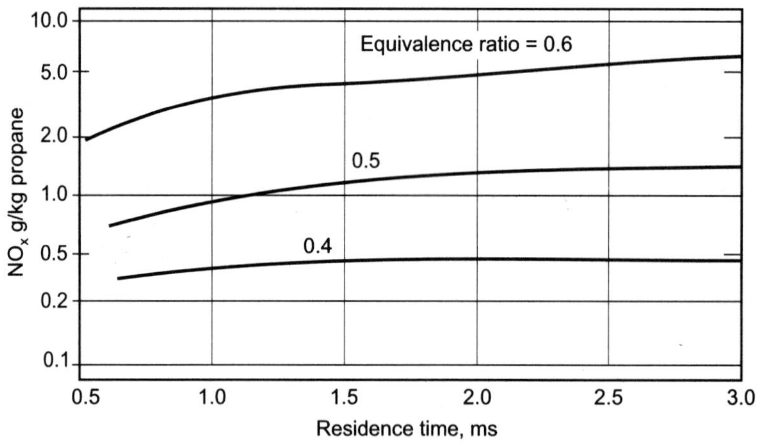

Fig. 9.11 NO_x Variation with Residence Time (Anderson, 1975)

One method to diminish temperature of the flame and consequent NO_x formation in stationary engines is by injecting water to extract combustion heat. Water or steam may be directed into the flame in the form of a spray from nozzles located at the upstream end of the combustor or through holes combined with the fuel nozzle. The water can also be introduced upstream of the combustion liner into the air stream prior to entry through the swirlers. If steam is to be used for controlling NO_x, it may be introduced at the compressor discharge, in the upstream region of the airflow or into the combustion zone. The effectiveness of this procedure has been demonstrated for gas and liquid fuels on the General Electric MS7001F gas turbine. Water injection provides better results when NO_x emission rates are at a peak due to high operating condition pressures and temperatures. A larger quantity of steam is generally required than water to achieve the same level of reduction in NO_x emission.

But this procedure has disadvantages in the form of higher fuel consumption of about 2 to 3 per cent, and also adds to the initial cost of the system by $10 to 15 per kilowatt. CO and UHC emissions may also be expected to rise as a consequence of reduced combustor performance. The purity of the water introduced into the combustor must also be taken into account to inhibit corrosion of the turbine blades. Oscillations in combustion chamber pressure have also been known to increase with the injection of water.

NO_x molecules can also be chemically reduced to nitrogen and water vapor by introducing ammonia on a catalytic bed, and this is the underlying principle behind the selective catalytic reduction (SCR) method. Either alone or in combination with other techniques, SCR is a popular choice for NO_x reduction. In a combined cycle unit the ammonia injection system and catalyst bed may reside between heat exchanger circuits at a point where temperature is within the window from 560 to 670 K for maximum efficiency. The method needs a control system to deliver the proper quantity of ammonia.

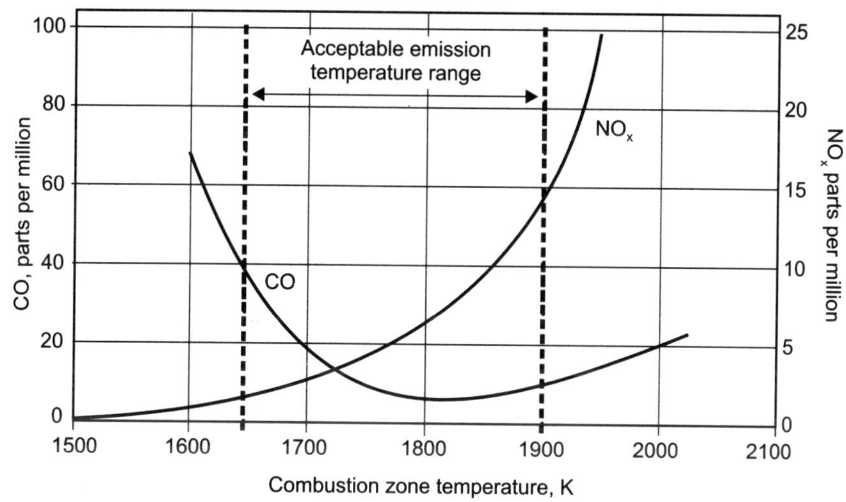

Fig. 9.12 Combustion Zone Temperature and CO, NO_x (Lefebvre, 1999)

The SCR procedure can introduce some potentially complicating factors. If the $NO_x - NH_3$ reaction does not proceed to total completion, a small amount of NH_3 slips through the catalyst untouched, eventually escaping to the atmosphere as a hazardous air pollutant. When natural gas is the sole fuel, ammonium sulfate [$(NH_4)_2SO_4$] and bisulfate [$(NH_4)HSO_4$] formation is not of concern, but problems can arise when dual fuel units are switched from gas to fuel oil due to the presence of sulfur. Reaction of ammonia with sulfur tri-oxide (SO_3) in the flue gas produces the sulfate and bisulfate compounds, which can foul and corrode turbine blades and other downstream components in heat exchangers. Problems of storage of ammonia in concentrations above certain levels at the site are also experienced due to toxicity issues. In spite of the demerits, the method finds favor among many turbine manufacturers.

Since flame temperature plays an overwhelming role in the production of pollutants, most emissions control mechanisms focus on constraining it within set limits. As indicated in Figure 9.12, CO levels are low when combustion chamber temperatures exceed 1650 K, while NO_x production rate becomes objectionable beyond 1900 K. Hence, the range of 1650 to 1900 K must be achieved for all power settings of the engine. Thus, a larger amount of air is required at the entrance to the combustor liner at full power to reduce the flame temperature. As engine load decreases, more of this air must be directed towards the downstream end to dilute the burned gases and maintain the temperature above the low limit.

9.15 REVIEW QUESTIONS

9.15.1 How does clearance between the rotor and the stator affect performance of a gas turbine's compressor and turbine sections?

9.15.2 What design aspects of a power system for a ship's propulsion system contribute to easing maintenance and repair tasks?

9.15.3 What are the advantages and disadvantages of a gas turbine compared with a diesel engine on board a ship?

9.15.4 Discuss some major constraints and requirements for the use of gas turbines in land vehicles.

9.15.5 What are the principal methods used to prevent corrosion of steel in the marine environment? What is the primary objective in these methods?

9.15.6 Compare the effects of liquid and gas fuels on performance of a gas turbine.

9.15.7 How does air temperature at the inlet to the compressor affect power produced and heat consumption of a gas turbine?

9.15.8 Why is air extracted from the compressor, and what is the effect on the performance of the gas turbine?

9.15.9 How does a thermal system differ from a pneumatic system in controlling tip clearances in the turbine section?

9.15.10 Mention some methods to control production of soot in the combustor.

9.15.11 How does firing temperature and residence time in the combustor affect formation of objectionable exhaust gas components?

9.16 EXAMPLE PROBLEMS

Example Problem # 9.16.1: A 2.35 in. long cantilever beam consists of a strip of spring steel 0.2245 in. wide and 0.0135 in. thick. Calculate the spring stiffness constant of the beam.

Solution: The spring constant of a cantilever beam is $3EI/l^3$. The moment of inertia of the cross section is:

$$I = (1/12)bh^3 = [0.2245 \times (0.0135)^3]/12 = 4.603 \times 10^{-8} \text{ in}^4.$$

Assume Young's Modulus = 30×10^6 lb/in^2

Bending constant $EI = 30 \times 10^6 \times 4.603 \times 10^{-8} = 1.381$ lb-in^2

So spring constant $k = 3EI/l^3$ lb/in.

$$= 3 \times 1.381/2.35^3 = 0.3192 \text{ lb/in.}$$

Example Problem # 9.16.2: Assume the strip of spring steel in Example Problem # 9.16.1 carries a weight of 0.0025 lb at its free end. What is the natural frequency of vibration of the strip?

Solution: Let density of steel be 0.28 lb/in^3.

Mass per inch of strip $\mu = (0.2245 \times 0.0135 \times 0.28)/386 = 2.1985 \times 10^{-6}$ lbs-in^{-1}-sec^2.

So total mass = $(0.0025 + 2.35 \times 2.1985 \times 10^{-6})/386 = 6.490 \times 10^{-6}$ lbs-in^{-1}-sec^2.

Natural frequency: $\omega^2 = k/m = 0.3192/6.490 \times 10^{-6} = 49184$

Hence $\omega = 35.31$ cycles/sec.

Example Problem # 9.16.3: A rectangular bar spins about an axis perpendicular to the bar. The center of gravity of the bar is 7.285 in. from the spin axis. If the spin speed is 3250 rpm, beam is 5.5 in.

long of section area 1.235 in^2 and material density is 0.28 lb/in^3, determine the centrifugal force on the bar.

Solution: Weight of the bar = 1.235 × 5.5 × 0.28 = 1.9019 lb.

Centrifugal force = $(W/g)\omega^2 r$ = (1.9019/32.2) × $(2\pi \times 3250/60)^2$ × 7.285/12

= 4149.18 lb.

Example Problem # 9.16.4: A 2750 horsepower diesel engine weighing 47350 lb. has 8 cylinders of size 9.375 in. bore, 9.5 in. stroke. The engine is 6.23 ft wide and 26.2 ft long. Calculate the weight and floor area required for the same total output from another engine rated at 248 horsepower each, and each cylinder is 3.65 ft wide, 5.61 ft long and weighs 2812 lbs.

Solution: The number of smaller cylinders required will be 2750/248 = 11. Total weight will be 11 × 2812 = 30932 lb. Floor area = 11 × 3.65 × 5.61 = 225.24 ft^2, compared with 6.23 × 26.2 = 163.23 ft^2 for the single engine.

Example Problem # 9.16.5: Six 12-cylinder direct drive diesel engines power a freight ship. Cylinder bore is 30 in., develops 11200 horsepower and weighs 117 lb/bhp. The engine is 24.5 ft tall, 8.35 ft wide and 41.3 ft long. Ten 8 cylinder engine of 14.5 in. bore is to be substituted, with reduction gear. Weight of the gearbox is 18 per cent as much as the engine and adds 22 per cent to the floor space of the engine. If the replacement engine experiences same stress level, estimate the new dimensions and floor area required.

Solution: For the same output at the same piston speed and mean effective pressure, the number of 14.5 in. bore cylinders required will be 8 × $(30/14.5)^2$ = 34.24. To account for losses in the gears, use 10, 14.5 in. cylinder engines, or a total of 80 cylinders of 14.5 in. bore. Total engine weight is:

$$11200 \times 117 \times (14.5/30) \times (80/72) = 703,733 \text{ lbs.}$$

This compares with 11200 × 117 = 1,310,400 lb. for the single engine. To this must be added 20% for the gears, so the total weight of the engines and gears will be 703,733 × 1.18 = 830,405 lb. The height of the engines will be: 24.5 × (14.5/30) = 11.84 ft. The floor area covered by the engines plus gears will be: 22 × (80/72) = 24.44% greater.

9.17 EXERCISE PROBLEMS

Exercise Problem # 9.17.1: A 2.25 in. diameter, 6.25 in. long bar spins about an axis normal to the bar. Find the center of gravity of the bar from the spin axis if the spin speed is 3722 rpm, beam material density is 0.22 lb./in^3, and centrifugal force on the bar is 13890 lb.

Exercise Problem # 9.17.2: A 11.25 in. long cantilever beam is 1.255 in. wide. Calculate the thickness if the beam has stiffness = 125 lb./in

Exercise Problem # 9.17.3: If the beam in Exercise Problem # 9.17.2 has natural frequency of vibration = 32.5 cycles/sec, what weight must be attached at its free end?

Exercise Problem # 9.17.4: A 39500 lb. engine develops per cylinder 335 horsepower, has 6 cylinders of size 6.25 in. bore, 6.85 in. stroke. The engine is 4.27 ft wide and 11.25 ft long. Calculate the weight and floor area required for the same total output from another engine rated at 437 horsepower each, and each cylinder is 4.17 ft. wide, 10.83 ft. long and weighs 4215 lbs.

Exercise Problem # 9.17.5: A round bar spins about an axis perpendicular to the bar. The center of gravity of the bar is 8.255 in. from the spin axis. If the spin speed is 3425 rpm, beam is 6.21 in. long, material density is 0.28 lb/in^3, determine the radius of the bar when centrifugal force is 8257 lb. on the bar.

REFERENCES & BIBLIOGRAPHY

Bird, J, Grabe, W., "Humidity Effects on Gas Turbine Performance", ASME paper # 91-GT-329, New York, 1991.

Boyce, M. P., "Handbook for Cogeneration and Combined Cycle Power Plants", ASME Press, New York, 2002.

Broekhaus, K. O., Rand, M. J., "The Design and Layout of the Propulsion Plant of the RNLN Air Defence and Command Frigate", ASME paper # GT-2002-30673, New York, 2002.

Caguiat, D. E., "Rolls Royce/Allison 501-K Gas Turbine Anti-Fouling Compressor Coatings Evaluation", ASME paper # GT-2002-30261, New York, 2002.

Driscoll, M., McFetridge, E., Arseneau, W., "Evaluation of At Sea Tested LM2500 Rainbow Rotor Blade Coatings", ASME paper # GT-2002-30263, New York, 2002.

Harvey, E., Kingsley, J., Stauffer, M., "United States Navy Integrated Power System Gas Turbine Generator Set Test Experience", ASME paper # GT-2002-30260, New York, 2002.

Hibner, D. H., "Dynamic Response of Viscous Damped Multi-Shaft Jet Engines", Journal of Aircraft, Volume 12, No. 4, 1975.

Hoeft, R. F., "Heavy Duty Gas Turbine Operating and Maintenance Considerations", GER-3620B, General Electric I & PS, Schenectady, NY, 1993.

Kaya, H., "Catalytic Combustion Technologies", Journal of Gas Turbine Society of Japan, volume 25, number 98, page 48–51, Tokyo, 1997.

Koschier, A. V., Mauch, H. R., "Advantages of the LV100 as a Power Producer in a Hybrid Propulsion System for Future Fighting Vehicles", ASME paper # 99-gt-416, New York, 1999.

Kurz, R., Fozi, A. A., "Acceptance Criteria for Gas Compression Systems", ASME paper # GT-2002-30282, New York, 2002.

Nakazawa, N., Ogita, H., Takahashi, M., Yoshizawa, T., Mori, Y., "Radial Turbine Development for the 100 kW Automotive Ceramic Gas Turbine", ASME paper # 96-GT-366, New York, 1996.

Rangwala, A. S. "Reciprocating Machinery Dynamics: Design and Analysis", Marcel Dekker, New York, 2001.

Rangwala, A. S., "Dynamic Analysis of Turbo-machinery", McGraw-Hill, New York, 2004.

Stalder, J. P., "Gas Turbine Compressor Washing State of the Art: Field Experiences", ASME paper # 98-GT-420, New York, 1998.

Thompson, B. D., Badgley, R., "Application of an Advanced Hybrid Rotor Dynamics Model to the Complete Structure of a Marine Gas Turbine Engine", ASME paper # 88-GT-123, New York, 1988.

Thompson, B. D., Wainscott, B., "Systematic Evaluation of US Navy LM2500 Gas Turbine Condition", ASME paper # 00-GT-667, New York, 2000.

Wadia, A. R., Wolf, D. P., Haaser, F. G., "Aerodynamic Design and Testing of an Axial Flow Compressor with Pressure Ratio of 23.3:1 for the LM2500+ Gas Turbine", ASME paper # 1999-GT-210, New York, 1999.

Yoshida, Y., Oyakawa, K., Aizawa, Y., Kaya, H., "A High Temperature Catalytic Combustor with Starting Burner", ASME paper # 00-GT-087, New York, 2000.

Part III

Component and System Failure Modes

10

CHAPTER

FAILURE ROOT CAUSE

10.1 INTRODUCTION

Turbine engines are afflicted by troubles arising from sources as varied as ingestion of external debris at the compressor's intake side to internal aerodynamic interference between the stationary vanes and rotating blades as the flow of air or gases passes over their surface. Not all incidents result in catastrophic breakdowns, but the potential for operation to diverge significantly from its intended course can lead to deterioration in performance in the form of increase in specific fuel consumption or power output. The consequent increase in fuel costs over sustained periods of time may then force the plant operator to schedule a shutdown to address the problem.

External factors acting as root source of failures are many. A number of airports are located in areas where bird sanctuaries are within flying range of the birds. When the birds happen to traverse the path along which aircrafts takeoff and land, the risk of a few birds getting sucked in due to the powerful suction created by the engines is high. Debris in the form of pieces of a tire's rubber tread, metal nuts or bolts and branches or twigs lying on the runway are other hazards. If chunks of metal were to separate from any of the rotating parts as a consequence, the unbalance will lead to higher than normal vibrations. High ambient temperature in a desert environment can cause lubricant oil to lose much of its viscosity and lead to premature failure of the bearings.

Within the engine, blockage of air flowing through the annulus of the compressor can cause the flow to surge and even reverse in direction. A compressor stall can be severe enough to produce severe vibrations in the whole engine. Large dynamic stress amplitudes will result, and metal fatigue can lead to structural failures. A fire in the combustor section if left unchecked can engulf an entire aircraft. Centrifugal forces act on rotating blades in the compressor and in the turbine. In combination with thermal fatigue in the turbine section, the blades may not be capable of accepting the imposed loads and fail with severe consequences.

10.2 DISK BURST FROM OVER-SPEED

Among all failures, consequential effects are the greatest when a rotating disk fails due to a turbine running into over-speed regime. All of the disks in a turbine's rotating assembly are designed to withstand a certain level of circumferential, or hoop, direction stress created by the centrifugal effects of the attached blades at the rim, as well as the disk's own weight. Beyond the safe maximum speed, the disk's material will lose its ability to withstand the stress. It is worth mentioning that centrifugal force varies with the square of the speed, hence the circumferential stresses in the disk will rise exponentially with the rotor speed.

When the disk has a central bore, circumferential stresses reach a peak at the inner surface of the bore. The bore at the center is required if other parts such as a vent tube goes through the bore. Thus, when the material capability to accept the centrifugal force is reached, the disk will burst throughout its geometry. Fragments will then fly outwards. The intensity of the ejected projectiles will be strong enough to not only destroy the turbine engine but also destroy the complete installation. Needless to say, humans happening to be in the vicinity at the time of the incident will not be spared.

The root cause of a turbine going into over-speed condition can be traced to multiple factors occurring simultaneously. An abrupt shedding of a large electrical load on a power generator can tend the prime mover, either steam or gas turbine, to increase in speed. Speed controllers are invariably used to ensure the turbine unit's operating speed is maintained within a small percentage range of 3600 rpm for 60 Hz electrical power or 3000 rpm for 50 Hz power. On occasions, however, the speed controller fails to perform its assigned function, mostly due to corrosion and wear of many of its mechanical links and springs. Malfunctioning of steam admission valves to the high-pressure section is another factor. Thus, when the speed controller and the admitting valves fail to operate when a large chunk of electrical load is shed, the situation for the turbine unit to run into the over-speed regime is setup.

Another possible situation can occur during startup of the turbine-generator set. Prior to reaching regular operating speed, some anomaly within may cause the startup to be aborted. However, the large physical size and associated inertia of the rotating system together with absence of driven load may preclude the rotor from slowing down. In combination with steam admission valves, or fuel injection system in a gas turbine driven unit, the rotor keeps accelerating into over-speed condition. If the maximum safe speed is breached, any of the numerous disks in the turbine with the lowest burst speed will fail with cataclysmic results.

10.3 DEFICIENCIES IN LUBRICATION SYSTEM

A failure in the lube oil system of a turbine may not cause as much damage as from a burst disk, but problems associated with the bearings not receiving their regular supply of lubricant are more frequently encountered. During regular operation the lubricant film produces enough hydrodynamic pressure to lift the rotor from the stationary journal bearing's surface. A regular flow of oil through the annular clearance between the bearing surface and the journal levitates the rotor, dissipates heat built up due to friction losses and flushes out any debris in the form of tiny metal chips that may separate from the mating surfaces.

When the flow of oil is disrupted, the hydrodynamic uplift on the rotor is lost. Metal to metal contact between the journal and the bearing will lead to grinding. Chunks of Babbitt from the bearing's surface will dislodge and lead to severe scoring of both the normally smooth surfaces. The normal concentric location of the rotor and the stator will also be disturbed. Radial clearances in modern steam and gas turbines at the rotating blade tips are held within tight limits to prevent leakage, thus maximizing power output as also efficiency. In the absence of levitation in the bearings, the blade tips will rub against the stationary casing, possibly chipping and even breaking the blades. Widespread damage may also be expected in the inter-stage seal areas. Since the turbines operate at very high speeds, the rotor will turn several turns before it can be made to come to a complete stop. The rubbing process will then extend to inflict damage in the machine internally.

Causes for the lube oil supply to stop flowing are many. Any of the components in the lubricant processing system, such as the filtration, magnetic chip detector or cooling units, may fail. Pumping devices, plumbing fixtures, angle and T-joints and other accessories may spring a leak. If the leak is left unchecked, over a period of time the reservoir level may drop to unsustainable levels. Oil supply to the bearings will then be hampered.

Large steam turbine installations have a number of journal bearings, with each section having bearings at the two extremities of the rotor. The bearing located farthest away from the oil processing unit is likely to feel the first effects of any breakdowns. Oil flow and pressure indicators may be placed near this bearing to inform the control room of any anomalies in operation of the oil supply system.

10.4 MALFUNCTION IN CONTROL VALVES AND ACTUATORS

The design of most control valves, manipulators and actuators tends to be complex, requiring a number of small parts, linkages, springs, diaphragms and other elements. Assembled inside the valve body or housing, the parts nevertheless deteriorate from corrosion, fretting and wear. For example, a quick-acting valve may suffer from fretting at the valve seat as the valve disk repeatedly slams against the seating area. Where the fluid medium handled by the device is at an elevated temperature, wearing effects will rapidly cause the part to degrade to an extent its functionality is impaired.

Suspended contaminants such as sand or iron oxide particles can cause erosion of metallic parts in a valve. Thus, the cage of a pressure reducing valve with a large number of precisely drilled holes may have some of the material between the holes torn off as millions of particles flow over the surfaces. The piston inside the cage will not be able to accurately control the flow volume passing through the valve. As a consequence, the control room may set the flow at 40 per cent but the actual flow going through the valve may be 60 per cent due to the damaged cage. The control personnel are then at a loss to decide when some other gauge in the turbine system is indicating flow of gas at 60 per cent but the errant valve is set at 40 per cent.

Effective control of processes with defective valves and manipulators in a turbine system is hampered when deficiencies prevent full closure and permit leakage of the fluid medium. Even a tiny leakage of natural gas past a defective valve is not acceptable if the gas were to escape to the outside. In spite of best efforts to detect presence of the gas with sophisticated sensing instruments, failures do occur. Thus, a burning cigarette or backfiring from an automobile engine in such an environment can cause the entire installation to go up in smoke and fire in a matter of minutes.

Pressure relief valves are critical items when it comes to safety issues in boiler installations. Temperature and pressure of the steam discharged from the super-heater section are extremely high. When load demand on the electrical generator drops, steam requirements of the turbine will also fall. Both the pressure and temperature at the super-heater outlet will then climb, perhaps past the maximum sustainable limits. The pressure relief safety valve is then required to release a lot of the steam, accompanied by a deafening noise. If this valve were to fail to operate, explosion in the boiler accompanied by fire will be the direct consequence.

10.5 IN-FLIGHT SHUTDOWN OF ENGINE

In-flight shutdown of an aircraft engine can pose an extremely difficult situation for the crew, especially if ignition cannot be obtained after repeated efforts to restart the engine. An aircraft engine can cease to operate during flight due to a number of reasons. Excessively rich or lean mixture of fuel and air can be caused by sudden and large fluctuations in outside air pressure or temperature, which can result in the combustor to lose the flame. In the event the engine cannot be restarted, the crew may be forced to alter its assigned flight pattern and land at the nearest airport.

Most modern aircrafts are designed to be capable of flying and staying aloft with one engine inoperable. Thus, the twin-engine Boeing 777 on a transatlantic flight can keep operating, perhaps at a reduced pace, at least for a period of time even on a single engine.

Excessive rotor vibrations, exhaust gas temperature at power turbine exit straying beyond its upper bound, compressor discharge pressure falling to below normal and lubricant oil pressure or flow volume decreasing to unacceptable levels are some other factors that may cause the engine's microprocessor based controller to order the fuel delivery to the combustor to be shut off. The control system will even intimate to the crew the reason for the engine ceasing operation.

The probability of occurrence of an engine shutting down during the flight due to sudden increase in vibrations and other operating parameters is generally low. When the engine is operating at cruise conditions, the thrust produced is only a fraction of that required during takeoff, since only aerodynamic drag on the aircraft has to be contended with. Thus, the risk associated with any of the prime operating indicators going out of their safe limits is also low. Only in the event of factors such as improper routine maintenance practices can the probability of the engine shutting down increase. Here, too, the engine's microprocessor may be able to give advanced warning to the crew.

10.6 COMPRESSOR SURGE AND STALL

A full stall in a compressor is accompanied by a loud bang and high vibrations. In an aircraft engine the vibrations may be strong enough to shake the aircraft's structure. At the turbine exit, the exhaust gas temperature will also climb. After a stall incident, ground personnel will need to inspect the rotor blade tips with a bore-scope. In the absence of action from the crew such as shifting down the throttle, the control system will bring down the rotor speed and power output.

A surge in pressure in the compressor section of a gas turbine is known to occur under certain circumstances. The pressure tends to surge as the air leaves the stationary vanes, just prior to entering the rotating cascade of blades. When the angle of attack deviates substantially from the theoretically correct value, flow eddies develop at the leading edge of the rotating blade. As the deviation in the attack angle progressively becomes worse, the turbulent regime also tends to increase in severity to a point where most of the passage between adjacent airfoils is blocked. The air behind the turbulence will then tend to increase in pressure as it struggles to get past the turbulent blockage. In extreme cases the flow may even tend to reverse in direction. Blockage may even be set up by ingestion of a foreign object such as leaves getting trapped in the cascades.

A stalling compressor is generally difficult to distinguish from pressure surge since many of the characteristics are not readily identifiable from sensor signals. If a stall is considered to be a more intense form of pressure surge, a full blown stall may be deemed as reversal in the direction of air flow. Consider, for example, pressure taps placed in holes drilled in the casing in the space between adjacent rows of stationary and rotating airfoils. Spikes in the pressure signal at certain frequencies may indicate one form, while a broad fronted spectrum in a band of frequencies may be depicting another form of pressure surge.

10.7 COMBUSTOR FIRE

The primary role of the liner inside the combustor of gas turbines is to contain the flames from ignition of fuel. As the fuel, liquid or gaseous, burns in the presence of oxygen arriving from the compressor's diffuser section, the products of combustion rapidly increase in temperature.

Within the flame zone the temperatures soar to levels far in excess of the melting point of the liner substrate. Protection for the liner to withstand the thermal onslaught is derived from two sources. Application of thermal barrier coating physically isolates the liner surfaces from direct exposure to the hot gas stream. In addition, cooler air bled from the compressor is made to flow over both surfaces of the combustor liner. Holes are provided in the liner to permit some of the coolant to enter into the hot stream. By drilling the holes at an angle, the cooler air is made to flow along the liner walls, which once again aids in separating the liner from the hot gases.

The liners are designed such that they can withstand the imposed thermal load. Stress due to thermal conditions is created when a component exposed to elevated temperatures is restrained from growing thermally. Also, thermal stresses are induced when temperature gradients are present, even when permitted to grow. To meet the need for freely increasing in dimensions, the combustor liner is mechanically supported as lightly as possible. This is achieved with stops to restrain physical motion in the three principal directions.

Annular liner configurations favored for aircraft engines rely on a series of circular rings made of sheet metal. The rings have interlocking joints at both ends. As the temperature rises from the cold condition at start of ignition to operational levels, the joints gradually open up and let the liners grow in length, thus avoiding buildup of thermal stresses.

In spite of the protective measures, the liners are subject to failure arising from thermal induced fatigue. One reason is lack of uniformity of temperature distribution in the gas stream's cross-section. Thus, some regions of the liner are exposed to higher temperature levels than others, leading to thermal gradients and consequent thermal stress. As small cracks develop and grow in length with increasing number of operating hours, parts of the liner might separate. Containment of the flame zone within its boundaries is jeopardized. The flames may even leap out of the main engine's casing and spread to surrounding parts in the wing section. This sets up a dangerous situation for the safety of the entire aircraft. Timely action on the part of the crew to douse the flames through discharge of extinguisher cylinders in then necessary.

10.8 VIBRATIONS FROM ROTOR UNBALANCE

Lack of symmetry in the distribution of mass in a rotating assembly about the spin axis creates unbalance in the system. The asymmetry may arise from a number of sources. Machining errors of individual disks may leave certain areas with greater axial thickness than others. Blades attached at the rim of the disk may not be equal in weight, or may have unequal location of center of gravity from the shaft axis. During assembly of mating disks in the turbine module, a mismatch in the centerline of the disks will lead to unbalance. Thermal bow setting into a turbine rotor is another example. Besides rotor unbalance, vibrations may also be caused by misalignment and aerodynamic excitation.

Centrifugal force is created as the unbalanced mass spins around the centerline. In the absence of an equivalent mass 180º away, centrifugal force is developed, which has to be carried through the rotating assembly and eventually reacted at the supporting bearings. The rotating force vector produces dynamic stresses in the load carrying components. Besides the normal loads carried by the string of components in the path, imposition of the additional load compromises on the structural ability to withstand and still keep operating.

Fatigue is induced as a direct consequence of the dynamic stress pattern in the material of the component, together with the number of variations in the stress. The fatigue limit of a part depends on the static and dynamic loads and the number of cycles of applied dynamic load. Dynamic stresses produced, however, play a leading role. The endurance strength of the material is based on its tensile strength, in conjunction with factors such as surface finish, manner of load application, stress concentration and effects of corrosion. Thus, after a certain number of load cycles at a specific level of maximum total stress, the material loses ability to withstand further loadings and the component experiences rupture.

10.9 THERMAL BOW IN ROTOR

After a certain number of hours of operation, steam and industrial gas turbines need to be shut down for scheduled maintenance. Unforeseen circumstances may even force the operators to cease operation. As part of the shutting down sequence, it is imperative the rotor be put on turning gear as the decelerating rotor speed comes in the range of 5 to 8 rpm.

Turning the rotor on the gear may continue for 3 to 5 hours in order to avoid a set bow in the rotor. The reason is due to the uneven cooling of the rotor's surfaces that can lead to non-uniform contraction. For instance, when a steam turbine is going through shut down, the lower surface of the rotor is exposed to cooler air blowing in through the underside of the rotor while warmer currents flow over the upper surface. In the absence of a slow roll the result would be unequal thermal distortion on the two sides. A permanent bow may also be induced in the rotor.

A thermal bow effectively induces a large amount of unbalance in the rotor. Power generation turbine rotors tend to have large thickness through the radius. In combination with the long span between the bearing supports, the result is to cause the rotor to sag in the mid-span section. Putting the rotor on a slow roll permits even temperature distribution and hence more uniform shrinkage on all sides of the rotor.

Starting an engine with a bowed rotor condition is a difficult proposition since vibrations will be high. If top speed is at all reachable, the dynamic activity will be large enough to cause failure of attached accessories and piping. Operation would then have to be abandoned.

Various remedies have been tried to fix a bowed rotor, but have met at best limited degree of success. Hanging the rotor vertically from one end relies on the weight of the rotor itself bringing it back to shape. Sand blasting on the high side attempts to reduce the offset. Shaving off material would compromise on the ability to restrict leakage.

10.10 ENSURING PROPER FUEL SUPPLY

Dissolved and suspended contaminants invariably flow into the combustor with the injected fuel. If not filtered and separated, oily, gritty sand and ferric oxide particles may flow into a natural gas fired boiler. The contaminants may stick to the walls of pipeline bringing in the gas and stationary components.

If left unchecked, over a period of time the deposits may become heavy enough to restrict the flow, while stationary components such as valves and accessories may be prevented from functioning. The components may even experience erosion as millions of particles flow over their surfaces.

Industrial and power generation gas turbines require an elaborate filtration system to separate the offending solid particles. Fixed vanes inside cyclonic filters cause the incoming gas flow to be whirled around the cylindrical housing mounted vertically. The whirling motion causes the large solid particles to flow outwards under the action of centrifugal force, and stick to the walls of the vessel. As the gas flows upwards, it passes through a bank of urethane type of filtering elements that trap tinier particles. Removal of the particles is essential since the natural gas has to enter the combustor through finely drilled holes at the tip of the fuel injector, practically entering and mixing with the air stream in the form of a haze.

Sour gas refers to the presence of sulfur compounds in natural gas. As the gas burns in the presence of oxygen, it produces H_2S and SO_2, products that are particularly offensive to the sense of smell. Sour gas requires treatment and conversion to a sweet form before it is acceptable for use in the gas turbine.

Dissolved impurities are a major concern in liquid fuels. Chemical compounds of sodium, potassium, calcium and phosphorous are present in the stream. When treated with special agents, the impurities chemically combine to form granules, which may then be easily washed off. Sulfur, on the other hand, is more difficult to remove, requiring a more complex process. Even minute levels of sulfur leads to formation of chromium sulfate as the hot gases flow over the nickel-chrome substrate of the blades. This causes wavy patterns on the surface of the airfoil, which increases losses as the pressurized gas flows over the surfaces.

10.11 HYDRAULIC OILS AND FLUIDS

Hydraulic fluids are required for actuators to perform in a gas turbine system. Lubricant oil is required for the main bearings. Servo-motors operate the speed control mechanism. Fluid for accessories and gear box have their own special requirements to ensure mating gear surfaces experience the least level of wear and spall.

Industrial and power generation gas turbines have variable vanes at the inlet to the compressor to preclude occurrence of stall during startup, and require an actuating mechanism. Aviation and aero-derivative engines, on the other hand, have variable stator vanes in the first six to seven stages in the high-pressure compressor. Power generation turbines operate at an essentially fixed speed, thus needing variable setting only at the inlet. Multiple stages in the high-pressure compressor require this ability since the rotor operates in a much wider speed band. The variable vanes perform the twin tasks of directing the flow of air while also varying the available cross-sectional flow area.

Individual vanes are attached to the compressor casing through bored holes in the case, with the protruding stem fastened with a nut. The vane setting is manipulated through a link that connects the stem with actuating rings that go around the compressor casing. The actuating ring turns around the casing by small amounts and causes the smaller links to shift the vanes through the desired angle. This action effectively directs the flow of air into the leading edge of the rotating blades at the right angle of incidence. The actuating ring in turn receives the force to turn through a hydraulic cylinder, with its piston imparting the necessary force.

It is essential the various fluids, oils and other mediums are qualified to perform their assigned job. To this extent, turbine manufacturers require the different oils to possess a number of characteristics that are within specific upper and lower bounds. Thus, viscosity, density, flash point, moisture content, heat value and a host of similar other parameters must meet their target if the turbine installation is to work at its peak capability. In the absence of the fluid meeting any of the technical requirements, the manufacturer's performance and related warranties may not remain valid.

10.12 NEED FOR CONCENTRICITY BETWEEN ROTOR AND STATOR

Performance of a turbine is affected in the absence of coincidence of the true centerlines of the stator and the rotor. When the two centerlines do not match, there will be too much radial space on one side and too little 180º away.

On the compressor side, excess radial clearance between the blade tips and the stator affects performance in two ways. Instead of flowing over the airfoil surface, when some of the air escapes past the blade tip, the pressurization obtained from the stage will decline. This decrement will be compounded as the flow goes through successive stages and show up as a loss at the compressor discharge that may not be easily ignored. The other effect will be in the form of reduced stall margin. As the dead space above the blade tip increases, increased turbulence occurs as the flow randomly traverses from above one blade tip to the next. Vortices are likely to be shed with formation of eddies in the enlarged space. The result is increased resistance to the flow from the upstream to smoothly flow through the cascade and consequent reduction in the overall stall margin. As a rule, the radial clearance must be between 1.5 to 3.5 per cent of the camber of the blade section at the tip. There has to be some minimum clearance to avoid rubbing of the blade tips against the casing in the event there is excessive unbalance and consequent bending in the rotor. The higher percentage is required for adequate overall performance.

Performance on the turbine side is also affected in multiple ways when the radial clearance is not at optimum levels. Some of the hot and pressurized gases will flow over the blade tips without doing useful work. The reduced work output will be recorded by the torque sensor. To offset the loss, the automatic control system will order additional fuel into the combustor. But the extra fuel will also increase the gas temperature at the turbine's exhaust. The rotating blade tips are provided with knife edges that cut into the honeycomb segments attached to the turbine casing. The feature helps to seal and substantially reduce the leakage of the hot gases.

During initial manufacture the tolerance on both axial and radial clearances between the rotating and stationary modules are set at levels that will ensure maximum performance in modern turbines. On the compressor side, the blade tips are ground to a diameter that will obtain the best fit with the mating stator on all the stages.

10.13 LACK OF ALIGNMENT WITH DRIVEN MACHINERY

The alignment of the rotors in a large steam turbine installation among the high, intermediate and low-pressure sections must be set to close tolerances, as also with the generator rotor. The two halves of the coupling on adjacent rotor must match both in the horizontal and vertical planes. Besides linear alignment, angular misalignment between two mating rotors must receive even greater attention.

Alignment in the vertical plane requires insertion of shims of varying size between the bearing pedestal and the mount block. For shifting in the horizontal plane, attachments to the bearing pedestal help to obtain precise location. The alignment process is carried when the rotor is cold and also when it is hot. Once the alignment is obtained, anchor bolts on the bearing housing are tightened.

Ramifications of misalignment between adjacent rotors can be many. Angular misalignment will cause the centerlines of the mating rotors to move along a conical path, causing vibrations in the bearings to increase, perhaps even to unacceptable levels when the amount of misalignment is high enough. Angular misalignment will also induce fluctuating overturning moments in both rotors. Operation of the turbine may have to be discontinued to rectify the situation.

Large turbine installations are often installed on a rigid steel framework. The intent is to permit piping and accessories to go through on the underside. Routine maintenance chores also must be carried out with the arrangement. The steel framework must also be thoroughly grounded. Excess vibrations in one bearing may cause the mount block on the steel foundation to distort or shift, which in turn further aggravates the misalignment.

10.14 PRECAUTIONS DURING ENGINE ASSEMBLY

During the manufacturing cycle, turbine components go through several steps. After the final machining of a disk or section of a rotor, the component must be dynamically balanced. A thin disk may be balanced in a single plane, but a longer rotating shaft will require multi-plane balancing to ensure not just centrifugal forces but also the accompanying overturning moments are in a state of equilibrium.

A case in point is the long mid-shaft of high-bypass turbo-fan engines. The shaft goes through the high-pressure core module to transmit power from the low-pressure turbine to the fan and low-pressure compressor module. In certain engines, the shaft spins during engine cruise condition at a speed close to the first natural frequency of vibration. Maximum engine speed is in the super-critical range, well above the fundamental frequency. To balance the shaft, balance lands are provided at quarter, half and three-quarter length from one end. There is no place on the shaft to attach balance weights, so material is ground from the balance lands. The center land aids in balancing the fundamental mode, while the quarter and three-quarter lands are required for the second vibration mode. Since operation is in the super-critical regime, exceedingly tight balance specifications are mandated for the shaft.

Once a disk is brought to balance within requirements, the blades are attached at the rim of the disk. The bladed disk is then once again put on the balancing machine. At this point, instead of adding balance slugs, the location of heavier and lighter blades may only need to be exchanged to obtain proper balance. The balanced bladed disks are next assembled to form the turbine or the compressor module. The complete module is next dynamically balanced in two planes. In the next higher level of assembly, all of the rotating modules are assembled to form the complete rotor.

Many large power generation steam and gas turbine rotors are designed such that the first natural frequency of vibration is well below the fixed operating speed of 3000 or 3600 rpm. This fundamental mode has practically all of the dynamic activity in the bearings, with the shaft experiencing practically no flexural activity. Sometimes referred to as the bounce mode of operation, balancing of the rigid rotor requires attachment of the balance masses at the two ends of the rotor, preferably in the coupling since it is outside the turbine casing.

Aviation and aero-derivative engines, in contrast, have rotating systems that are far more flexible. During startup and acceleration, the engine goes through a series of natural frequencies of vibration. Individual vibration modes characteristically have significant dynamic activity in certain portions of the engine, both in the rotating and stationary structures. Machining, assembly and balancing requirements are more tightly set to ensure the engine operation is not impaired by excessive vibrations.

10.15 INADEQUATE THERMAL PROTECTION OF HOT PATH

Metal temperatures soar to exceedingly high levels in the combustor and turbine sections of gas turbines. Steam turbines are not exposed to elevated metal temperatures even in the initial stages of the high-pressure section, mostly because combustion of fuel takes place outside the turbine in the boiler.

The elevated metal temperatures expose the parts to special risks associated with thermal fatigue. In spite of the parts treated by expensive processes such as thermal barrier coats and internal coolant passages built inside the parts, thermal fatigue limits useful operational life of airfoils or otherwise distorts their geometry permanently.

A good example of the latter is the stationary vanes at the inlet to the turbine section. The coated vanes have a complex network of passages to cool the leading and trailing edges as also the suction and pressure surfaces. At the inner and outer radii the vanes are firmly attached to the inner and outer annulus. This provides rigidity and resistance against thermal distortion. The mid-span region of the airfoil, however, lacks in rigidity, especially at the leading and trailing edges. As the engine accumulates operating hours, the mid-span region tends to distort from the influence of the hot gases flowing over the vane. At both edges thermal bow occurs, and prevents the vane from receiving the gas flow at the leading edge and directing the gas flow into the rotating blade cascade at the correct angle of incidence.

The thermal bow of the vanes directly impacts upon the performance and power output of the turbine. The vanes cannot be brought to their original shape through repair, so the solution to the problem lies in replacing the vanes.

Problems are also encountered in the cooling passages of airfoils. Out of necessity, the cooling passages are ducts of small diameter. Sometimes the entrance to the ducts may be obstructed by extraneous material, rendering the duct mostly useless. Oxidation along the walls of the ducts reduces the flow area, as also the amount of coolant reaching the hot surfaces. Where the airfoil meets the platform near the trailing edge of the airfoil is the most difficult to cool due to the lack of material in the region, but unfortunately this area also develops maximum stress from centrifugal effects. Stress concentration also is high due to the sharply changing geometry.

Chipping of the coated material exposes the metal substrate to the hot gases. The uneven temperature distribution in the surrounding region leads to high thermal stresses, in the process reducing the available life of the part.

10.16 CORROSION OF COMPONENTS

Exposure to the hostile environment of turbine components rapidly increases the rate of corrosion. Even when not directly exposed to salt water, marine air carries with it compounds of sodium, magnesium and other elements that corrodes turbine parts when operating at sea. Corrosion also is a problem in areas where steel mills and petroleum refineries are in the vicinity due to the presence of gases and solid particles. Thus, a turbine unit operating within a petro-chemical complex is likely to experience higher levels of corrosion in its system.

Exposure to corrosive elements gradually removes the smooth finish of parts and causes the surface to become uneven. If at the same time erosion is also a problem, nicks and scratches will compound the problem and increase the stress concentration in the local region. Aside from enhanced friction losses when air or products of combustion flow over the part's surfaces, the useful life of the product is also reduced.

10.17 ENGINE STARTUP AND SHUTDOWN RELATED PROBLEMS

Turbine engines require a number pre-start checks to ensure the control system issues the command to start. Availability of lube oil at the right pressure and temperature and fuel pressure at the injector are rime parameters that come under the influence of the automatic control system of the turbine. In the case of steam turbines, pressure and temperature of steam available at the high-pressure section's intake valves must be in the proper range.

If the unit has been on turning gear, some designs have a single motor for the turning gear and for starting. As the rotor picks up speed, most all of the instrumentation readings must be closely monitored. Due to the large physical size and inertia of the rotors, the starting motor must produce adequate torque to bring up to a speed point where continued operation is sustained by the internal power produced in the engine.

Aviation engines rely on air motors for the start, with pressurized air delivered by the auxiliary power unit or obtained from ground sources. The high-pressure rotor first starts turning, reaching to a point when enough air is brought into the engine before the low-pressure rotor also starts to turn. At a convenient point in speed, fuel is injected into the combustor, accompanied by sparks at the tip of the igniter. A successful start occurs when a stable flame is achieved in the combustor.

Power generation and industrial gas turbines follow a similar trend in the matter of fuel injection and ignition. If the flame is lost prior to reaching operating speed, the ratio of fuel and air not coming up to stoichiometric requirement must be checked if all other instrumentation indicators are in the acceptable range.

10.18 REVIEW QUESTIONS

10.18.1 What is the prime reason for a turbine disk to burst, and what precautions must be taken to avoid the failure?

10.18.2 What are the possible effects of bearing lubricant oil being above its normal range?

10.18.3 Explain the circumstances when control valves malfunction.

10.18.4 What are the principal reasons of excessive vibrations in the bearings?

10.18.5 Why is it essential to use the right grade of hydraulic fluid for an actuating mechanism of a turbine?

10.18.6 Under what circumstances will an aircraft engine shutdown during a flight?

10.18.7 When will a gas turbine combustor catch fire? What steps may be taken to extinguish the fire?

10.18.8 What are the chief indicators of a turbine engine running into a stall?

10.18.9 How does stalling of a compressor affect the overall health of the engine?

10.18.10 Explain the effects on operation of a unit when the rotor is not adequately running centered with the stator case.

10.18.11 When do the cooling channels inside a turbine airfoil restrict the flow of coolant? What is its effect on the airfoil?

10.19 EXAMPLE PROBLEMS

Example Problem # 10.19.1: An overhung disk on a shaft is supported in two bearings placed 3.715 in. and 9.328 in. from the disk centerline. The disk has 8.75 in. radius and 3.725 in. width. The disk material has density $\rho = 0.285$ lb./in^3, and the shaft material has $E = 27.83 \times 10^6$ lb./in^2. Ignoring the weight of the shaft, calculate the reactions at the bearings due to the weight of the disk.

Solution: Weight of the disk: W = Volume × density = $\pi R^2 l\rho$

$$= \pi \times 8.75^2 \times 3.725 \times 0.285 = 255.22 \text{ lb.}$$

Assume x is distance along shaft, $x_A = 0$, $x_B = 9.328 - 3.715 = 5.613$ in.

and $x_{Disk} = 9.328$ in.

For static equilibrium, $(R_A + R_B - 255.22) = 0$, or $R_A = 255.22 - R_B$.

Also: $\Sigma M = 0$. Taking moments about A: $255.22 \times 9.328 - R_B \times 5.613 = 0$

Hence: $R_B = 424.14$ lb. ↑, and $R_A = 255.22 - 424.14 = 168.92$ lb. ↓

Example Problem # 10.19.2: In Example Problem # 10.19.1, obtain an expression for bending moment in the shaft.

Solution: In the shaft segment between the bearings, consider a section located at a distance x from bearing A ($x < 5.613$).

Moment $M(x) = R_A \times x = 168.92x$ in-lb. clockwise.

Hence, $M_A = 0$

$M_B = 168.92 \times 5.613 = 948.15$ in-lb. clockwise.

In the shaft segment between bearing B and the disk ($x > 5.613$):

$$M(x) = -R_B \times (x - 5.613)$$

$M_B = -[424.14 \times 3.715] = 1575.68$ in-lb. counter-clockwise.

$$M_{Disk} = [424.14 \times 3.715 - 168.92 \times 9.328)] = 0$$

Example Problem # 10.19.3: In Example Problem # 10.19.2 assume the shaft is spinning at 5400 rpm, and the disk has an unbalance of 0.525 oz. at the rim. Determine the rotating unbalance force on the disk.

Solution: The unbalance in the rotating disk creates a centrifugal force in the radial direction that also rotates with the disk. This force is given by: $F = (W/g) \times \omega^2 r$, where Wr provides the amount of unbalance in the form of weight W placed at a distance r from the axis of rotation, ω is the rotor speed, and g is the gravitational constant.

$$\text{Centrifugal force } F = [0.525 \times (2 \times 3.14159 \times 5400/60)^2]/(386.0 \times 16)$$
$$= 27.18 \text{ lb.}$$

Example Problem # 10.19.4: In Example Problem # 10.19.3 calculate the maximum reactions at the two bearings.

Solution: In Example Problem # 10.19.1, replace the weight of the disk by the sum of weight and the unbalance force.

Weight of the disk + unbalance force $W_T = 255.22 + 27.18 = 282.40$ lb.

Assume x is distance along shaft, $x_A = 0$, $x_B = 9.328 - 3.715 = 5.613$ in.

and $x_{Disk} = 9.328$ in.

For static equilibrium, $(R_A + R_B - 282.40) = 0$, or $R_A = 282.40 - R_B$.

Also: $\Sigma M = 0$. Taking moments about A: $282.40 \times 9.328 - R_B \times 5.613 = 0$

Hence: $R_B = 469.32$ lb. ↑, and $R_A = 282.40 - 469.32 = 186.91$ lb. ↓

Example Problem # 10.19.5: The rotor of Example Problems # 10.19.1 to 10.19.4 is placed inside a stator casing. If the maximum deflection of the shaft during operation is 0.0085 in., determine the diameter of the casing when the minimum radial clearance between the rotor and the stator is required to be 0.0035 in. on the radius and the rotor is spinning at the rated speed.

Solution: Total deflection due to rotating unbalance and dead weight is 0.0085 in.

Minimum radial clearance = 0.0035 in.

Hence radius of the casing:

$$\begin{aligned} R_{casing} &= \text{(disk radius)} + \text{(minimum radial clearance)} + \text{(radial deflection)} \\ &= 8.750 + 0.0035 + 0.0085 \\ &= 8.762 \text{ in.} \end{aligned}$$

10.20 EXERCISE PROBLEMS

Exercise Problem # 10.20.1: A disk mounted on a shaft is supported in two bearings with the disk in between, placed 8.265 in. and 11.428 in. from the disk centerline. The disk is of 7.875 in. radius and 1.725 in. width. The disk material has density $\rho = 0.283$ lb./in^3, and the shaft material has $E = 27.83 \times 10^6$ lb./in^2. Ignoring the weight of the shaft, calculate the reactions at the bearings due to the weight of the disk.

Exercise Problem # 10.20.2: In Exercise Problem # 10.20.1, obtain an expression for bending moment in the shaft.

Exercise Problem # 10.20.3: In Exercise Problem # 10.20.2 assume the shaft is spinning at 9550 rpm, and the disk has an unbalance of 0.0525 lb. at the rim. Determine the rotating unbalance force on the disk.

Exercise Problem # 10.20.4: In Exercise Problem # 10.20.3 calculate the maximum reactions at the two bearings.

Exercise Problem # 10.20.5: The rotor of Exercise Problem # 10.20.1 to 10.20.4 is placed inside a stator casing. If the maximum deflection of the shaft during operation is 0.00345 in., determine the diameter of the casing when the minimum radial clearance between the rotor and the stator is required to be 0.00175 in. on the radius and the rotor is spinning at the rated speed.

REFERENCES & BIBLIOGRAPHY

Catlow, R., “Blade Tip Clearance Control Apparatus”, US Patent 5,211,534, 1993.

Childs, D., “Turbo-machinery Rotor Dynamics”, John Wiley & Sons, New York, 1993.

Childs, D., Kleynhans, G., “Experimental Rotor Dynamic and Leakage Results for Short (L/D = 1/6) Honeycomb and Smooth Annular Pressure Seals”, Proceedings, Institute of Mechanical Engineers, 5th International Conference On Vibrations in Rotating Machinery, London, 1992.

Eder, M., “Engine Fuel System Design Issues”, Aviation Fuels with Improved Fire Safety: A Proceedings, pages 61–64, National Academy Press, Washington, DC, 1997.

Gaffin, W. O., "JT9D-70/59 Improved High Pressure Turbine Active Clearance Control System", NASA CR-159661, 1979.

Greathead, S., Bostow, P., "Investigations into Load Dependent Vibrations of High Pressure Rotor on Large Turbo-Generators", Proceedings, Institute of Mechanical Engineers, Conference on Vibrations in Rotating Machinery, Cambridge, page 279–286, 1976.

Hendrick,G. J., Dabinis, J. S., Feulner, M. R., "Analysis of Instability Inception in High-Speed Multi-Stage Axial Flow Compressors", ASME Journal of Turbo-machinery, volume 119, page 714–722, New York, 1997.

Holmes, R., Box, S., "On the Use of Squeeze Film Dampers in Rotor Support Structures", Machine Vibration, volume 1, page 71–92, 1992.

Kawecki, E. J., "Thermal Response Turbine Shroud Study", Air Force Aero Propulsion Laboratory Technical Report AFAPL-TR-79-2087, 1979.

Koul, A. K., Immarigeon, J. P., Dainty, R. V., Patnaik, P. C., "Degradation of High Performance Aero Engine Turbine Blades", Proceedings, ASM Materials Congress, pages 69–74, Pittsburgh, 1993.

Marquette, O. R., Childs, D. W., San Andres, L., "Eccentricity Effects on Rotor Dynamic Coefficients of Plain Annular Seals: Theory versus Experiment", ASME, Journal of Tribology, volume 119, pages 443–448, New York, 1997.

Olsson, W. J., Martin, R. L., "B747/JT9Dflight Loads and Their Effect on Engine Running Clearances and Performance Deterioration: Nacelle Aerodynamic and Inertial Loads (NAIL)/JT9D Jet Engine Diagnostics Programs", NASA CR-165573, 1982.

Shemeld, D., "A History of Development in Rotor Dynamics – A Manufacturers Viewpoint", Rotor Dynamic Instability problems in High Performance Turbo-machinery, NASA Report # CP 2443, pages 1–18, 1986.

Wiseman, M. W., Guo, T., "An Investigation of Life Extending Control Techniques for Gas Turbine Engines", Proceedings of the American Control Conference, Arlington, VA, June 25–27, 2001.

11 CHAPTER

ROLE OF THERMODYNAMICS AND AERODYNAMICS

11.1 INTRODUCTION

A prime characteristic of all heat engines is the reliance on the power of hot gas to perform useful work. Thermodynamic efficiency of heat engines is related to the use of high temperatures, which is the cornerstone in the development of the Brayton cycle. Applying the concept to rotating engines, improvements were introduced in steam turbines for power generation appearing in the late 1800s and to gas turbines in the first quarter of 1900s.

Pressurization of the air-fuel mixture prior to combustion in the early stages of engine development was also recognized as another contributing factor, which was first put into practice in the form of a turbocharger for piston driven engines. Advances in aerodynamics theory brought the realization that turbulence can cause substantial loss of power at the tip of the propellers of a conventional piston driven engine beyond set limits. This was coupled by the observation that the weight to power ratio of reciprocating engines increases exponentially as the size and speed capability of the airplane increases. Piston engines also have the disadvantage of the reciprocating components imposing inertia loads on the aircraft's frame structure. These limitations provided the incentive to develop turbine engines to power larger transport aircrafts flying at faster speeds.

A constant-volume engine relies on a series of events occurring at specific points cyclically. The operation of a reciprocating engine is based on the constant-volume principal using a closed cycle. In contrast, the key advantage of the 'constant pressure' method used in gas turbines is the ability to handle high mass flows and consequent high power output. All of the events of intake, compression, ignition, expansion, and exhaust take place simultaneously in the constant pressure Brayton cycle. The events of an open cycle take place in different components of the engine. The air drawn in is compressed, its pressure and temperature raised and its volume decreased in the compressor section. Fuel injected in the combustion chamber burns with the aid of oxygen at a nearly constant pressure, while experiencing a substantial increase in temperature and volume. The increased volume causes the velocity of the flow to dramatically increase through the essentially unvarying engine area. The flow next proceeds to the turbine wheel, where it expands to further increase in volume and accompanied by drop in pressure and temperature.

In using air as the working fluid for a gas turbine, its three primary properties of pressure, volume and temperature are undergoing continuous variation. For a given mass of air, the value of any property can be determined if the other two values are known from Boyle's and Charle's laws. From Boyle's law, pv = constant, where absolute temperature, T, of a mass of air is constant, p, is absolute pressure, and v is volume. And from Charle's law, if the pressure of a mass of air is held constant, its volume is directly proportional to its absolute temperature, or v/T = constant. The combination of Boyle's and Charle's laws yields the general gas law, or:

$$pv/T = R, \text{ constant} \tag{11.1}$$

11.2 LIMITS ON COMPRESSION RATIO FOR GAS TURBINE

In evaluating the performance of an ideal cycle for a gas turbine where all forms of losses in the components are not taken into consideration, two specific parameters that control the work output and efficiency stand out: compression ratio and maximum cycle temperature. This may be observed from Figure 8.1. Commonly referred to as the Joule or Brayton cycle, the ideal cycle is represented by points 1-2-3-4, where the steady flow of energy is expressed as:

$$Q = (h_2 - h_1) + (V_2^2 - V_1^2)/2 + W \tag{11.2}$$

per unit of mass flow, where Q is the heat transferred, h_1 and h_2 are enthalpy at given points of the cycle, V_1 and V_2 are flow velocities and W is work done (*ref*: Rogers, et al, 1994). The same condition is applicable to the major components in the assembly. Thus, in the compressor:

$$W_{12} = -(h_2 - h_1) = -c_p (T_2 - T_1) \tag{11.3}$$

In the combustion chamber:

$$Q_{23} = (h_3 - h_2) = c_p (T_3 - T_2) \tag{11.4}$$

In the turbine:

$$W_{34} = (h_3 - h_4) = c_p (T_3 - T_4) \tag{11.5}$$

By definition, cycle efficiency is:

$$\eta = (\text{net work output})/(\text{heat put in})$$
$$= [c_p (T_3 - T_4) - c_p (T_2 - T_1)]/[c_p (T_3 - T_2)]$$

where c_p is specific heat at constant pressure. For isentropic processes, $T_2/T_1 = r^{(\gamma - 1)/\gamma} = T_3/T_4$, where r = pressure ratio $p_2/p_1 = p_3/p_4$. If the ratio of specific heats c_p/c_v is denoted by γ, the cycle efficiency then takes the form

$$\eta = 1 - \left(\frac{1}{r}\right)^{\left(\frac{\gamma - 1}{\gamma}\right)} \tag{11.6}$$

The useful output that a given plant is capable of delivering, W, is of interest in establishing its size. The difference between the output from the turbine and work required to drive the compressor is:

$$W = [c_p (T_3 - T_4) - c_p (T_2 - T_1)]$$

Or

$$\frac{W}{c_p T_1} = t\left(1 - \frac{1}{r^{(\gamma - 1)/\gamma}}\right) - \left(r^{(\gamma - 1)/\gamma} - 1\right) \tag{11.7}$$

where $t = T_3/T_1$. T_1 is the normal atmospheric temperature that does not change much.

11.3 DIFFERENCES IN IDEAL AND ACTUAL CYCLE DESIGN

A number of cycle designs are possible for a gas turbine where the processes of multi-stage compression and expansion, reheating, heat exchanging, and inter-cooling are available. Prior to entering the combustion chamber, air can be preheated using the energy of escaping hot gases in a heat exchanger. An inter-cooler between compressor stages reduces the volume of air while bringing down its temperature, and this increases the efficiency of the compressor. Besides maximizing the generation of power and overall thermal efficiency, the objective in configuring and assembling of the components is to control part load operation at a different speed. Addition of the components requires extra piping, sensors and control valves, but the payoff comes from flexibility of operation and savings from reduced fuel consumption and exhaust emissions.

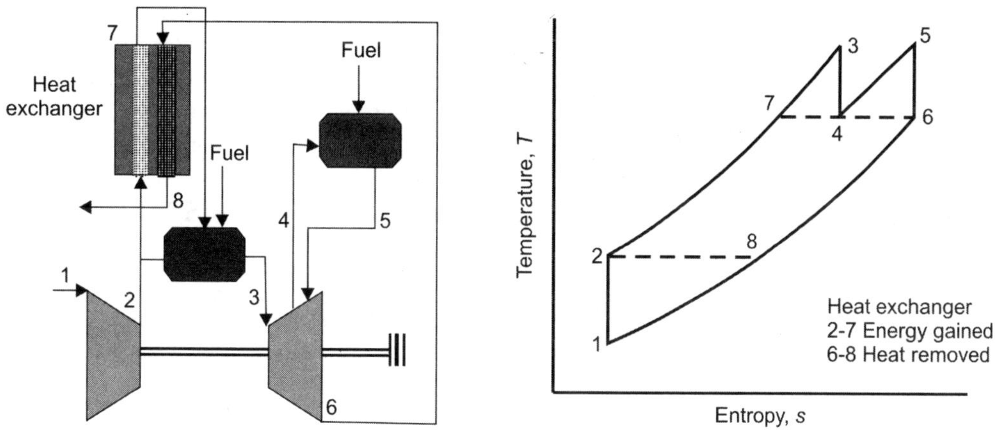

Fig. 11.1 Combined Reheat and Heat Exchanger in Gas Turbine Cycle

When the heat exchanger is combined with the reheat element in the cycle, the efficiency is improved. The elevated temperature of the turbine's exhaust is beneficially employed in the heat exchanger, and the additional heat provided does not offset the increased work output. The schematic diagram of the combined reheat and heat exchanger in a gas turbine is shown in Figure 11.1.

Dividing the compression process into low and high-pressure groups and passing the airflow through an inter-cooler between the two has some of the characteristics of reheat cycle. The work output is maximized if the air is inter-cooled to T_1, and when the compression ratios in the high and low-pressure units are equal.

11.4 MERITS AND DEMERITS OF CENTRIFUGAL COMPRESSOR

Indicative of its major advantage, the centrifugal compressor has essentially two components, an impeller rotating at high speed and a diffuser manifold. Pressurization of as much as 6:1 can be obtained with just a single stage. Air is drawn into the eye, or inducer, of the impeller, and whirled around by the vanes. Centrifugal force drives the air away from the spin axis to the rim, or the exducer. Centripetal acceleration is present due to pressure head, so static pressure also increases from the eye to the tip. The air is ejected at a high velocity into fixed diverging passages of the diffuser where it is decelerated, accompanied by a rise in static pressure.

But the entrance area at the inducer is limited, so the amount of air admitted is also limited. For small engines such as for helicopter engines and for auxiliary power units, the smaller flow suffices. Lager transport aircraft engines and for power generation turbines require a far greater amount of air to be admitted, so the radial compressor cannot be gainfully employed. At compression ratios exceeding

7:1, friction losses increase rapidly and cause efficiency to drop, indicating another demerit of the centrifugal compressor.

Friction is responsible for some loss of stagnation pressure in the diffuser. Losses are also experienced as the flow turns around the edges of the moving vanes in the clearance space between the impeller and the casing. This loss may be minimized by keeping the clearance small or by a shroud attached to the vanes. Shrouds are used in process compressors but not in aviation related impellers.

Flow through the expanding space of the diffuser does not entail any work to be done on the air, and energy absorption is limited to the flow between the inlet and outlet of the impeller. If the air enters the eye axially, the initial angular momentum of the flow is zero. To permit air to enter smoothly, the vane section at the eye is curved. When air exits at the impeller tip at absolute velocity v_2, it has a tangential component v_{w2} and a radial component v_{r2}. Ideally, flow velocity v_2 is equal to impeller tip speed u, but inertia effects trap the air between the vanes and carry it in the impeller. This leads to a 'slip' effect.

11.5 CENTRIFUGAL COMPRESSOR OPERATING CHARACTERISTICS

Compressor performance is typically specified in the form of curves of discharge pressure and temperature plotted against mass flow at selected rotor speeds. Other variables such as pressure and temperature at the entry point and physical characteristics of air also play a role. The dimensional analysis reduces the number of variables, which leads to formation of a lesser number of dimensionless groups. The method permits depicting the operation of a compressor with two sets of plots.

Note that density $\rho = 1/v$, hence $RT = p/\rho$, and the dimension RT takes the form $ML^{-1}T^{-2}/ML^{-3} = L^2T^{-2}$, which is the same as (velocity)^2. Density of the fluid may be deleted because $\rho = p/RT$. The other physical property, viscosity, will emerge in the form of Reynold's number. The flow at the inlet is generally throttled when a strong turbulent regime is present to decrease power requirements, leading to a drop in pressure and mass flow. As a consequence, the density and the Reynold's number are also lowered.

For a centrifugal compressor, the performance of a compressor calls for changing the delivery pressure and temperature with mass flow, rotor speed, and inlet conditions. The two terms $m\sqrt{T_{01}}/P_{01}$ and $N/\sqrt{T_{01}}$ correspond to non-dimensional mass flow and rotor speed.

Pressure and temperature curves against the mass flow and rotor speed in their non-dimensional format provide the most convenient method of describing a machine's capabilities. The mass flow may also be given as $m\sqrt{(RT)}/D^2p = \rho Av\sqrt{(RT)}/RTD^2p$, which is in proportion to flow Mach number M_f. Rotor speed is represented as $u/\sqrt{(RT)}$, which is proportional to rotor Mach number M_R.

Stagnation pressure and temperature ratios are calculated separately as a function of the non-dimensional mass flow for a specific speed point, and plotted to yield a family of curves. The isentropic efficiency may be obtained by replacing the temperature ratio with its derivative:

$$\eta_c = \frac{T'_{02} - T_{01}}{T_{02} - T_{01}} = \frac{(p_{02}/p_{01})^{(\gamma-1)/\gamma - 1}}{(T_{02}/T_{01}) - 1} \tag{11.8}$$

Values for isentropic efficiency as a function of the non-dimensional mass flow may then be derived and plotted from the pressure and temperature curves for selected values of rotor speed.

The predicted performance curve data of centrifugal compressors is sometimes marred because of surging. After the start and stabilization of a centrifugal compressor at a given speed and prior to turning on the discharge valve, the pressure ratio reaches a certain point because of the pressure head from the action of the impeller on the trapped air between the vanes. As the valve is opened and flow begins, the diffuser will also add to the pressure buildup. The pressure reaches a peak at some point, when the efficiency is a maximum. Further opening of the valve causes the mass flow to exceed the design point, and will result in a decrease in the pressure ratio. The airflow angles will not match the vane angles, the streams break away, and efficiency will rapidly fall off because most of the power is expended in overcoming friction.

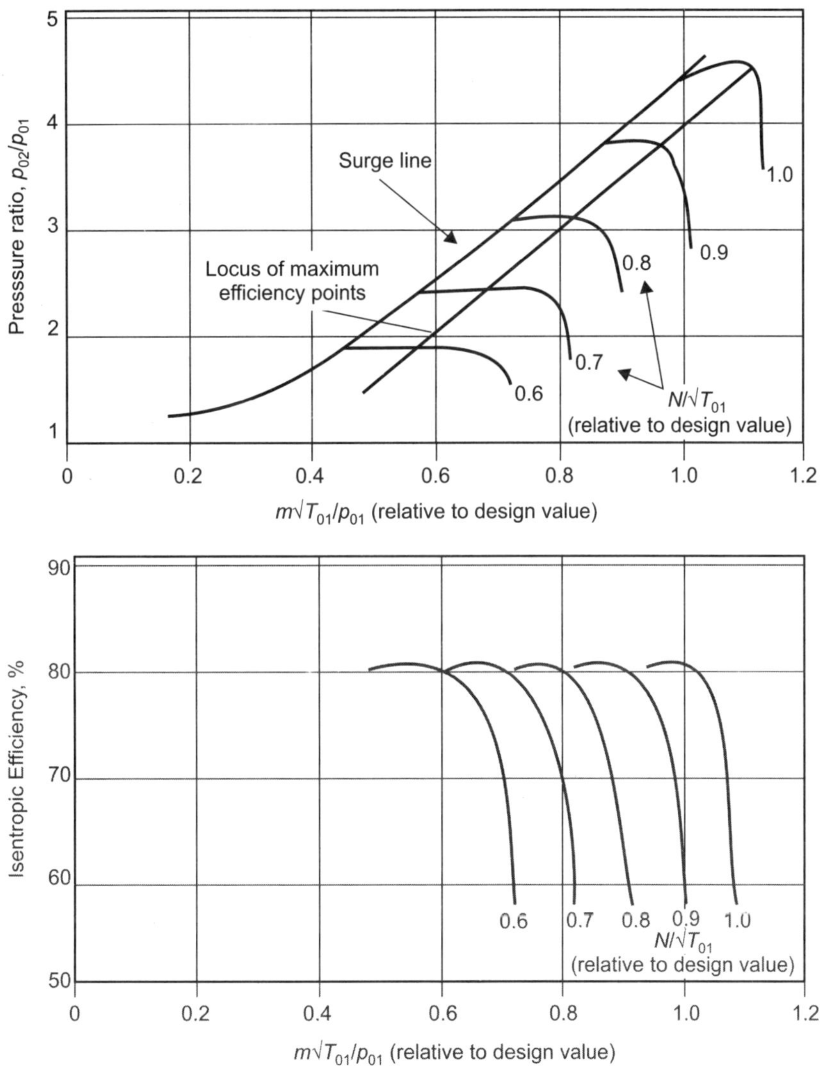

Fig. 11.2 Centrifugal Compressor Operating Characteristics

An obstacle appears in the case of a centrifugal compressor running at a fixed speed, with the flow rate exceeding that at maximum pressure. The extra mass flow at lower pressure causes density to fall and radial velocity component to rise. Since rotor speed is constant, total velocity and incidence angle at the diffuser leading edge must increase up to a certain point. Additional mass flow cannot be accommodated, and the compressor reaches the '*choked*' condition. At the choking point on the performance curve, further increase in mass flow cannot be obtained for the specified running speed of the unit. Similar other performance curves may be plotted for different speed points. The collection of curves provides the variation in the pressure ratios for a range of mass flows, as shown in Figure 11.2 using either the compression ratio or the isentropic efficiency. A set of curves for the predicted isentropic efficiency completes the performance data definition of the centrifugal compressor. The end points of individual curves are of interest. Surge conditions prevail at the minimum flow end point, while the choked condition is reached at the maximum flow end point of the curves. The contour connecting the surge points provides the '*surge line*' of the compressor. A locus of the points of maximum efficiency is also indicated for selecting the design point.

11.6 SURGING IN CENTRIFUGAL COMPRESSOR

Surging is marked by a sudden drop in delivery pressure, and accompanied with large aerodynamic pulsations, which disrupt the pressure/mass flow relationship away from the calculated performance characteristics of the unit. Prior to reaching the maximum pressure ratio point, the pressure on the delivery side must be lower to permit the flow to occur. If the pressure downstream of the compressor does not fall quickly enough, the air will reverse its direction and flow back, accompanied by a drop in the pressure ratio. At the same time the pressure on the delivery side will fall, so the compressor will resume its forward flow to repeat the cycle of events at a rapid pace. Note that the sequence of events may take place at any time besides the initial start. The tendency to surge predominates when on the downstream of the compressor the system is not in a position to consume the flow of air, particularly when the gas turbine is operating over a wide spectrum of operating conditions.

The likelihood of surging occurring in the diffuser passages, where frictional forces around the vanes retard the flow, is greater. Surge tendency is also observed to increase with the number of diffuser vanes, mostly because of the difficulty to evenly distribute the flow in all passages. If each impeller channel supplies air to multiple diffuser channels which discharge into a common outlet, the air can flow up one channel and flow back down another in the event of an onslaught of the surge regime. If the situation arises in just one pair of passages the discharge pressure decreases, and causes the surging to intensify. One remedy to reduce this tendency is to provide for a larger number of vanes in the impeller than in the diffuser, which helps to even out pressure and velocity differences between the impeller passages prior to the air reaching the diffuser.

11.7 ROTATING STALL

Rotating stalls in an otherwise stable operating regime are another source of unstable operation. Lack of uniformity in the flow pattern or geometry of the passages can disrupt the flow in one channel and deflect it into the following channel at a reduced angle of incidence and into a preceding channel at an increased incidence (Figure 11.3).

The non-uniform flow of air in the channels may be a consequence of a number of factors. For instance, blockage of a portion of the passage by ingested material, defect on a blade surface or some other form of obstruction may cause the flow to deviate from its intended flow path. The preceding channel will then stall, which reduces the incidence angle to the middle channel and permit it to recover.

The stall thus traverses from one passage to the next. Looked at from the eye of the impeller, the passing would appear in a direction opposite to that of the shaft's rotation. A rotating stall may contribute to surging, leads to aerodynamic pulsations, and eventual fatigue failures in the turbine engine.

Depending on the severity of the obstruction to the smooth flow, the intensity of the rotating stall may also be expected to vary. In general, when the flow passages run between smooth surfaces of blades guiding the flow will not experience the rotating stall phenomenon. When the surfaces suffer from erosion or corrosion, the likelihood of increased roughness in operation of the compressor unit may be expected.

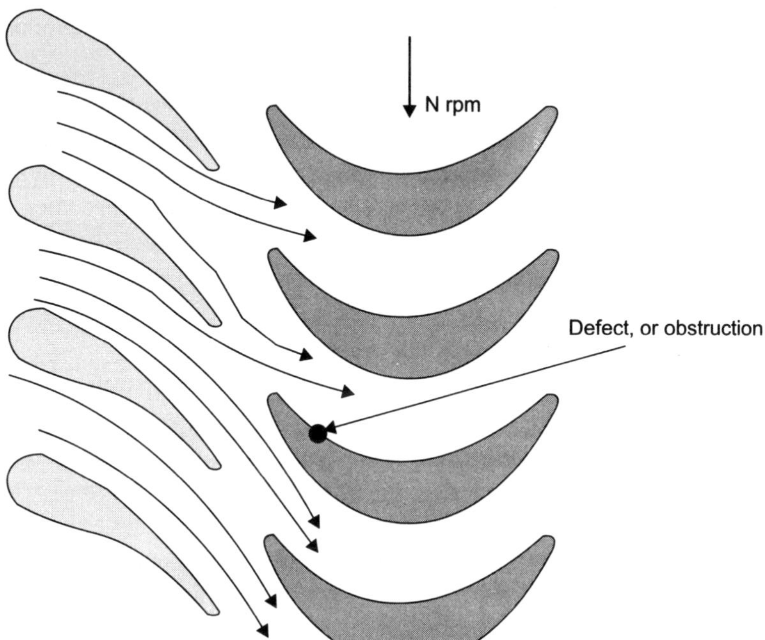

Fig. 11.3 Rotating Stall Event

11.8 FLOW COMPLEXITIES IN AXIAL COMPRESSION

The capacity to handle a considerably higher mass flow of air because of the increased frontal flow area makes the axial flow compressor suitable for gas turbines with large power output. The aerodynamics of individual axial stages has made tremendous strides, and so fewer stages are needed to achieve the desired level of compression. Modern gas turbines have axial compressors that provide overall pressure ratios approaching 40:1. Axial compressors also have the advantage of operating at a relatively high operating efficiency. With all the many fold advantages of centrifugal compressors, they are relegated to the lower end of the power spectrum, where the flow is not large enough to justify efficient operation with axial blading.

The design of axial compressors with entry flow velocity exceeding the sound of speed throughout the blade height has not been successful. But relative velocities exceeding the speed of sound in a portion of the rotating blades are employed in industrial and aviation engines, the compression process being referred to as 'transonic'. Pressurization occurs as a consequence of a series of diffusions that take place in both the rotating and stationary passages. The absolute velocity of the fluid medium between the rotating blades increases, but the relative velocity decreases with the effects of diffusion. The difficulties encountered are compounded with increased compression ratio, which emphasizes the need to restrict the amount of diffusion in each stage.

Aviation engines sometimes have axial compressors split into two or more spools driven by separate turbines. The airfoils do not possess the requisite strength to withstand damage from foreign objects, especially in the latter stages. Additionally, a large number of close fitting blades and vanes make this type of compressor complex and expensive to manufacture. Axial flow compressor designs find applications in situations where the demands of efficiency and power output predominate considerations of cost, simplicity, and flexibility of operation.

Air flowing through the annulus of an axial compressor performs the relatively more difficult task of going against an adverse pressure gradient. As the compression ratio increases, the danger of stall troubles rises correspondingly. Among other reasons, stalling is encountered in individual airfoils when the difference between the flow direction and the blade's angle of incidence becomes excessively large. The excessive variation leads to turbulence, and tends to act as a blockage to the flow of air. Stability of the flow is jeopardized by the very fact that the movement is taking place in the same direction as that of the increase in pressure. Flow reversals occur under the right conditions of mass flow and rotor speed, the conditions varying from those for the blade's design.

The airfoil sections are designed to maximize efficiency. In the first stage the larger tip radius increases peripheral speed, which poses a significant issue. Oftentimes compressor blades are required to operate over a transonic range, meaning that flow over a portion of the blade exceeds unity Mach number. Blade sections based on circular arcs, sometimes referred to as bi-convex blades, have been found to be effective in the transonic mode of operation. With even greater Mach numbers, parabolic based sections are required. At entry velocity below the critical Mach number, the performance of the blades does not exhibit much variation with the speed. Above the critical speed the losses mount rapidly to a point where an appreciable increase in pressure increment is not experienced, and the blade loses the capacity to provide diffusion for the flow. The Mach numbers are in the region of 0.7 to 0.85 at zero angle of incidence. Increased Mach numbers also reduce the range of incidence angle for which losses may be at an acceptable level.

Air is first imparted acceleration by the rotating blades in an axial compressor, and then decelerated in the stator passages, where the kinetic energy is converted to static pressure. A series of diffusions take place in both the rotor and stator passages. Absolute velocity of the working fluid is increased in the rotor, but the relative velocity will decrease. It is imperative for the flow area to diminish moderately in a diffusing flow. Thus, a single compressor stage can only deliver a small increment in the pressure, considerably less then what a turbine can advantageously use with its efficient pressure gradient, converging blade passages, and accelerating flow.

The compressor in an axial flow engine, and to a lesser extent the centrifugal engine, presents a number of challenging problems. The importance of a good design is indicated by noting that a high-bypass turbo-fan engine experiences 0.75 per cent improvement in specific fuel consumption for each 1 per cent gain in fan efficiency. Also, a gain of 1 per cent in the high-pressure compressor efficiency yields 0.5 per cent improvement in the specific fuel consumption.

The centrifugal compressor relies on imparting high velocity, or kinetic energy, as the flow of air passes through the passages between the airfoils of the rotating impeller. The kinetic energy is then converted to potential, or pressure, energy when the flow exits the impeller and passes through the expanding space of the diffuser. The principal advantages of a centrifugal compressor are low weight, ruggedness, simplicity of design, and high compression ratio per stage. Close to 80 per cent efficiency may be reached with the compression ratio in the range of 6 or 7 to 1. Above this ratio the efficiency falls sharply from friction and shock wave losses due to the attendant high velocity. In comparison, axial compressors have the advantage of developing high compression ratios and high efficiencies (see Figure 11.4).

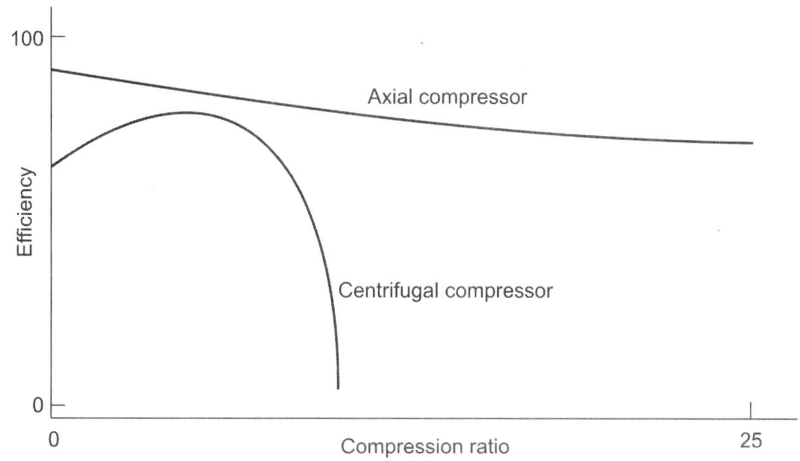

Fig. 11.4 Centrifugal and Axial Compressor Efficiency Comparison

11.9 STALL INCEPTION AND INTENSIFICATION

The task of aerodynamic designing of an axial compressor is complicated due to the twin requirements of minimizing losses and avoiding flow reversals for all operating conditions. Since the flow takes place against the pressure gradient, the tendency to reverse its direction is present at all times. In this respect, note that in the turbine section the flow takes place in the same direction as that of the expanding gases, flow reversal does not occur, and hence fewer stages are needed than in the compressor.

Stalling can take place in a few selected airfoils when the flow angle deviates significantly from the blade angle. Since the blades are closely spaced, each blade is impacted at the leading edge by the air stream passing through the preceding passage. A 'cascade' form of effect prevails as the air flows through the passages between adjacent blades, and has a major impact on the design and placement of the blades (Figure 11.5).

High on the list of difficulties encountered in designing axial compressors is the stall problem. When the angle at which the flow of air strikes the blades, or the angle of attack, becomes too small, the pressure zones reduce in value. On the other hand, if the angle is too large, the compressor will stall because the upper foil surface layer becomes turbulent and destroys the pressure zone. Note that the angle of incidence is of pre-determined value set during the design process. The angle of attack depends on engine speed and temperatures at compressor inlet and discharge.

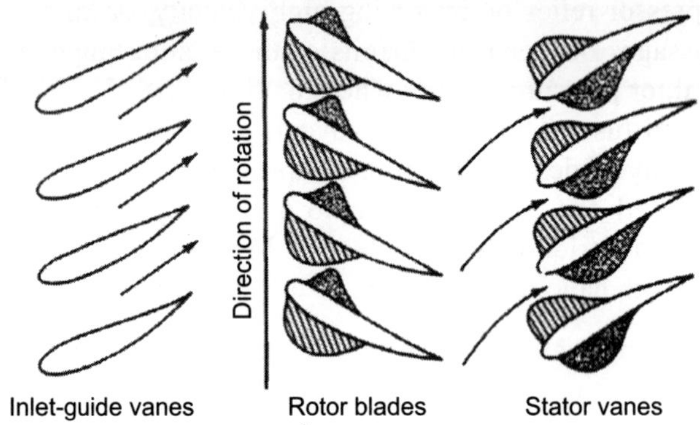

Fig. 11.5 Cascade Effect

Decreasing the flow velocity or increasing rotor speed tends to increase the angle of attack. More generally, any action that decreases flow velocity relative to rotor tip speed will increase the angle of attack and increase the tendency to stall. A number of causes can lead to such a situation. For example:

- The decrease in airflow may result from compressor discharge pressure becoming too high due to excessive fuel flow schedule during rapid acceleration. Or the inlet pressure may fall too low relative to the discharge pressure.
- The difference in discharge and inlet pressures may become too small because of high inlet temperatures or because of distortion at the inlet.
- During ground operations, choking conditions can lead to a stall. When engine speed falls below the design speed, the ratio of compression falls and excess volume of air is present in the rear stages. This leads to choking in the rear and consequent reduced airflow, which in turn decreases flow velocity in the front stages. The upshot is increased tendency to stall. In the absence of corrective action, the front stages will stall at low speeds.
- Excessive temperatures at the compressor inlet due to ram effect at high aircraft speeds are another cause of stalling. Higher air temperature reduces the density and lowers the compression ratio, and will also lead to choking in the rear stages. The choking-stall condition is similar to that caused at low engine speeds.

Corrective actions for the low-speed, high-temperature stalls at the front end include:

- Reducing the angle of attack at the front stages at lower engine speeds.
- Bleed some of the air in the middle and rear stages to increase flow rate at the front end at low speeds.
- Split the compressor into two sections or spools. The front spool speed may fall off more than the rear spool at lower speeds so that front speed matches the lower choked airflow.
- Use variable guide vanes for the front several stages. The angle of attack can be reset to lesser angles at low engine speeds.
- A variable area exhaust nozzle helps in reducing the compressor load during acceleration.

An indication of a mild stall is abnormal noise patterns, such as rumbling, buzzing, chugging, or choo-choo-ing. A rapid climb in the exhaust gas temperature, fluctuating speed, and high vibrations also occur because of pulsating flow in the compressor, accompanied by poor response to power lever movements. More severe stalls are marked by loud bangs, accompanied by flame, vapor, or smoke emitting at the engine's inlet or exhaust. Explosive stalls may cause engine failure. Recurring stalls generally require maintenance action.

11.10 COMBUSTION SYSTEM OPERATION

Physical and chemical aspects are embraced during combustion of fuel, and are the subject of physics and chemistry. Physical aspects include mass and heat transfer, thermodynamics, gas and fluid dynamics, while chemistry influences pollutant emission among the products of combustion, heat release rate and radiation properties of the flame at high temperatures. In aviation applications, the chemical process also impacts lean light-off and flameout limits at high altitudes.

The combustion process is characterized by steep temperature gradients and species concentration as it is executed in one or two milliseconds. Relative to the fresh air and fuel mixture, the burned gases are far higher in volume and temperature and lower in density, with the waves traveling at under 1 m/second. The deflagration regime of combustion of fuel is marked by a luminescent flame front that may be viewed as an interface between the burned gases and the unburned mixture.

Aerodynamic evaluation of a combustion system is required to achieve a stable flow pattern throughout. Inside the combustion liner considerable flow re-circulation is essential to stabilize the flame, and maximum benefit must be derived from the cooling air along the walls. Mixing in the combustion and dilution areas is of significance. Proper mixing in the primary zone enhances the rate of burning, while also minimizing soot and nitric oxide formation.

Current emission regulations call for combustion efficiency in excess of 99 per cent. Failure to achieve this level is objectionable partly because fuel is wasted, but more so because it is manifested in the form of excessive pollutants in the environment. For aircraft engine combustors, a design hallmark calls for it to be large enough to ensure an adequate level of combustion efficiency during engine restart at the highest altitude at which re-light is required. After a flameout during flight, combustion efficiency in the 75 to 80 per cent range is justified, because with the engine wind-milling the temperature and pressure of ambient air are low enough to affect the stability of the flame.

The design of the combustion chamber is one of reaching the best compromise between many conflicting requirements, which likely will vary considerably with different applications. Aviation and ground-based gas turbine combustion systems differ in several respects. Aircraft gas turbines rely solely on liquid petroleum distillates. Depending on availability and price, industrial and power generation turbines need to be configured for both gas and liquid fuels. Military naval vessels are often required to carry their own fuel for their mission, resulting in widespread use of marine diesel as fuel. Aero-derivate gas turbine engines for warships are also subject to strict emission limits while in harbor, similar to industrial engines.

Liquid fuel is admitted at the front end of the burner in the form of vaporized spray from specially designed nozzles or in a pre-vaporized condition from vaporizing tubes. The chamber's geometry causes air near the nozzle to stay close to the front wall for cooling. Air also enters through holes in the liner, which mixes rapidly with the fuel to form a combustible mixture, with more air admitted through additional holes. A high temperature is obtainable from the combustion when the mixture of air and fuel is nearly stoichiometric.

Turbulent flow conditions throughout the combustor enhance thorough mixing for improved combustion efficiency, as also for more uniform temperature distribution. Combustion of gaseous fuels does not entail vaporization and breaking down of heavy hydrocarbons into lighter fractions, but the need for intimate mixing of molecules of the hydrocarbons and oxygen and for the chemical reactions remain the same.

The air is introduced in stages because the overall air-fuel ratio is around 100:1 while the stoichiometric ratio is about 15:1. The stages may be identified as the primary stage where 15 to 20 per cent of the air is admitted around the fuel jet to ensure high temperatures. Fuel is injected in the same direction as the air stream. Air enters through curved radial vanes, referred to as swirl vanes, to produce a vortex that induces a region of low pressure along the axis of the chamber. The vanes, assembled in the form of a cup, are attached to the flame tube. About 30 per cent air enters through holes in the flame tube in the secondary zone to complete the combustion. This air must enter at the proper location to avoid local chilling of the flame, which can potentially reduce the reaction rate. In the tertiary or dilution zone, the remaining air combines to cool the gases to a temperature that suits requirements at the inlet to the turbine.

Incomplete combustion can occur when the flame region is locally chilled where the secondary air is admitted. The reaction rate may then fall to a point when some decomposed fuel remains in a partially burned state. The temperature at the combustor exit will fall below expected levels. The problem may be overcome with devices that increase turbulence and circulate secondary air into the burning gas regime. Combustion efficiency is enhanced as a consequence, but the increased turbulence will also lead to greater pressure losses due to friction at the wall.

11.11 INEFFICIENCIES IN COMBUSTION PROCESS

Combustion chambers of gas turbines are afflicted by not so well known but nevertheless significant problems related to startup and ignition, fuel injection nozzle, use of non-standard fuels and cooling of the flame tube.

Combustion of fuel is self-sustaining and proceeds continuously during normal operation. But during startup of a gas turbine, the mixture needs to be ignited. Methods for starting an engine from rest vary with size and type of turbine units. In aircraft applications the engine may be started using compressed air obtained from the aircraft's own auxiliary power unit or a ground based cart. An air starter motor located in the engine's accessories module and operating through a speed reduction gear is required to turn the rotor. Turning only the engine's core section suffices for the purpose. Directly passing the air over turbine blades has also been used in the past. Heavier industrial gas turbine and steam turbine trains require substantial amount of power produced in an electric motor, diesel engine or steam turbine for the initial start. Newer units are designed with generator coils wound such that it also acts as a starter motor. Once the rotor speeds up to a certain point, the airflow from the compressor is capable of burning the fuel continuously following injection of fuel and turning on the ignition. Suitably placed igniter plugs near the primary combustion zone readily spread the flame around after it initially occurs in annular combustors; inter-connecting cross tubes between cans perform the same task in can type of combustors. The ignition system then cuts off once the engine operation stabilizes to control erosion of the igniter's electrodes.

11.12 FUEL COMBUSTION RELATED PROBLEMS

In aviation engines liquid fuel is forced under high-pressure through an orifice to produce a conical spray of fine droplets in the primary zone of the combustor. The 'atomized' fuel passes through a conical chamber through tangential ports in the fuel injector nozzle to impart swirling motion to the spray. Some fuel is also directed axially through the core of the vortex. The spray sheet breaks up under controlled pressure to produce droplets varying in size from 50 to 100 microns. Smaller droplets outside the size range do not penetrate into the stream, while larger ones take longer time to evaporate.

Aviation gas turbine engine starting have the additional requirement when an in-flight shutdown occurs at altitude and the light in the combustor is lost. The mixture of air and fuel must be capable of re-igniting as the engine continues to windmill. Ignition of fuel at start is sensitive to pressure inside the chamber. The difficulty of igniting increases as the pressure reduces, which also points to problems encountered when re-igniting following an in-flight shutdown at altitude. Re-lighting of the engine at the aircraft's peak altitude may thus be considered a benchmark. As a precaution to avoid flameout when an aircraft is taking off or coming in for landing during adverse weather when substantial amounts of water may be ingested into the engine, turning on the ignition system is usually recommended. Engine operation during ascend is at maximum speed and mostly at flight idle during landing.

Combustion may not even initiate when the mass flow is too large. In addition, further difficulties are encountered during startup and shutdown periods. For instance, the mixture of air and fuel may be too rich as the engine accelerates and the fuel throttle is activated, since the flow of air takes time to build up to the necessary level, leading to an unacceptably low ratio of air and fuel. One possibility for rectifying the situation is to measure the variation in fuel flow rate and control it with the engine's microprocessor to avoid loss of flame during the transient periods, and thus also prevent rapid temperature changes in the turbine. The defined stable flame region in the diagram is also affected by pressure in the chamber. Chemical reactions slow down as the pressure decreases, which tend to restrict the flame's stability region. Aircraft engines also have to contend with reduced pressure of air exiting the compressor when operating at peak altitudes.

A cheaper form of fuel, obtained from crude oil after lighter and more expensive fractions are extracted, is the residual oil. But it suffers from some severe disadvantages due to the presence of vanadium, alkali metals and ash. The corrosive nature of these extraneous materials does not permit use of residual oil at operating temperatures above 900 K.

At such reduced cycle temperatures, however, the thermal efficiency is unacceptably low, except under certain operational circumstances. Some other related issues of excessive viscosity, tendency to polymerize to tar when overheated and high carbon content may be resolved without much difficulty. Alkali metals can be removed, and fuel additives made from compounds of magnesium can eliminate the effects of vanadium. Removal of sulfur entails further processing during refining. But the additional processes of washing with water, centrifuging the mixture and need for chemicals dissipate the initial cost benefits of the cheaper residual oil.

11.13 DIFFICULTIES IN FUEL INJECTION

The liquid fuel nozzle is required to deliver a nearly stoichiometric mixture of air and fuel uniformly in the primary zone, with the fuel flow adjusted to accommodate engine operation from idle to full load conditions. But problems arise in a simple burner since fuel flow rate varies with the square of supply pressure. To overcome this, two fuel manifolds are used, one supplying fuel to the orifice at the center for low load operation and a larger annular orifice concentric with the one at the center at higher loads. The total flow in the two stages fulfills the engine's fuel requirements without calling for excessive pressure and without sacrificing atomization qualities and spray pattern at low fuel flow.

Industrial gas turbines using natural gas sometimes have to depend on liquid fuel if the gas supply is interrupted. Dual fuel burner designs have separate concentric annuli through which the gas and liquid fuels flow, with an additional annular passage for steam or water injection to control exhaust emissions. Manufacturers also use other fuel injector designs. The rotary design, for instance, flings the fuel from a spinning cup into the air stream; a high tip speed ensures proper atomization, but the fuel pressure is low.

Aircraft engines of necessity require expensive liquid fuels. Aviation gas turbine engines successfully replaced piston engines following World War II mainly because of the advantages of power, weight and speed, in spite of their lower thermal efficiency. Natural gas also is expensive, but finds extensive use in stationary gas turbines since it does not contain sulfur and other contaminants, and unlike liquid fuels does not require atomization or vaporizing. Although somewhat unreliable in terms of availability, natural gas fired turbines still have the prospect of conversion to gas obtained from coal.

The functional nature of a combustion liner imposes on it high temperature levels and steep thermal gradients. Mechanical strength of nickel and cobalt based alloys used for the liner deteriorates considerably when temperatures exceed 1100 K, hence means must be designed to relieve the heat buildup. With gas turbines employing higher operating pressures and temperatures to improve performance and power, the need for cooling the combustion liner becomes even more acute. At the same time the liner is required to possess a minimum number of operating hours (*ref*: Lefebvre, 1999).

The temperature of the liner increases due to the combination of heating by radiation and convection from the internal hot gas flow and cooling due to radiation to the outer case and convection to the air in the annulus. Depending on the volume, pressure, temperature and chemical composition of the gases flowing through it, as also the dimensions and shape of the component, a considerable amount of heat is radiated on the liner. The size and number of hot and glowing soot particles formed during the combustion also control the intensity of radiation.

Film cooling of the inner surface of the liner is useful in achieving additional extraction of heat, and is accomplished by injecting air along the wall axially through slots and holes machined in the liner. The small diameter, closely spaced holes may be provided by laser drilling or by the EDM method. Since the turbulent hot gases gradually eliminate this film, the hole and slot pattern is repeated at specific intervals along the length. Cooling air is supplied through rings rolled or machined in the liner, through stacked rings with holes sized to deliver adequate amounts of cooling air and through corrugated spacers attached between overlapping segments of the liner. Depending on the design and method of attachment, the rings also serve to provide stiffness to the liner. Cooling efficiency in the liner is enhanced by improving the heat transfer coefficient on the coolant side and by increasing surface roughness of the heat transferring areas.

11.14 EXHAUST GAS EMISSIONS

All combustion processes generate exhaust gases composed of carbon monoxide (CO), carbon dioxide (CO_2), water vapor (H_2O), oxides of nitrogen (NO_x), oxygen, nitrogen, unburned hydrocarbons (UHC) and particulate matter in the form of carbon. Because of their impact on people's health and the environment, government agencies have passed aggressive rules to control emission of pollutants from turbines. Carbon monoxide is poisonous even in small quantities, unburned hydrocarbons are objectionable because of toxicity, NO_x produces acid rain, and smoke and soot lead to visibility concerns. Of considerable interest is the formation of ozone in the troposphere above the earth by the pollutants. Recent research indicates a number of respiratory problems are a direct consequence of excessive pollutants in the air.

With conventional combustors, any modifications to alleviate generation of smoke and oxides of nitrogen (NO_x, most notably nitrogen oxide and nitrogen dioxide) will invariably result in increased levels of emission of carbon monoxide (CO) and unburned hydrocarbons, with the reverse also being true. Flame temperature plays an overwhelming role in the production of pollutants, most emissions control mechanisms focus on constraining it within set limits. CO levels are low when combustion chamber temperatures exceed 1650 K, while NO_x production rate becomes objectionable beyond 1900 K. Hence, the range of 1650 to 1900 K must be achieved for all power settings of the engine.

Thus, a larger amount of air is required at the entrance to the combustor liner at full power to reduce the flame temperature. As engine load decreases, more of this air must be directed towards the downstream end to dilute the burned gases and maintain the temperature above the low limit.

Regulating the amount of air entering the primary combustion zone through a variable geometry mechanism has been successful to a considerable extent in overcoming this problem. Larger quantities of air are admitted at higher pressures to obtain more complete combustion and to minimize the incidence of soot and NO_x formation. When the pressure is low, the primary airflow is restricted to raise the air-fuel ratio and to reduce the velocity, thereby improving combustion efficiency and lowering emission of CO and unburned hydrocarbons.

The concept of staged combustion attempts to achieve the same objectives by using two or more separate zones, each designed specifically to optimize certain features of combustion. The process may take the form of axial or radial staging. In the lightly loaded primary first zone, high combustion efficiency and minimized CO and unburned hydrocarbon production are achieved by operating at a relatively high equivalence ratio ϕ of around 0.8, where equivalence ratio is defined by the actual fuel ratio divided by the stoichiometric fuel ratio.

Carbon deposits accumulate on fuel nozzles, which get worse with increasing temperatures and pressures, and after a point they begin to burn off. Increasing the fuel/air ratio accelerates the deposits, mostly due to reduced oxygen availability to completely burn the fuel. Fuel characteristics also affect the carbon deposits and burner performance.

11.15 POWER EXTRACTION IN TURBINE

The primary function of the turbine is to extract power from the pressure and kinetic energy of hot gases coming out of the combustor. A large portion of this power, in the neighborhood of 75 per cent, is expended in driving the compressor, with the remainder producing the thrust in an aero-engine or electrical power in an industrial power generation turbine. The drive power to compress the air is as high as 100,000 horsepower in larger engines. The load imposed on a single rotating blade can be gauged from the amount of power, 250 to 300 horsepower, as the hot gases flow over the airfoil, roughly the same as that produced in an eight cylinder automobile engine. But the space needed and the weight of a single turbine blade is much less than that of an automobile engine.

The operating mechanism of an axial turbine does not differ substantially from that of an axial compressor. While a compressor adds energy to the airflow by increasing the pressure, the turbine conversely absorbs energy from the gases and converts it into mechanical power for the shaft. Compressor blades are designed to have an increasing flow cross-section between adjacent blades in the downstream direction. The effects of diffusion cause conversion of its kinetic energy into pressure during compression, which leads to reduced flow. A reversed effect occurs in the turbine, where the flow cross-section between adjacent blades narrows in the downstream direction, thereby creating a nozzle effect to cause the flow to accelerate and perform useful work.

Gas enters the row of nozzle blades at static pressure and temperature p_1 and T_1 and velocity v_1, is expanded to p_2 and T_2. The gases approach the stator with velocity v_1 at an angle α_1 from the axial direction, and leave with an increased velocity v_2 at an angle α_2. The gases leaving the stator then pass to the rotor, where combining vector v_2 with blade speed u gives velocity relative to the blade c_2 at an angle β_2 from the axial direction. After being deflected, and some further expansion, in the rotor blade passage, the gas departs at p_3 and T_3 with relative velocity c_3 at an angle β_3. Assuming that the design is such that the axial velocity v_a is constant, then the absolute velocity of the gas is reduced in the rotor to v_3 at an angle α_3. α_3 is referred to as the 'swirl angle'.

In a constant pressure, or impulse, turbine the expansion occurs in nozzle guide vanes, where all of the gas potential energy is converted into kinetic energy (Figure 11.6). The area at the inlet and at the exit between the blades is equal. Exchange of momentum takes place as the gas impinges on the rotor blades at constant pressure, essentially as in a water wheel. In a reaction turbine, on the other hand, the gas expands both in the nozzle vanes and through the blades of the rotating wheel. Hence, the blades also feature a narrowing cross-section of the flow passage between adjacent blades for further acceleration of the flow. Analogous to the lifting of an airplane's wing, acceleration of gases generates an aerodynamic force that is imparted on the blades, causing the wheel to turn. The reaction turbine offers better efficiency than the impulse turbine, but the latter produces higher power output, which may permit reduction in the number of stages.

The design of stator and rotor blade shapes calls for accurate knowledge of losses encountered as the gas stream flows over their surface. Friction losses associated with the flow of gas in a turbine arise from growth of boundary layer over the blade profile and the inner and outer walls of the annulus. Losses are also incurred from secondary flows as the boundary layer is turned by an adjacent curved surface. In addition, gas spilling over the blade tips in the clearance region interacts with the outer wall boundary layer, leading to a loss in useful work output.

In the laying out of a blade section to scale, it is necessary to first establish the pitch and the chord. The 'opening' determination then calls for adjustments in the blade profile to fit and blend with the requisite gas outlet angles α_2 and β_2. The process is repeated at a number of radii between the root and the tip. The procedure leads to a complete definition of the blade shape. The selection of pitch and chord dimensions is based on a number of factors. The pitch to chord ratio (s/c) affects the blade loss coefficient. Blade height h is already calculated, so the effect of chord on the aspect ratio (h/c) must be

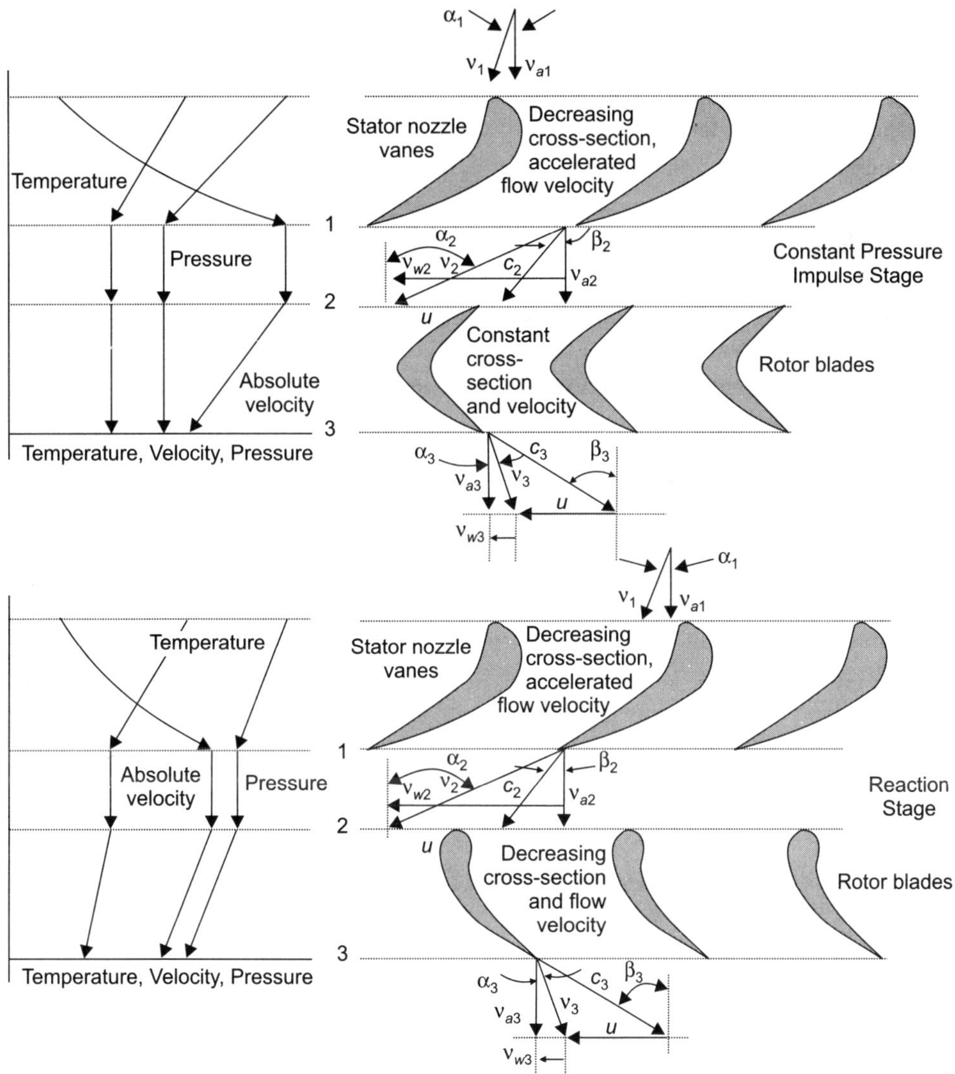

Fig. 11.6 Turbine Stages: Impulse (upper), Reaction (lower)

taken into account. The rotating blade chord has a major influence on blade stresses. Finally, the blade pitch affects the stresses at the blade attachment area with the disk.

11.16 REVIEW QUESTIONS

11.16.1 What do role compression ratio and turbine inlet temperature play in the performance of a gas turbine?

11.16.2 When is it more advantageous to use a centrifugal compressor? When is it not advantageous to use a centrifugal compressor?

11.16.3 What are the primary reasons for a compressor to experience rotating stall?

11.16.4 To what extent and why do the efficiency curves for axial and radial flow compressors differ?

11.16.5 How do indications of a mild stall in a compressor differ from a severe one?

11.16.6 List some representative values for the overall and stoichiometric ratio of air-fuel ratios. Why is the difference so large?

11.16.7 Discuss the many problems experienced in the combustor of a gas turbine.

11.16.8 What are the methods commonly employed to protect the liner from hot gases?

11.16.9 What design characteristics of the combustor liner aid in reducing mechanical and thermal stresses?

11.16.10 Discuss the role of firing temperature in the combustor on the production of objectionable exhaust gas components.

11.16.11 Mention some losses that are encountered as the flow goes over the curved surfaces of stator vanes and rotor blades.

11.17 EXAMPLE PROBLEMS

Example Problem # 11.17.1: An air-standard Brayton cycle admits air at 97° F and 12.65 lb./in². The compression ratio is set at 24.75, and maximum permissible cycle temperature is to be 2378 R. Calculate the pressure and temperature at each point 1 through 4 of the cycle per pound of air mass.

Solution: Intake pressure p_1 = 12.65 lb./in², temperature $T_1 = 460 + 97 = 557$ R

Compression ratio $r = p_2/p_1 = 24.75$, or $p_2 = 24.75 \times 12.65 = 313.09$ lb./in²

$T_2 = T_1 \times r^{(\gamma-1)/\gamma} = 557 \times 24.75^{0.281} = 1370.43$ R, where $\gamma = 1.39$

$p_3 = p_2 = 313.09$ lb./in² and $T_3 = T_{max} = 2378$ R

$T_4 = T_3 \times (p_4/p_3)^{(\gamma-1)/\gamma} = 2378 \times (1/24.75)^{0.281} = 966.51$ R

$p_4 = p_1 = 12.65$ lb./in²

Example Problem # 11.17.2: In *Example Problem # 11.17.1*, determine the work to be done in the compressor.

Solution: From the first law form of analysis, work to be done in the compressor is:

$W_{12} = -(h_2 - h_1) = -c_p(T_2 - T_1) = -0.24 \times (1370.43 - 557) = -195.22$ BTU/lb., where c_p is assumed 0.24 in the compressor and in the turbine sections.

Example Problem # 11.17.3: In Example Problem # 11.17.1, determine the work done in the turbine section.

Solution: From the first law form of analysis, work done in the turbine is:

$W_{34} = (h_3 - h_4) = c_p(T_3 - T_4) = 0.24 \times (2378 - 966.51) = 338.76$ BTU/lb.

Example Problem # 11.17.4: In *Example Problems # 11.17.2 and # 11.17.3*, determine the efficiency of cycle per pound of air mass.

Solution: Net work done: $W_{net} = W_{12} + W_{34} = -195.22 + 338.76 = 143.53$ BTU/lb.

Heat input: $Q_{23} = (h_3 - h_2) = c_p(T_3 - T_2) = 0.24 \times (2378 - 1370.43) = 241.82$ BTU/lb.

Efficiency: $\eta_{th} = W_{net}/Q_{23} = 143.53/241.82 = 0.5936$ or 59.36%

Example Problem # 11.17.5: In *Example Problems # 11.17.1 through # 11.17.4*, verify the efficiency of the cycle from the expression for isentropic efficiency.

Solution: From equation (11.6) isentropic efficiency:

$$\eta = 1-\left(\frac{1}{r}\right)^{\left(\frac{\gamma-1}{\gamma}\right)} = 1-(1/24.75)^{0.281} = 0.5936$$

11.18 EXERCISE PROBLEMS

Exercise Problem # 11.18.1: Air enters a compressor inlet at 89° F and pressure 11.72 lb./in^2. If at the discharge end pressure is 321.58 lb./in^2, calculate the isentropic efficiency.

Exercise Problem # 11.18.2: In Exercise Problem # 11.18.1 calculate the temperature at the discharge end if $\gamma = 1.39$.

Exercise Problem # 11.18.3: In Exercise Problem # 11.18.1 and 11.18.2 determine the work done in the compressor per pound of air.

Exercise Problem # 11.18.4: In Exercise Problems # 11.18.1 to 11.18.3, temperature at turbine entry is 2438 R. Calculate the pressure and temperature at the turbine exit per pound of air mass.

Exercise Problem # 11.18.5: In Exercise Problems # 11.18.1 to 11.18.4, determine the work done in the turbine section.

REFERENCES & BIBLIOGRAPHY

Ainley, D. G., "Internal Air Cooling for Turbine Blades – A General Design Survey", Aeronautical Research Council, R & M 3013 (HMSO), 1957.

Burghardt, M. D., "Engineering Thermodynamics with Applications", Harper and Row, New York, 1993.

Houghton, E. L., Brock, A. E., "Tables for Compressible Flow of Dry Air", Arnold, 1970.

Keenan, J. H., Kaye, J., "Gas Tables", John Wiley & Sons, New York, 1948.

Kempster, A., Czech, N., "Protection against Oxidation of Internal Coating Passages in Turbine Blades and Vanes", presented at Power Gen Conference, 1998.

Koul, A. K., Immarigeon, J. P., Dainty, R. V., Patnaik, P. C., "Degradation of High Performance Aero Engine Turbine Blades", Proceedings, ASM Materials Congress, pages 69–74, Pittsburgh, 1993.

Liepmann, H. W., Roshko, A, "Elements of Gas Dynamics", John Wiley & Sons, New York, 1957.

McQuiggan, G., "Design for High Reliability and Availability in Combustion Turbines", ASME paper # 96-GT-510, New York, 1996.

Rogers, G. F. C., Mayhew, Y. R., "Engineering Thermodynamics, Work and Heat Transfer", 5th edition, Longman, London, 1994.

Rogers, G. F. C., Mayhew, Y. R., "Thermodynamics and Transport Properties of Fluids", Blackwell, London, 1995.

Saravanamuttoo, H., Rogers, G., Cohen, H., "Gas Turbine Theory", Pearson Education, England, 2001.

Sawyer, J. W., ed., "Gas Turbine Engineering Handbook", Turbo-machinery International Publications, Atlanta, GA, 1985.

Schilling, W. F., "Low Pressure Plasma Sprayed Coatings for Industrial Gas Turbines", NATO Advanced Workshop, Coatings for Heat Engines, Italy, April, 1984.

12
CHAPTER

ENGINE VIBRATION

12.1 INTRODUCTION

By itself vibratory motion is not objectionable. A bob of a pendulum swinging back and forth on a rigid rod typifies the situation. A shaft whirling while spinning in a casing is another instance, unless the bend in the shaft causes it to rub against the case.

For further illustration, consider a footbridge spanning a stream. Made from wood, the little structure has length, width, weight, fixity conditions at the ends, and elasticity modulus of the material of the bridge. In addition, the bridge also has another parameter called natural frequency of vibration. Now say a troop of soldiers comes marching in lockstep. If the rhythmic frequency of pounding from the marching soldiers were to coincide with the bridge's natural frequency of vibration, resonant conditions prevail and the bridge span will move up and down. Once again, the dynamic motion of the bridge deck is not objectionable. What is objectionable, however, is the dynamic stress produced in the material of the bridge as the deck heaves up and down. In combination with static stress developed from the weight of the bridge and of the soldiers, if the endurance limit is reached, the bridge will collapse, oftentimes with disastrous consequences. The structure may very well be capable of withstanding the static weight imposed, but the dynamic stresses are of far greater consequence. The phenomenon also explains why the troop sergeant will order the soldiers to break step as they approach the bridge, then march once again in formation after they cross the bridge.

Concerning bridges, there is the well-known incident about the bridge spanning the Tacoma Narrows straits in the state of Washington. The structure was erected in the early 1930's in the suspension form. Some 10 years later, one fine morning the bridge abruptly collapsed, with about 35 vehicles plunging into the waters down below. Being a relatively new structure, the authorities were surprised. Adequate protection had been provided against corrosion from the marine air environment. After further investigation, the failure was attributed to wind blowing from the sea at the rate of 5 to 6 gusts per minute. The strong winds caught the underside of the bridge deck, causing one section of the structure to twist elastically. Finite element type of analysis and computers were not available at the time of design, but later evaluations indicated the torsional natural frequency of vibration was in the same range as the howling winds that are characteristic in the region. The two-lane bridge was not massive enough, thus becoming sensitive enough to the dynamic excitation from the wind. Together with lack of damping against twisting in the system, the bridge failed.

12.2 OPERATIONAL CAUSES OF MACHINERY VIBRATIONS

Rotating unbalance force is not the only culprit that leads to vibrations and failures in turbine engines. A host of other factors can also produce excessive vibrations in the shaft, compressor and turbine airfoils, supporting structures and outer casing. Predominant among them are the wake produced in the flow stream and the associated blade passing frequency and misalignment in the bearings or coupling. Corrosion in tiny crevices between mating rotating parts can cause the gap to widen and lead to the rotating assembly to lose balance.

At steady state operating conditions, a well-balanced shaft should not experience easily detectable vibrations in the bearings. Net radial forces are created when the fluid flowing through the machine are not balanced around the periphery. Stress buildup from zero to maximum as the rotor goes from standstill to operating speed and reversing back to zero when shutdown is a prime example that holds the possibility of failure in a shaft.

Imbalance in the rotating structure may be due to operating circumstances as well as during initial manufacture and assembly. If a rotating part were to rub against a stationary component, metal chips fall off from the contact area. As the machine runs for many operating hours, more chips gradually dislodge. On the radially opposite side a similar pattern of chips may not be falling off, thus resulting in unbalance. Radially directed centrifugal force is consequently set up, the magnitude of which is $(W/g)\ \omega^2 r$, where W/g represents the unbalance mass at radial location r and ω is angular speed of the rotor. Since the shaft is spinning, the force vector is also turning at the same speed.

During the machining phase of disks, some material may be left on the high side when the axis of the machining center does not coincide with the true spin axis of the engine. Similar problems are encountered during assembly of rotating disks, spacers and other related parts. Mating parts may not have coinciding axes of spin, and must also be coincidental with the total machine's true axis of spin.

12.3 JOURNAL BEARING FAILURE

Failures in journal bearings occur mostly when starved from lubricant oil. Disruption in the flow of clean lube oil at the right pressure, temperature and flow velocity can prevent the bearing from levitating the rotor. In the absence of the annular region filled adequately with the lubricant, the required level of hydrodynamic forces are not developed.

The net result is the rotor bottoms out and rubs against the bearing surface. Metal chips are dislodged from both contacting surfaces that are not flushed out of the annular space due to the restricted oil flow. The particles further grind the surfaces and open up the clearance space. The increased eccentricity then causes rubbing at the inter-stage seals and at the blade tips. Vibrations in the turbine rotor increase, and eventually the turbine is forced to shut down.

Temperature of the lubricant oil also plays a significant role. An increase of, say 60° F, from 180° F to 240° F, can cause the viscosity of turbine oil to fall by as much as 90 per cent. Friction in the bearing raises the temperature of the oil. A major task of the lubricant is to carry away the heat associated with the friction. After draining back to the oil reservoir, it is essential the lube oil is cooled to a level that permits safe operation in the bearings. Failure in the cooling system should be rectified on a priority basis.

12.4 LIMITS IN ROLLING ELEMENT BEARING LIFE

The life of a bearing is defined as the total number of revolutions, or the number of hours at a fixed speed, of bearing operation required for the failure criteria to develop. Under ideal conditions the first evidence of fatigue failure will consist of a spalling of the load carrying surfaces. The Timken Company goes by the failure criterion of pitting or spall of an area of 0.01 in^2. In testing a group of bearings, the objective is to establish the median and the L_{10}, or rated, life. The L_{10} parameter implies that 90 per cent of identical bearings operating at a constant speed and load will complete or exceed the test before the failure criterion develops. When a number of batches of bearings are under test, the median life is usually between four and five times the L_{10} life.

The concept of probable survival of a batch of bearings also needs examination. If a machine uses N bearings with each bearing having the same reliability R, then the reliability of all the bearings is $(R)^N$. The distribution of bearing failures can be approximated by the Weibull procedure. By making adjustments to the Weibull parameters (*ref:* Mischke, 1965), the distribution of bearing failures takes the form:

$$R = \exp\left[-\left(\frac{L}{6.84L_{10}}\right)^{1.17}\right] \tag{12.1}$$

Thus, if a certain application requiring reliability of 98 per cent for the bearing is to last for 2500 hours, this establishes its rated life.

12.5 FAULTS IN ROLLING ELEMENT BEARING

The failed bearings shown in Figure 12.1 are removed from a number of machines. Spectroscopy, a widely applied technique, provides a quantitative analysis to determine the concentration of as many as 20 different elements that are commonly used in the manufacture of rolling element bearings and in the bearing seals. Iron, aluminum, chromium, copper, tin, lead, silver, titanium and nickel are detectable and reported in parts per million. Oil samples are obtained immediately after return from the bearings. Elements used as lubricant additives, such as calcium, barium, zinc, phosphorus, magnesium, boron and molybdenum are also included in evaluation of the lubricant oil. The presence of only 2 ppm of copper may be sufficient to warrant maintenance action. The source of the copper in this case is the bronze bearing cage. Sometimes even a slight increase or presence of an element can be cause for alarm.

Concentration trends are routinely monitored to indicate if a continuing wear condition exists, the rate of wear, and as a consequence, the urgency of the problem. The oil samples may be from many different types of environment. An effective program is dependent upon manual or computer based interpretation of the analytical data, and is based on wear metal guidelines for the type of equipment.

Since the spectrometer can differentiate between metals present in an oil sample, the readings can be used advantageously in a number of ways. For example, if only iron and aluminum are present, the job is simplified and the inspection can be restricted to components made of the metals.

Damaged seal, shield or cage during mounting

Fatigue failure from overload, dirt or misalignment

Extended flaking from prolonged use after initial fatigue failure

Fig. 12.1 Examples of Bearing Failure (Morton, 1967)

Knowing the relative concentration of the elements will further narrow down their possible source. Increase in levels of nickel and silver is indicative of bearing wear, and a relatively simple replacement of the bearing liners may suffice. Higher levels of silicon in conjunction with a corresponding increase in iron, aluminum and chromium is probably caused by dirt ingestion, and replacement of air filter and oil change may be the only maintenance actions required. A trend showing the presence of boron in a water-cooled system would indicate a coolant leak. If left unchecked, the coolant combines with combustion products and forms harmful acids that corrode metal.

12.6 VIBRATION ASSOCIATED WITH SHAFT TWISTING

Torsional vibration in a rotating shaft assembly is an oscillatory angular motion causing twisting. The oscillatory motion is superimposed on the steady rotational motion of the shaft. Even though the vibration cannot be detected without special measuring equipment, its amplitude can be destructive. The motion by itself is not much of concern, but the torsional stresses affect the structural integrity and life of the components. Thus, the angle of twist in the shaft sets the limit of torsional vibration.

Torsional vibratory motions can produce stress reversals that cause metal fatigue. Components tolerate less stress reversals than steady stress. In addition, stress concentration factors associated with machine components decrease the effectiveness of load bearing materials.

In modeling the system for torsional analysis of a rotating system, the support bearings do not play any role, the bearings actually facilitate the rotor to spin about the axis. The system has low damping, and is essentially limited to internal material damping of the shaft. The damping is in contrast to lateral vibrations of a rotor system where significant damping is derived from the bearings and support structure.

Large amplitudes of torsional vibration can be present in the rotor of a machine with little or no vibration induced in the stationary structure. If a harmonic exciting torque is present in the system, either in the driving element or in the reactive torque of the driven element, then the possibility of a destructive failure of the shafting elements of the rotating system due to torsional fatigue exists.

When the frequency of any harmonic component of the exciting torque is equal to one of the natural torsional frequencies of the rotor, then a condition of resonance exists. Coincidence of the natural and exciting torque frequencies occurs at the critical speed. Also, for each natural frequency there is a series of critical speeds corresponding to the various harmonics of the exciting torque.

For example, systems with gearing tooth contact force between mating gear elements impose lateral forces on the bearings and moment reactions on the gear case. If the system is in a state of torsional vibration, then cyclic force reactions are superimposed on these steady forces. If torsional vibration amplitude is large enough to unload the gear teeth, then clattering noise ensues as a result of the backlash in the gearing, providing a warning of the presence of torsional vibrations. This large amplitude vibration, if allowed to continue, can damage tooth contact surfaces. Under such circumstances the cyclic component of the load on the bearings provides some damping as a result of squeeze film action.

Initially considering a simple two inertia system connected by a shaft with torsional flexibility will highlight the basic concepts in torsional analysis. The simple two-inertia system is a convenient starting point for making estimates, particularly where the driving and driven elements of a machinery system may each be approximately represented by a lumped inertia (Figure 12.2). A mass-less shaft with torsional stiffness K_{12} connects two disks with polar mass moments of inertia J_1 and J_2. Angular rotations of the disks are denoted by angles θ_1 and θ_2. A sinusoidal exciting torque equal to $T_1\text{Sin}(\omega t)$ is applied to disk 1. Solution for the natural frequency of vibration is based on the governing differential equations of vibration. Viscous damping torque, proportional to the rate of change of the angle of twist, may be present in the shaft connecting the disks or at either of the two disks. The natural frequency expression without damping is:

$$\omega_n = \sqrt{\frac{K_{12}}{J_1 J_2 /(J_1 + J_2)}} \tag{12.2}$$

If damping is introduced in the form of torque equal to the viscous damping coefficient B_{12} times the difference in angular velocities of the disks ($d\psi/dt = d\theta_1/dt - d\theta_2/dt$), a steady state solution yields a relationship for the amplification factor of the angle of twist. The damping ratio takes the form:

$$\xi = \frac{B_{12}}{2\omega_n J_1 J_2 /(J_1 + J_2)} \tag{12.3}$$

where the damping ratio is a single function of δ (the logarithmic decrement) and $\xi = \delta/2\pi$. As an indication of damping in a system, logarithmic decrement represents the ratio of measurement of two successive peaks of vibration. But this damping is generally too low to be of practical significance in torsional vibrations of a shaft.

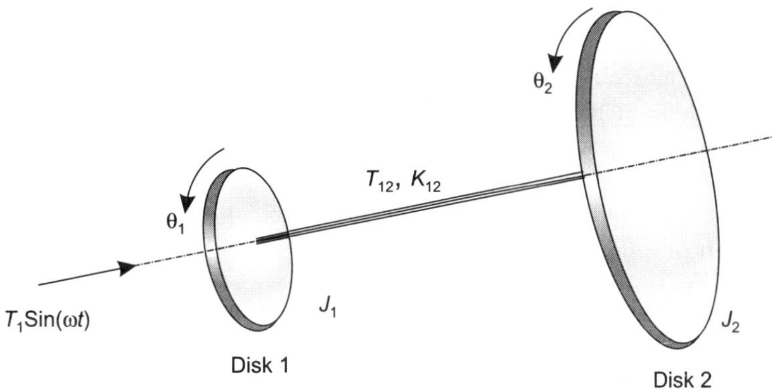

Fig. 12.2 A Simple Two-Inertia System in Torsion

12.7 TORSIONAL VIBRATIONS IN PROPELLER-DRIVEN AIRCRAFT TURBINE ENGINE

The propeller blades and rotor of a turbo-prop engine are subjected to a unique source of dynamic excitation. As the rotor goes through one full turn, a wake is shed in front of the leading edge of the aircraft's wing as air flows over its surface. Individual propeller blades encounter the wake twice during a single turn. The blades are hence subjected to second excitation order. In other words, if the rotor is spinning at 1625 rpm, the propeller blades will experience 3250 times variation in the pressure profile of the air flowing past it.

In the design phase of the propellers, it will be necessary for the manufacturers to ensure any of the natural frequencies of vibration of the blades do not interact with the second multiple in the operating speed regime of the rotor. If the engine operates in a given speed band, an acceptable blade design will require all of the natural frequencies to be away from the associated frequency range. Section 12.20 provides a striking example of the impact of the second excitation order, 2P, on the propeller blade's dynamic characteristics. The Campbell diagram of Figure 12.8 shows the first flatwise mode of vibration intersecting the 2P curve in the high-idle to cruise speed range of the engine. It is surmised the engine and propeller blade manufacturers decided that the condition was deemed acceptable since the engine does not linger for any length of time as the engine speed increases from the high-idle to cruise conditions, together with the tacit approval of government agencies.

Propeller blades are oftentimes made from the best quality wood, since seasoned wood has good damping characteristics. Metal propeller blades, in contrast, do not offer internal damping characteristics, except for low level damping associated with the molecules rubbing against each other. Certain other polymers bonded to internal metal spars have also been used, mostly because of the difficulty to procure the right grade of seasoned woods.

Also of interest is the effect of the 2P excitation on the shaft to which the propeller blades are attached. The shaft must not only be able to endure excitation of the first mode but also from the second mode. Failure from high-cycle fatigue is going to be of interest, where the total maximum stress and number of cycles, or *S-N*, data must be used in the evaluation. Hence, the shaft material must withstand number of cycles due to 1P and also due to 2P excitation.

It is interesting to note that a strong interaction oftentimes exist in the dynamic characteristics of propeller blades and of the driving shaft. The interaction may be strong enough to prevent identifying the primary source of dynamic activity, whether the blades are causing the shaft to vibrate or the other way round. The propeller assembly is overhung at the end of the rotor. When the plane of the propeller rotates because of the slope of the shaft's centerline, incremental axial velocities are induced. Thus, whirl in the propeller could be recognized as a direct consequence of shaft whirl. At high air speeds, destabilizing forces are set up in the propeller assembly, and can overcome damping forces to cause destructive whirling of the system. Thus, the whirling propeller blades may also be responsible for causing the engine shaft to deflect and set up dangerous vibrations.

12.8 FLEXURAL ROTOR VIBRATION

Transverse, or lateral direction, vibrations of a shaft are analytically more complex than torsional and axial vibrations in several respects. The deformation of the rotor is due to a combination of bending and shearing, an inherently more complicated elastic motion than simple stretching or twisting. Shaft bending is not confined to a single plane, and motion must be looked at in two orthogonal planes. Within each plane it is necessary to characterize the dynamic motion in the form of linear displacements and angular rotation. Gyroscopic and other asymmetric effects cause the motion of the shaft in the horizontal and vertical directions to be coupled.

To top it all, the location and dynamic characteristics of the support bearings play a substantial role in transverse direction vibrations. This calls for development of detailed models of the bearings and their support structure. Despite these complications it is possible to develop equations of motion for transverse vibration of the rotor that are analogous to torsional vibrations.

In addition to the fundamental, or first, critical speed of a long rotor with distributed mass, there will be an infinite number of critical speeds. Each critical speed has a characteristic mode shape. To understand this, consider a rotor of uniform cross section. Linn et al (1951) calculated the effect of bearing flexibility for the first four critical speeds of a uniform un-damped rotor. The solution is derived from the differential equation of the elastic curve of the whirling shaft. The first critical speed on rigid bearings is given by:

$$\omega_1 = \pi^2 \times \sqrt{\frac{gEI}{WL^3}} \tag{12.4}$$

where: g = acceleration due to gravity
E = shaft material modulus of elasticity
I = moment of inertia of shaft cross section
W = weight of rotor
L = length of rotor.

When the shaft is supported on flexible bearing supports, both the stator and the rotor are treated as flexible elements. Damping in the system, c_b, will be associated with the bearing support only. If the rotor is infinitely stiff, the un-damped critical speed is given by $\omega_{nb} = \sqrt{(K_s/m)}$, where K_s is support stiffness and the new damping ratio $\xi_b = c_b/(2m\omega_{nb})$.

The location of the bearings along the length of a rotor will have a major effect on shaft response to unbalance. For instance, if the bearings were placed closer to each other and away from the ends to the nodal points of the first mode of the free-free rotor, supporting structure damping will not be effective. Excessive response to unbalance will be experienced when operating near the first critical speed.

12.9 OPERATION AT SHAFT CRITICAL SPEED

The first step in the layout of a rotor-bearing system is to calculate the un-damped critical speeds. In the example shown in Figure 12.3, the natural frequency is calculated to be 3,890 rpm for the first mode and 7,580 rpm for the second mode. Of special significance is the response to rotor unbalance, especially since machine operation takes place through the first critical speed.

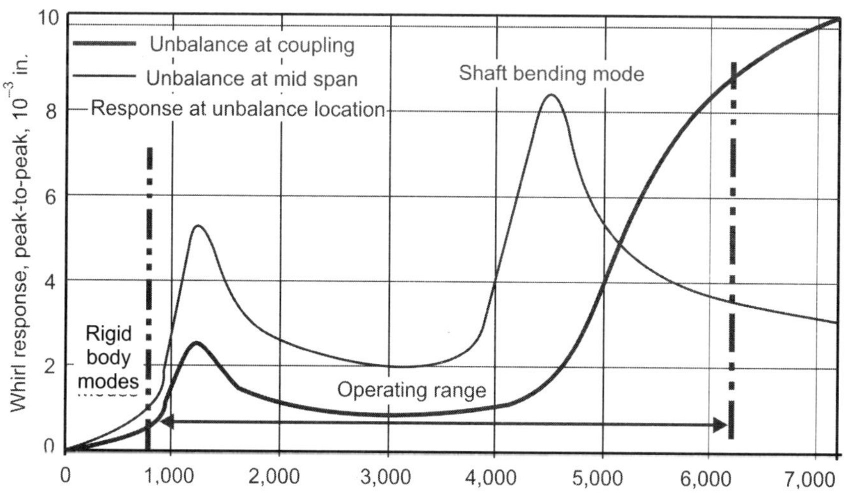

Fig. 12.3 Calculated Rotor Response due to Unbalance

The shape of the modes of vibration dictates that unbalance at mid span portion of the rotor will be of interest for the first mode. Unbalance located at the ¼ and ¾ span length from either bearing will result in high deflections in the rotor when operation of the machine is near the second mode of vibration. Un-balance or misalignment at the flexible coupling will also be sensitive in the second mode. Coupling unbalance, however, will have little effect in the first mode. If operation were required through the second critical speed on this machine, then special attention would be needed to balance the coupling. Unbalance here could cause excessive whirl at the overhung coupling, together with high dynamic loads in the bearing at the higher operating speed. Rotor response at the location of unbalance is shown in Figure 12.3.

12.10 INADEQUACIES IN BALANCING OF ROTOR

A rotor is in dynamic balance if it rotates about its principal axis without wobbling, assuming that there is no dynamic motion due to the elasticity of the rotor and the lubricant in the bearings. When the journal axis of rotation does not coincide with its principal axis, a net radial force acts on the rotor due to centrifugal acceleration. The process of effecting essential coincidence between the principal inertia axis of the rotor and the shaft axis is called dynamic balancing.

The balancing procedure for correction balance of a rotor after it is placed in service may be considerably different from the initial process employed during its manufacture. Commercial balancing machines using either the soft bearing or the hard bearing concept are used to balance individual rotating components and the total assembly. Field balancing of the machine in place avoids the time consuming job of removing the rotor to be reworked and re-balanced at the factory. The object of trim balancing is to add correction weights in one or more planes along the length of the rotor to maintain acceptable levels of vibration.

At low speeds, a shaft supported in flexible bearings does not bend much. But as the lowest natural frequency of vibration is approached, the bending distortions cannot be ignored. The theory of balancing is based on the premise of canceling out motions in principal modes, as many of them as necessary, by the use of compensating weights. The stiffness of the bearings is assumed to be independent of frequency, since they are essentially massless springs.

The single-plane, or one-shot, balancing method offers the advantage of simple and quick application on machines in service. The process requires a trial mass to be placed at a known radial location on the rotor. Dividing the change in the measured response by the magnitude of the trial weight provides a useful vector quantity called an influence coefficient at the specific speed. The magnitude of the vector represents the amplitude of vibration, with an associated phase angle between the forcing function and the displacement. Phase measurements can be obtained from directly reading non-contact or displacement probes, velocity transducers or accelerometers. An electronic key-phasor or strobe light is also used to obtain the phase angle on the shaft. The application of a single mass in an assumed linear system theoretically will nullify any dynamic motion in the shaft running at that speed. But in the case of a long rotor with a number of disks, other points may experience higher vibrations. At other speeds also high vibrations may be encountered, representing a lack of balance.

12.11 EFFECTS OF UNBALANCE IN ROTOR

To understand the effects of unbalance on a flexible rotor, consider a single mass disk mounted on an elastic shaft. The mass center of the disk is displaced from the center of the elastic shaft, resulting in an unbalanced rotor. The total rotor effective unbalance is equal to the product of the disk's mass and the unbalance eccentricity. The rotating unbalance force on the disk causes the shaft centerline to precess, or orbit, about the geometric centerline. When the bearing supports are rigid, the motion may be referred to as synchronous precession with a circular orbit. If this rotor is mounted on flexible bearing supports, total motion of the disk includes motion of the end supports.

Next, consider two cases of a rotor in soft bearings. In one case the rotor has an unbalance mass at its mid plane, causing the rotor to be statically unbalanced. As the shaft turns, centrifugal force on the mass will cause the centerline of the shaft to travel in a circular path. In the other case the rotor has two unbalance masses at opposite ends, and are offset 180° to each other. The two unbalance weights will offset each other statically. Also, as the shaft rotates, centrifugal force on the masses will cancel each other, however, an overturning moment will be present due to the axial offset of the masses. The resulting trajectory of the shaft's centerline will be in the form of two cones, and the apex of the cones will be at mid plane due to the symmetric nature of the rotor and the unbalances. When the two forms of unbalance are combined, the apex of the cones moves away from the center of gravity. An apex will not be present because the shaft will move in a complex combination of motions. A random combination of axial and radial magnitudes of vibration and phase angle location of the static and couple unbalance will lead to a condition referred to as dynamic unbalance.

Flexibility in the supports of the rotor will add to the motion of the rotor in the lateral directions. When support flexibility at the two ends is not symmetric, the lateral motion will also have differing characteristics, if tacitly assumed that the rotor is essentially rigid, with the dynamic activity confined to the supports and no flexure present in the shaft.

12.12 SELF-EXCITED VIBRATIONS IN ROTOR ASSEMBLY

Self-excited vibrations in a system begin by themselves without any warning, initiating spontaneously. The amplitude increases until some nonlinear effect limits any further increase. The excitation source for these vibrations is a uniform source of power associated with some inherent mechanism within the machine system. The destabilizing force sustaining the motion is derived from and controlled by the motion itself, and the force disappears when the motion stops.

Whirling and whipping instability arises from many sources. Internal rotor damping arises from hysteretic whirl. A material related characteristic, internal friction between the molecules results in hysteresis. Bending in the shaft due to some radial deflection sets up elastic restoring forces in a direction parallel and opposed to the direction of deflection. Friction in the shaft material also causes a phase shift in the developed stress, which results in a tangential force component. This sets up the condition for the shaft to whirl.

Besides internal friction, loose mating components also are responsible for this phenomenon. Bolted joints, shrink fitted assemblies and rabbets are prime examples. If the bolts are not tightened to the proper torque setting, or adequate interference fit does not exist, relative motion between the components will provide cause for hysteretic whirl in the system. Hysteretic whirl occurs at shaft speeds above the first critical speed. But once whirl sets in, the induced critical whirl frequency is about half the onset speed.

Many other examples of unstable rotor vibrations can be cited. Fluid trapped in a hollow rotor can act as a high-speed centrifuge to set up destabilizing forces. Oil from bearing sumps and steam condensate may inadvertently collect and be trapped in the cavity of the rotor. When the shaft is operating with some nominal radial deflection, the fluid flies out, but does not remain in one fixed location. The surface of the shaft carries along some fluid with it in the direction of rotation. This advance sets up a force component in the tangential direction; the force is the source of instability because it induces forward whirl. The threshold speed for instability will be above the critical rotor speed.

Yet another source of excessive rotor vibration may arise when transmission of torque in high-speed rotors is accompanied by a rotational twist in the shaft. If that twist results in an angular deflection of the torque vector, then the torque will have a small radial deflection, which is nearly proportional to the radial deflection. The radial torque vector induces bending in the shaft and a deflection that is normal to the radial deflection. When this tangential component is large enough to overcome the energy dissipating forces in the system's damping mechanism, the threshold of instability is crossed and the shaft whirls.

12.13 SUB-SYNCHRONOUS VIBRATIONS

The stiffness and damping characteristics of the fluid film play a major role in the development of whirl at a frequency that does not match with shaft speed. As the speed increases and eccentricity ratio gradually becomes smaller, the cross coupling stiffness and damping terms become about equal. The speed ratio at which whirl will develop is given by $k_{r\theta}/c\omega$, where $k_{r\theta}$ is the film stiffness in the cross coupled radial and tangential directions. For plain journal bearings this ratio is about one-half. For a flexible rotor the threshold speed at which the instability occurs is approximated by $\omega_n/(k_{r\theta}/c\omega)$, where ω_n is the rotor critical speed. Thus, a flexible rotor supported in plain journal bearings usually experiences sub-synchronous unstable motion when shaft speed is greater than twice the lowest rotor critical speed.

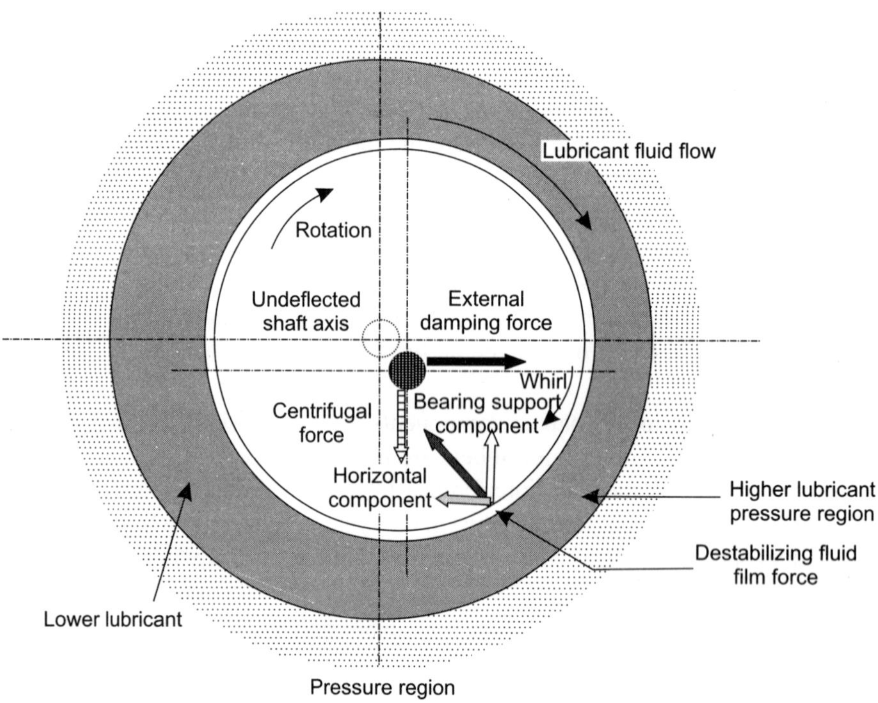

Fig. 12.4 Bearing Lubricant Film Causing Shaft Whirl (Ehrich, 1992)

Referring to Figure 12.4, a shaft rotating in a journal bearing filled with a lubricant has a nominal radial deflection. Because of the shaft's eccentricity and the average fluid velocity (about half that of the shaft surface), the lubricant develops pressure, which is not symmetric about the line of radial deflection. Also, fluid pressure on the upstream side of the clearance region is higher than on the downstream side. This causes the resultant hydrodynamic force on the shaft to have a tangential component in the direction of rotation of the shaft, tending to whirl the shaft in the same direction. When this force exceeds the stabilizing damping forces, unstable operation will be the consequence. The stiffness and damping characteristics of the fluid film play a major role in determining the development of whirl. As the speed increases and eccentricity ratio gradually becomes smaller, the cross coupling stiffness and damping terms become about equal.

12.14 RUBBING AT BLADE TIP AND INTER-STAGE SEALS

Steam and gas turbines require close control of radial clearance between the tip of the blades and the casing. The objective behind it is to prevent leakage of steam past the rotating blades without permitting the blades to rub against the case, and thus maximize operating efficiency.

In many cases, however, it is difficult to maintain concentricity between the rotor and the stator, with the result that the tip clearance is not uniform along the full circumference. This can happen when the rotor lacks adequate stiffness, or due to larger than normal shaft span between the supporting bearings, causing the rotor to sag downwards. Overhung stages past a bearing support (aircraft jet engines are an example) also tend to exaggerate this situation. From aerodynamic considerations the result of the unequal space around the circumference is to set up a tangential force at the blade tip in the direction of rotation of the rotor. As the clearance distribution becomes more pronounced due to changes in speed or power setting, tangential forces may become large enough to overcome the energy dissipating forces of the damper, with the rotor whirling as a consequence. To overcome the problem, during the design stages bearing and shaft stiffness must be adequately provided for to avoid excessive radial deflection of the shaft.

Rubbing of a shaft against the stator case takes place in an environment where there is no oil or other fluid lubricant, so the rubbing is essentially dry. The dry rub sets up conditions for the shaft to whirl. Dry rub can also take place in a bearing that is starved of the lubricant, from loss of clearance in a journal bearing, from loss of radial clearance in a labyrinth seal.

Coulomb friction due to rub between the rotating and static parts introduces a tangential force component equal to the product of the radial force and the coefficient of friction. The force induces the shaft to whirl, and the associated centrifugal force produces greater radial deflection, which in turn causes an even greater radial force and whirl producing tangential friction force. Note that the friction force will be in the opposite direction of shaft rotation, thus causing the shaft to whirl backward.

A similar situation is encountered in an axial thrust bearing, where due to the slope of the shaft contact takes place between the rotating and stationary disks at a pivot point. Remedies to alleviate the shaft whirl due to rubbing will range from stiffening of the shaft to reduce radial deflection to improving assembly techniques for maintaining concentricity between rotor and stator. If the rub persists during operation, lubrication of the contact surfaces should alleviate the problem. If lubrication is not possible, another remedy would be to use an abrade-able lining material on the stator to permit disengagement between the interfering components.

12.15 GEARED AND BRANCHED SYSTEMS

When two or more drivers are geared to a common load, hydraulic or electrical type of couplings may be placed between the drivers and the gear set. The couplings serve a dual-purpose role, one of isolating the gears from driver induced vibrations, the other as a clutch for connection and disconnection of load between the components. In many such coupling designs the vibration isolation may be complete enough to permit termination of the driver end of the unit at the coupling. The damping effect upon the vibration in the driver end of the system is small and may be disregarded in the amplitude calculations.

Hydraulic or electrical couplings are not employed in some applications calling for two identical driver units. Then the modes of vibration are of two types. In the opposed phase mode, the drivers vibrate against each other, with a node at the gear. The modes are calculated for a single branch in the usual manner, with the model terminating at the gear. The condition for a natural frequency is that shaft twist amplitude be set to zero at the gear. In the in-phase modes, both drivers vibrate in the same direction against the driven machinery. To calculate these frequencies the inertia and stiffness constants of the driver side of one branch are doubled; the calculations are then performed for only one driver. The condition for a natural frequency is zero residual torque at the end.

The two drivers can be phased such that certain like-phase modes can be eliminated. If the drivers are connected with clutches, these phasing possibilities cannot be utilized. In the general case of non-identical branches the calculation scheme proceeds as follows. Reduce the system to a 1:1 gear ratio. Call the two branches *a* and *b*. Perform the sequence calculation for the *a* branch, with initial twist angle equal to unity, and for the *b* branch the initial amplitude set to an unknown value *x*. At the junction equate the amplitudes and find *x*. With this numerical value of *x* substituted, the torques in the two branches and the torque of the gear are added; the sequence calculation is then continued up to the last mass. Figure 12.5 shows a schematic diagram of geared set of rotors.

The branch may consist of a single member elastically connected to the system. Examples of such a branch are a flywheel with flexible spokes or an armature with a flexible spider. Assume J to be the mass moment of inertia of the flywheel's rim and k the elastic constant of the connection. Then the flexibly mounted flywheel is equivalent to a rigid flywheel of equivalent moment of inertia $J' = J/[1 - J\omega^2/k]$.

For an un-damped rotor with isotropic supports, the orbits of the natural modes of whirl are circular relative to the fixed reference frame. Because of the axial symmetry, the order of the system equations may be reduced to half. The point matrix at a station and field transfer matrix between stations then reduce from 8 × 8 to 4 × 4 matrices.

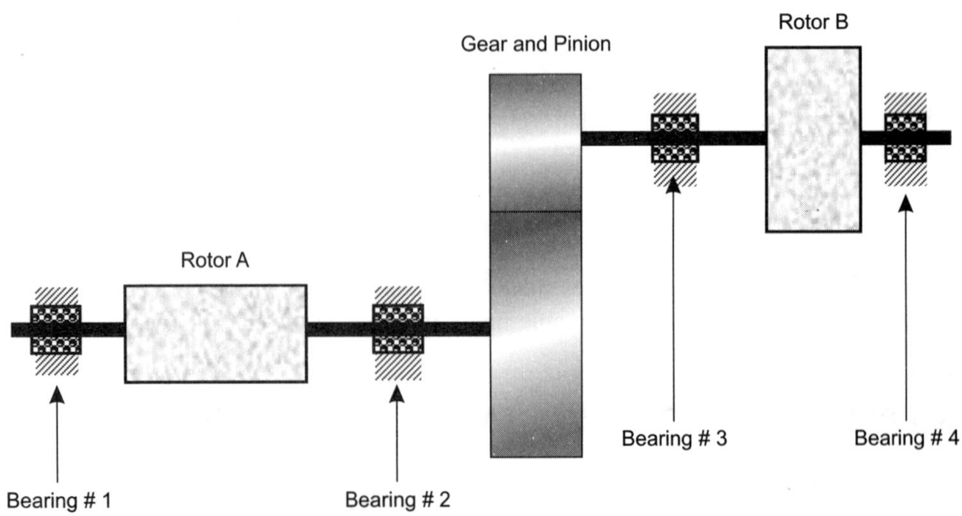

Fig. 12.5 Two-Shaft Rotating System

Natural frequencies of vibration of a geared system of two rotors are dependent on the gear ratio. The mathematical procedure calls for replacing the system with an equivalent system. The inertia and elastic constants on the driven side of the unit are multiplied by the factor $1/n^2$, in which the speed of the driver unit is a multiple of n, where n is the ratio of the speeds of the driver unit and the driven unit. Then calculate the system natural frequencies as if the gears do not exist. Torques and deflections thus obtained on the driven side of the substitute system, when multiplied by n and $1/n$ respectively, are equal to those in the actual geared system. Alternatively, the driven side may be used as the reference, in which case the driver constants are multiplied by the factor n^2.

12.16 AIRFOIL DYNAMIC CHARACTERISTICS

An in-depth study of the dynamic characteristics of a turbine blade is necessary in evaluating its capability to withstand the assigned loads. What may appear at first sight a relatively simple task of interpreting the modal characteristics of the blade is oftentimes hampered by the geometric configuration; the root structure is thick, with a complex fir tree form for attachment to the disk; airfoil trailing edge is thin; cross section profiles are varying at different radial heights; the blade is twisted and leaned.

Initial blade geometry is controlled by aerodynamic and performance considerations. Within the engine, however, it must meet fluid flow and structural criteria. Chord-wise motion of the airfoil is coupled with the flap, or perpendicular to the chord, motion, primarily because of the twist. Total bending or torsional deformation is not encountered in the blade. Varying degrees of the two movements exist in each mode of vibration as a direct consequence of the coupling between the degrees of freedom. Aerodynamic and mechanical loads also sometimes tend to dramatically change the blade's profile. The blade's design may also go through modifications, such as replacing a shrouded configuration with a wide chord or large sweep blade without the shroud.

In addition to finite element methods, individual blades need to be tested using modal analysis with impulse, white noise and sinusoidal excitation applied at the leading edge and at the trailing edge. Optical measurement systems based on Electronic Speckle Pattern Interferometry may also be employed to determine natural frequencies, and even more important, mode shapes. Very often each test program produces modes that may show small differences to totally new shapes, leading to ambiguities that cannot be easily resolved. Interpretation and classification of modes in the familiar flap, edge, torsional and stripe modes of vibration is not always feasible. Finite element analysis may help to sort the problem. Elements with one or two mid-side nodes and an appropriate master degrees of freedom can help in the identification, and may also reveal new modes among those measured. The blade may be assumed fully fixed at the root.

As a rule, finite element procedures are reliable, and often used in the design process for optimization studies. In rare cases a combination of analytical and test may be necessitated. Not so negligible differences can sometimes be explained by the method of excitation, accelerometer mass, height location for analysis or inappropriate boundary conditions. Closely spaced modes can combine to appear as a natural mode, but in reality combine elements of the closely spaced modes. Instead of a cursory look at measured or calculated results, it pays to be attentive to the details.

Suppression of flutter in an airfoil is the prime intent for adding an elastic support in the form of a shroud. Flutter is more commonly encountered in long and slender low-pressure turbine and compressor blades, which lack adequate torsional stiffness. When operating at speed the protrusions on the row of blades lock to form an almost continuous ring to couple the blades, resulting in a stiffer assembly. Since the protrusions are in contact as the blades go through a certain amount of untwist under centrifugal action, the shrouds rub against each other. The extent of the rub and the consequent damping has been examined extensively to determine suitable boundary conditions, with assumptions ranging from a fully locked to a freely slipping condition.

Z-shaped shrouds on gas turbine blades also perform the function of sealing while maintaining a pre-load along the circumference during operation. As the shroud contact edges wear, a change in the dynamic characteristics of the rotating assembly may be expected. Coupled modes in a tight shroud condition degenerate into first flap and first edge-wise cantilever type modes at lower frequencies when the shroud slackens.

12.17 BLADE FAILURE FROM LOW-CYCLE FATIGUE

Material fatigue sets in in the airfoils of a turbine when the engine is subjected to a number of loading and unloading cycles. A paper clip bent back and forth a few times illustrates the case of breakage due to fatigue. The airfoils also go through analogous variations in deflection and stress patterns when the turbine is started and shut down.

Prior to an engine running up in speed, the centrifugal force due to the mass of the airfoil is absent. As the engine speed picks up, the centrifugal force at various sections along its span increases. Thus, if the weight of the portion of the airfoil is W and radial height of the centroid of the portion from the engine centerline is r, centrifugal force on the cross-section is $(W/g)\omega^2 r$, where ω represents rotor speed. At the tip there is no mass, while maximum centrifugal stress will be encountered at the base of the airfoil, where it meets the platform. At the end of the mission, the engine is shut down and the centrifugal force reverts back to zero. Thus, a turbine engine going from start to top speed and back to shutoff causes the airfoils to experience one complete cycle from zero to maximum centrifugal stress and back to zero.

Blades in the turbine section also experience thermal stresses due to elevated metal temperatures. Thermal stresses develop when temperature gradients are present in the substrate. After ignition in the combustor the blade temperatures rapidly climb up, producing thermal stresses. As the engine settles into steady state conditions, the temperatures soak in throughout the airfoil geometry. Thermal gradients then disappear, as also the associated thermal stresses. During shutdown the reverse scenario occurs, except for one significant difference – the stresses change from tensile during heating to compressive during cooling.

The combination of centrifugal and thermal stresses constitute one load cycle. After an engine accumulates a certain number of load cycles, the airfoil material will lose its ability to withstand further imposition of loads. The damage incurred from each cycle adds up till the low-cycle fatigue limit is reached, and the component has to be replaced.

In aviation and aero-derivative engines, the emphasis on low weight results in the airfoils having substantially lower life capability as opposed to their industrial and power generating engine counterparts, where weight of the parts is not of consequence. Also of interest is the lack of option to replace the airfoils before the failure mechanism sets in, the part has to be replaced before rupture occurs. This points to the need to accurately predict the life of the rotating parts. No turbine operator would want to discard a turbine blade with some life remaining, and made from expensive super alloys.

12.18 VIBRATIONS AT BLADE PASSING FREQUENCY

A gas turbine's rotating components are subjected to resonant vibration and flutter phenomena. Resonant vibrations result from a coincidence in the blade's natural frequency and exciting force frequency. Exciting forces arise from uneven pressure distributions around obstructions in the gas flow path. For example, a prescribed number of support frame struts, nozzle vanes, combustor burners and bleed ports will result in a corresponding number of wakes in the gas stream past those obstructions. Rotating stall in a compressor, rubbing contact between rotor and stator and meshing of gears are examples of non-integral orders of obstructions in the gas stream. Stationary components such as nozzle vanes and support struts, on the other hand, will experience excitation caused by the rotating blades or by compressor rotating stall. Dynamic stresses, thus, must be adequately taken into consideration in the design of vanes and struts, but the absence of centrifugal force poses less stringent requirements than for rotating blades.

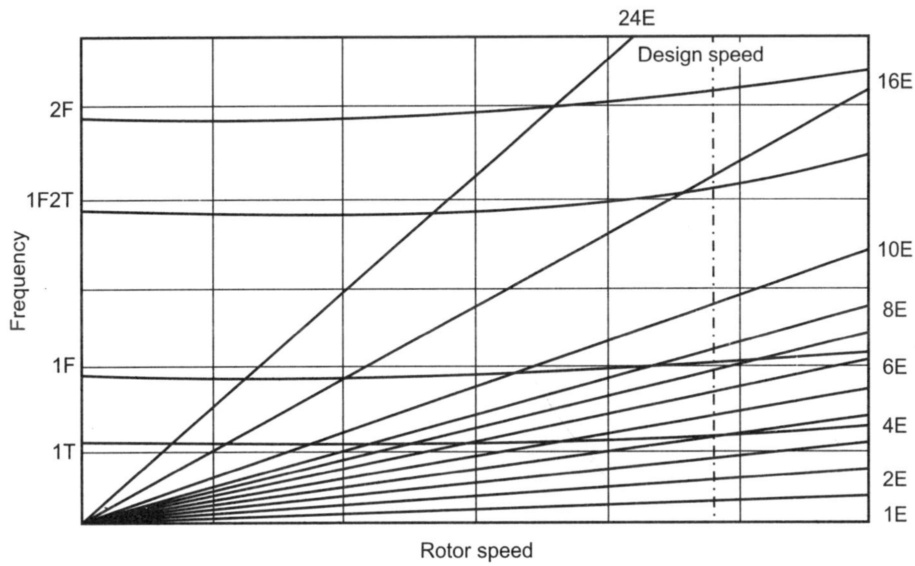

Fig. 12.6 Turbine Blade Campbell Diagram

When 8 support struts are present upstream in a flow path the turbine blades will encounter an equal number of wakes during one revolution of the rotor. Consequently, 8E, or 8 times rotor frequency, and its second harmonic will be of significance. The situation can be well summarized in the form of a Campbell diagram for identifying potential resonant situations, as shown in Figure 12.6. The figure graphically depicts the extent of interference between the blade's natural frequency and the different excitation frequencies for an engine operating at an essentially fixed speed.

Note that the blade's natural frequencies, especially the bending mode frequencies, increase with operating speed. This observation is consistent with the blade experiencing increased stiffness as a direct consequence of the higher centrifugal force as rotor speed increases. The example shows the blade's 1F2T (combined first flexure/second torsion) and 2F modes to be above the operating speed for 8E stimulus. The first flexural mode (1F) also has adequate margin for 16E excitation. However, the 1F mode for the 8E and the 1F2T mode for the 16E stimulus are too close to the design speed line, which may be considered unacceptable for the blade design. The 1F mode of vibration is next to the first torsional (1T) mode, but a higher frequency level also implies increased excitation levels. Blade tuning through modification of the blade's taper characteristics may permit setting of the problem modes to above the rotor speed. The 24E stimulus intersects all modes at a low rotor speed, and is not of consequence.

12.19 PROPELLER BLADE SEPARATION INCIDENT

In August 1995 an Embraer airplane on a scheduled passenger flight experienced the loss of a propeller blade from the left engine propeller while climbing though 18,000 feet at 160 knots. The aircraft crashed during an emergency landing half an hour after departing from Atlanta, Georgia. The aircraft was destroyed by ground forces and subsequent fire, with the pilot and seven passengers succumbing to fatal injuries.

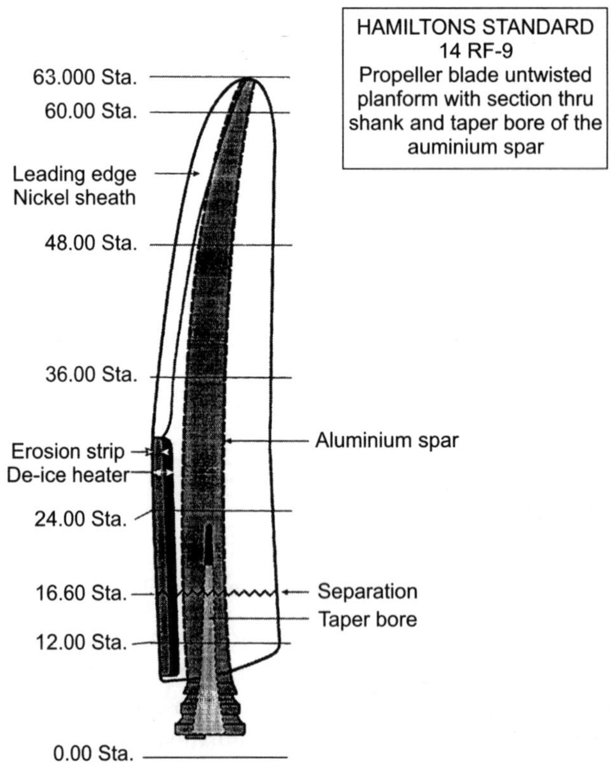

Fig. 12.7 Propeller Blade Design (NTSB/AAR-96/06, 1996)

Government authorities determined the probable cause of the accident as in-flight fatigue fracture and separation of the propeller (*ref*: NTSB/AAR-96/06, 1996). The resulting distortion in the engine's nacelle caused excessive drag, loss of lift on the wing and diminished control of the airplane. The fracture was attributed to a fatigue crack from multiple corrosion pits in the propeller. The manufacturer, Hamilton Standard, provides a family of composite propeller blades for use on turboprop airplanes. Figure 12.7 illustrates the design of the blade. A central solid forged aluminum alloy spar is used to carry the primary load. The airfoil shape of the blade is formed with glass fiber filled epoxy and foam, bonded to the spar with an adhesive. A tapered hole is bored in the center of the spar from the inboard end for installation of balance weights, while providing some reduction in weight. Earlier production runs of this design called for shot peening of the taper bore, but was deemed unnecessary and was discontinued. But the manufacturer's statistical data from field service experience indicate that blades without shot peening of the bores are susceptible to earlier corrosion and cracking. The blade involved in this accident was not shot peened.

It is also interesting to note that prior to certification the United State's Federal Aviation Administration required the airplane to be capable of successfully completing a flight during which likely structural damage may occur as a result of a propeller blade impact. However, the agency exempted the airplane manufacturer from complying with this requirement on condition that all practical precautions are taken in the design phase to take account of all features of the propeller to reduce the hazard, which might arise from the failure of a hub or a blade. Embraer's analysis indicated that the nacelle would not withstand the loss of a half or a full blade segment.

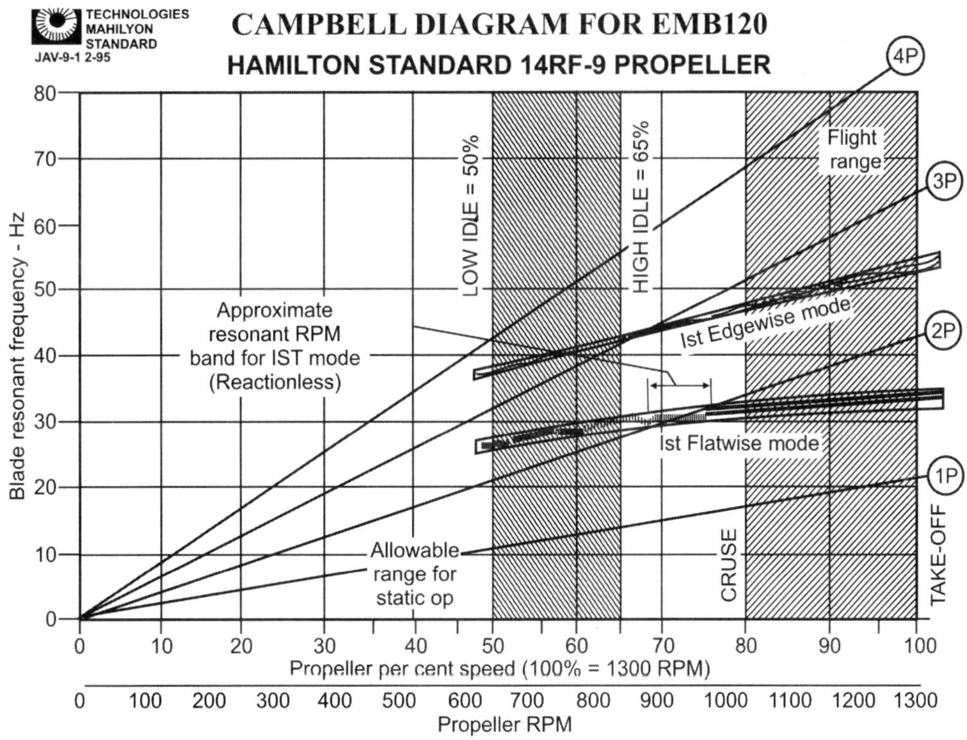

Fig. 12.8 Propeller Blade Campbell Diagram (NTSB/AAR-96/06, 1996)

To comply with airworthiness requirements the propeller manufacturer is required to demonstrate vibration characteristics of the assembly to ensure resonant frequencies responsible for producing critical vibration stresses do not occur within the normal operating range of use (*ref*: Rangwala, 2004). Each propeller must be shown to have vibration stresses that do not exceed values that have been shown by the propeller manufacturer to be safe for continuous operation. This must be demonstrated by measurement of stresses through direct testing, comparison with similar installations for which measurements have been made or service experience that proves the safety. A fatigue evaluation must be made and limits established for each metallic hub and blade. All reasonably foreseeable vibration load patterns must be considered. Fatigue limits must account for permissible service deterioration such as nicks, grooves, galling, bearing wear and variations in material properties. Variable pitch propellers (while engine is operating) must be subjected to a 100 hour test with the same power and speed settings when the blade experiences severe vibration characteristics. Flight tests are recommended if propeller diameter exceeds 13 feet.

Visual examination of the failed blade revealed a portion of the spar fracture was on a flat transverse plane, and contained crack arrest positions typical of fatigue cracking. Fracture faces were examined with a scanning electron microscope before the faces were cleaned. Near the origin area of the fracture a layer of heavy oxide deposits extending to a depth of 0.049 in. from the surface was observed. Fatigue crack initiated around at least two adjacent locations on the taper bore surface. Below the surface the cracks merged to form a single crack that propagated towards the face side of the blade, then progressed circumferentially around both sides of the taper bore. The crack spread 75 per cent of the spar's cross section. In areas beyond the terminus of the fatigue rough features with a matte appearance indicated typical overstress separation. After cleaning of the fracture surface additional examination revealed fatigue crack initiating from several corrosion pits in a line of pits extending over a distance of 0.070 in.

Prior to this accident, two other incidents of failures of the composite propeller blades from this manufacturer have been recorded. On both occasions the cracks originated from inside the tapered bore. In the first case analysis indicated forces induced from the rotation of the other three blades resulted in propeller assembly imbalance and loads on the engine's forward mount that exceeded the ultimate limits. This resulted in separation of the propeller and the reduction gearbox assembly from the airplane. The flight's crew was able to accomplish a safe landing, and injuries to passengers were avoided. In the second case causal findings were similar in nature. The three remaining blades and the fourth blade stub had moved to the feathered position, with the propeller and gearbox assembly remaining within the nacelle area, partially attached to the airframe. Laboratory examination of the failed blades indicated presence of corrosion pits in the tapered bore in both instances.

12.20 REVIEW QUESTIONS

12.20.1 Under what circumstances does resonance set in when a shaft is spinning in its end supports?

12.20.2 What is the role of bearings in torsional and flexural vibrations of a rotor-bearing system?

12.20.3 What is required when a turbine engine is to be run in the proximity of a critical speed?

12.20.4 How do manufacturing and assembly processes affect dynamic balance of a rotating disk?

12.20.5 Describe the various forces set up in the lubricant film of a journal bearing.

12.20.6 How does torsional vibration affect operation in a geared set of shafts?

12.20.7 What role do test measurements play in evaluating the suitability of a turbine airfoil for an application?

12.20.8 Discuss the impact of unbalance in a rotor on the dynamic behavior of a rotating machine.

12.20.9 What causes metals to rupture from fatigue?

12.20.10 What are the prime factors behind failures arising from low-cycle fatigue?

12.20.11 Explain the root cause of problems in airfoils associated with blade passing frequency.

12.21 EXAMPLE PROBLEMS

Example Problem # 12.21.1: A 1.785 in. diameter shaft is supported in rigid bearings set 47.25 in. apart. Shaft material $E = 28.355 \times 10^6$ lb./in^2 and density 0.278 lb./in^3. Calculate the fundamental natural frequency of vibration of the shaft.

Solution: From (12.4), the first critical speed on rigid bearings is: $\omega_1 = \pi^2 \times \sqrt{\dfrac{gEI}{WL^3}}$

Weight of shaft: $(\pi/4) \times d^2 \times l \times$ density $= (\pi/4) \times 1.785^2 \times 47.25 \times 0.278 = 32.854$ lb.

Moment of inertia: $(\pi/64) \times d^4 = (\pi/64) \times 1.785^4 = 0.4981$ in^4

Then first critical speed $= \pi^2 \times [(386 \times 28.355 \times 10^6 \times 0.4981)/(32.854 \times 47.25^3)]^{0.5}$

$= 124.53$ cycles/s.

Example Problem # 12.21.2: In *Example Problem # 12.21.1* calculate the torsional stiffness of the shaft if material shear modulus $G = 11.5 \times 10^6$ lb./in^2.

Solution: From Strength of Materials theory, $T/I_p = G\theta/L$

Polar moment of inertia $= (\pi/32) \times d^4 = (\pi/32) \times 1.785^4 = 0.9962$ in^4

Torsional stiffness $= T/\theta = GI_p/L = 11.5 \times 10^6 \times 0.9962/47.25 = 242453.57$ in-lb./radian

Example Problem # 12.21.3: In *Example Problems # 12.21.1 and 12.21.2* calculate the torsional natural frequency of the shaft.

Solution: The natural frequency from (12.2) is: $\omega_n = \sqrt{\dfrac{K_{12}}{J_1 J_2/(J_1 + J_2)}}$

Assume the weight of the shaft is concentrated at the two ends of the shaft. Hence, weights $W_1 = W_2 = 32.854/2 = 16.427$ lb.

Polar mass moment if inertia $J_1 = J_2 = 16.427 \times (1.785/2)^2/(2 \times 386) = 0.01695$

So torsional critical speed $= [242453.67/\{0.01695 \times 0.01695/(0.01695 + 0.01695)\}]^{1/2}$

$= 5348.70$ radians/s.

Example Problem # 12.21.4: A turbine engine's rotor support system requires a reliability of 99.75 per cent for the bearing to last for 2500 hours. Determine the rated life of the bearing.

Solution: Use $L = 2500$ and $R = 0.9975$ in equation 12.1.

$$0.9975 = \exp\left[-\left(\frac{2500}{6.84 L_{10}}\right)^{1.17}\right]$$

Ln $(0.9975) = -0.002503$, so $[2500/(6.84 \times L_{10})]^{1.17} = 0.002503$, or $L_{10} = 99985$ hrs.

Example Problem # 12.21.5: Experiments indicate that identical bearings acting under different radial load F_1 and F_2 and operating at speeds n_1 and n_2 have lives L_1 and L_2 according to the relation:

$$F_2 = F_1\left[\frac{n_1 L_1}{n_2 L_2}\right]^{1/a}$$

where $a = 3$ for ball bearings and 10/3 for roller bearings. A roller bearing can safely accept a load of 53.2 kN at 1712 rpm for an L_{10} life of 1485 hours. Determine its life if the load is reduced to 48.73 kN and the speed is increased to 1843 rpm.

Solution: In the equation, $F_1 = 53.2$ kN, $F_2 = 48.73$ kN, $n_1 = 1712$ rpm, $n_2 = 1843$ rpm, $L_1 = 1485$ hours and $a = 10/3$. Then $L_2 = 1848$ hours.

12.22 EXERCISE PROBLEMS

Exercise Problem # 12.22.1: The equations in *Example Problem # 12.21.4 and 12.21.5*: may be combined to obtain a specified levels of load, speed, life and reliability factor. Then:

$$F_2 = F_1\left[\frac{n_1 L_1}{n_2 L_2 (6.84)}\right]^{1/a} \frac{1}{[\ln(1/R)]^{1/1.7a}}$$

Determine the load rating F_2 if the bearing is to have reliability level of 95 per cent when $F_1 = 4.5$ kN, $F_2 = 3.75$ kN, $n_1 = 650$ rpm, $n_2 = 725$ rpm, $L_1 = 1400$ hours and $a = 10/3$.

Exercise Problem # 12.22.2: Data for a journal bearing is as follows: viscosity $\mu = 3.95 \times 10^{-6}$ reyn, speed $N = 1800$ rpm, radial load $W = 525$ lbs, radius $R = 0.875$ in., clearance $c = 0.0014$ in. and length $l = 1.625$ in. Determine the bearing's static pressure.

Exercise Problem # 12.22.3: A 6.75 in. diameter shaft supported in rigid bearings set 638.725 in. apart is required to have fundamental natural frequency of vibration at 524.53 cycles/s. Determine the shaft material E if density is 0.285 lb./in^3.

Exercise Problem # 12.22.4: Calculate the torsional stiffness of the shaft if material shear modulus $G = 11.5 \times 10^6$ lb./in^2, shaft inner diameter is 3.75 in., outer diameter is 6.875 in. and length is 348.735 in.

Exercise Problem # 12.22.5: A 5.785 in. diameter shaft is supported in rigid bearings set 249.55 in. apart. Shaft material $E = 11.35 \times 10^6$ lb./in^2 and density 0.145 lb./in^3. Calculate the fundamental natural frequency of vibration of the shaft.

REFERENCES & BIBLIOGRAPHY

Alford, J., "Protecting Turbo-Machinery from Self-Excited Rotor Whirl", ASME Journal of Engineering for Power, 87(4):333-334, New York, 1965.

American Petroleum Institute, "Centrifugal Compressors for Petroleum, Chemical and Gas Service Industries", API Standard 617, 6th edition, 1995.

American Society of Mechanical Engineers, "Compressors and Exhausters", Power Test Code # 10, 1965 (reaffirmed 1979), ANSI PTC # 10, 1979.

Begg, I. C., "Friction Induced Rotor Whirl – A Study in Stability", ASME, paper # 83-DET-10, New York, 1983.

Cataford, G. F., Lancee, R. P., "Oil Free Compression on a Natural Gas Pipeline", 86-GT-293, ASME, New York, June, 1986.

Choy, K. C., "Dynamic Analysis of Flexible Rotor Bearing Systems Using A Modal Approach", Ph. D. Thesis, University of Virginia, 1978.

Compressed Air and Gas Institute, "In-Service Standards for Centrifugal Compressors", Cleveland, OH, 1993.

Crede, C. E., Ruzicka, J. E., "Theory of Vibration Isolation", Shock & Vibration Handbook, 3rd edition., Mc-Graw Hill, New York, 1988.

Den Hartog, J. P., "Mechanical Vibrations", Dover Publications, New York, 1985.

Deutschman, A. D., Michels, W. J., Wilson, C. E. "Machine Design – Theory and Practice", MacMillan Publishing Co., New York, 1985.

Ehrich, F. F., O'Connor, J. J., "Stator Whirl with Rotors in Bearing Clearance", Journal of Engineering for Industry, 89(3):381 – 390, August, 1967.

Ehrich, F. F., "Handbook of Rotor Dynamics", Krieger Publishing Co., Malabar, FL, 1999.

Fowlie, D. W., Miles, D. D., "Vibration Problems with High Pressure Compressors", ASME paper # 85-Pet-28, New York, 1985.

Hildebrand, F. B., "Advanced Calculus for Applications", Prentice-Hall, New Jersey, 1986.

Houbold, J. C., Reed, W., H., "Propeller Nacelle Whirl Flutter", Institute of the Aerospace Sciences, 61 – 34, New York, January, 1981.

ISO 11340, "Mechanical Vibration – Methods and Criteria for the Mechanical Balancing of Flexible Rotors", 1994.

ISO 4867, ISO 4868, "Guidelines for Evaluation of Vibration in Merchant Ships", after 6954, 1993.

ISO 2631, "Balancing of Rotors", 1979.

ISO 6945, "Guidelines for Evaluation of Vibration in Merchant Ships", 1987.

ISO 1940 and ANSI S2.19, "Balancing Quality of Rotating Rigid Bodies", 1986.

ISO 2954, "Mechanical Vibration of Rotating and Reciprocating Machinery – Requirements for Instruments for Measuring Vibration Severity", 1984.

ISO/DIS 2953, "Balancing Machines – Description and Evaluation", 1984.

ISO/DIS 10816, "Measurement and Classification of Vibrations in Rotating Machines", ANSI counterpart S2.12 part 6-199X, 1993.

ISO/WD Standard 8579-2, "Vibrations in Enclosed Gear Reduction Units", 1981.

ISO 1940/1, "Mechanical Vibrations - Balance Quality Requirements of Rigid Rotors - Part I, Determination of Permissible Residual Unbalance", International Organization for Standardization, Geneva, Switzerland, 1986.

Kimball, A. L., "Internal Friction Theory of Shaft Whirling", General Electric Review, 27(4): 244, April, 1924.

Linn F. C., Prohl, M. A., "The Effect of Flexibility of Support upon the Critical Speeds of High Speed Rotors", Transactions, Society of Naval Architects and Marine Engineers, 59:536-553, 1951.

Mabic, II. II., Rcinholtz, C. F., "Mcchanisms and Dynamics of Machincry", John Wilcy & Sons, Ncw York, 1987.

Maslen, E. H., Sortore, C. K., Vazquez, J. A., Knopse, C. R., "Synchronous Response Estimation in Rotating Machinery", ASME paper # 00-GT-397, New York, 2000.

McCloskey, T. H., Adams, M. L., "Troubleshooting Power Plant Rotating Machinery Vibration Problems Using Computational Techniques: Case Histories", Institute of Mechanical Engineers, # C432/137, page 233 –244, London, 1992.

Military Standard, "Mechanical Vibration of Shipboard Equipment (Type 1 – Environmental and Type 2 – Internally Excited)," MIL-STD-167-1 (SHIPS), May, 1984.

Muszynska A., "Whirl and Whip – Rotor/Bearing Stability Problems", Journal of Sound & Vibration, 110(3):443-462, 1986.

Prohl, M. A., "A General Method for Calculating the Critical Speeds of Flexible Rotors", Journal of Applied Mechanics, page 142 – 147, 1945.

Rangwala, A. S., "Turbo-machinery Dynamics: Design and Operation", McGraw-Hill, New York, 2005.

Roark, R. J., Young, W. C., "Formulas for Stress and Strain", 5th edition, McGraw-Hill, New York, 1989.

Sawyer, T., "Gas Turbines – Volumes I, II and III", International Gas Turbine Institute, Atlanta, ASME, 1984.

Salamone, D. J., Gunter, E. J., "Synchronous Unbalance Response of an Overhung Rotor with Disk Skew", ASME paper # 79-GT-135, New York, 1979.

Schneider, H., "Balancing Technology", 4th edition, Carl Schenck, AG, Darmstadt, Germany, 1991.

Stadelbauer, D. G., "Fundamentals of Balancing", 3rd edition, Schenck Trebel Corp., Deer Park, NY., 1990.

Timoshenko, S., Young, D. H. and Weaver, W., "Vibration Problems in Engineering", John Wiley & Sons, Inc., New York, 1974.

13

CHAPTER

STEADY AND DYNAMIC STRESSES

13.1 INTRODUCTION

Mechanical and thermal stress buildup occurs in every load-carrying component of a turbine, be it due to thrust or torque developed, combustion of fuel or the flow of gases flowing over the surfaces. Some of the loads, and hence the associated stresses, are unidirectional and remain at a fairly constant level when the unit is operating at a steady state. More often than not, however, the loads tend to alternate in direction or magnitude or both, and consequently the stresses produced will be dynamic in nature. And as so often happens, the total stresses are a combination of both steady and dynamic components.

Uniform operation of the turbo-machine occurs in aero engines when an aircraft reaches cruise altitude and the autopilot takes over to maintain operation at a steady speed in a fixed direction. Depending on the distance between the departing and destination airports, the aircraft's power plants will keep operating at a consistent level for a number of hours till the cockpit crew manually activates the control mechanisms. On the other hand, power generation steam and gas turbines are more likely to be run at a steady pace continuously for thousands of hours. Except for shifts in power demands on the electrical generator arising from periodic variations during the day, the engine system and hence many of its components will not be subjected to substantial load changes and the related stresses.

Depending on the magnitude and frequency, alternating stresses exact far greater toll on the material characteristics of a part than steady state stresses. Even if the dynamic stresses are well below the yield or ultimate strength capacity, stresses that alternate from one direction to the other will lead to failure of the part after a certain number of load cycles. A steady and uniformly acting load can be applied for prolonged periods of time without the part being compromised even if the load produces stresses approaching the material's yield point.

13.2 MECHANICAL STRESSES IN SHAFT

A rotating shaft subjected to a fixed downward load and causing it to bend is a case in point for dynamic, or alternating, stress. For example, aerodynamic radial thrust is produced when a turbine rotor is not centered with the stator casing. So a downward load produces tensile stress in the upper fibers of the shaft and compressive stress on the lower surface fibers. The shaft may not whirl about the unbent axis, but as it turns 90°, the upper surface fibers reach the neutral axis and will have no stress. Turning the shaft through another 90° will bring it towards the bottom, and will now be seeing compressive stress.

Further turning of the shaft causes the upper fibers to once again cross the neutral axis and back to its beginning after one complete turn. Thus, even though the load on the shaft is in a fixed direction, the spinning shaft goes through one complete reversal from tension to compression state of stress.

Flexural bending of the shaft results from rotor unbalance. Since the unbalanced mass spins with the rotor, the centrifugal force vector is also rotating, and causes dynamic stress in the shaft. The bending vibrations are synchronous with the rotor speed, and the frequency of the alternating stress will match the spin speed.

Pulsating torque in the shaft is another source of dynamic stress. The propeller shaft of a turbo-prop engine is coupled with the power turbine shaft through a reduction gear. The propeller blades, gears and the turbine rotor have their own natural frequencies of vibration. At certain engine speeds, the propeller's vibration modes may interact with those of the gears and the turbine rotor to set up torque pulses. Torsional vibrations cannot be readily detected from outside in the turbine rotor since the bearings do not participate. The torque pulses are superimposed on the regular spinning motion of the rotor. But the torque pulses may be strong enough to produce high amplitudes of torsional stresses in the propeller shaft, gears and turbine rotor.

13.3 CENTRIFUGAL FORCE IN ROTATING BLADE

Compressor and turbine blades are subjected to centrifugal force due to their mass. The force, $Wr\omega^2/g$, where W is the weight of the airfoil section, r is radial location of the centroid of the airfoil section, ω is angular velocity and g is gravitational constant, is not uniform along the length of the blade since it depends on its section where the force is computed. Thus, at the blade's tip the force is zero, increasing gradually along the length and reaches a peak at the base where it meets the dovetail platform and where the total mass of the blade is to be taken into account.

Centrifugal force in the airfoils is a function of the square of the angular velocity of the rotor. Thus, at start of the unit the force is absent. As the machine picks up speed the centrifugal force climbs up. As an aircraft's turbo-fan engine accelerates to ground idle, flight idle, cruise and finally to takeoff speed, the centrifugal force in all of the airfoils will correspondingly increase. As the aircraft settles into its cruising altitude, the speed of the rotor once again reduces. Hence, when the rotor speed is not substantially varying, the centrifugal force in the rotating blades will remain at a fixed level. Power generation and industrial gas turbines need to operate at a fixed speed over sustained periods of time to enable the generator to produce electrical signal at the required frequency. Therefore, the centrifugal force in the blades will remain at a constant level, and the stress developing as a consequence at the various cross-sections along the height may be identified as steady stress.

There is another interesting aspect of the centrifugal force in a rotating blade that needs to be looked at closely, and that is its effect on the natural frequencies of vibration of the airfoils. To understand the phenomenon, consider a piano wire with one end fixed. The other end is wound around a pin. During tuning of the piano's musical notes, the tuner turns the pin to adjust the tension in the wire. As the wire is struck with the piano keys, the musical note is compared with the note of that produced by a tuning fork. The musical note from the piano wire changes because its natural frequency of vibration changes with the tension as the pin is turned. A similar situation exists in the case of a turbine blade, with one exception: the tension in the piano wire is uniform along its length, but the centrifugal force in the blades varies from zero at the blade tip to a maximum at the root.

The effects of the centrifugal force may be observed on a Campbell diagram of the blade's dynamic characteristics of the type shown in Figure 13.1. At start the engine speed gradually increases, the related centrifugal force is low and there is negligible effect on all of the natural frequencies of vibration of the airfoil. Acceleration of the engine towards higher speed ranges increases the internal centrifugal force, and causes the natural frequencies of vibration of all modes to increase in magnitude and the frequency curves to steadily climb up.

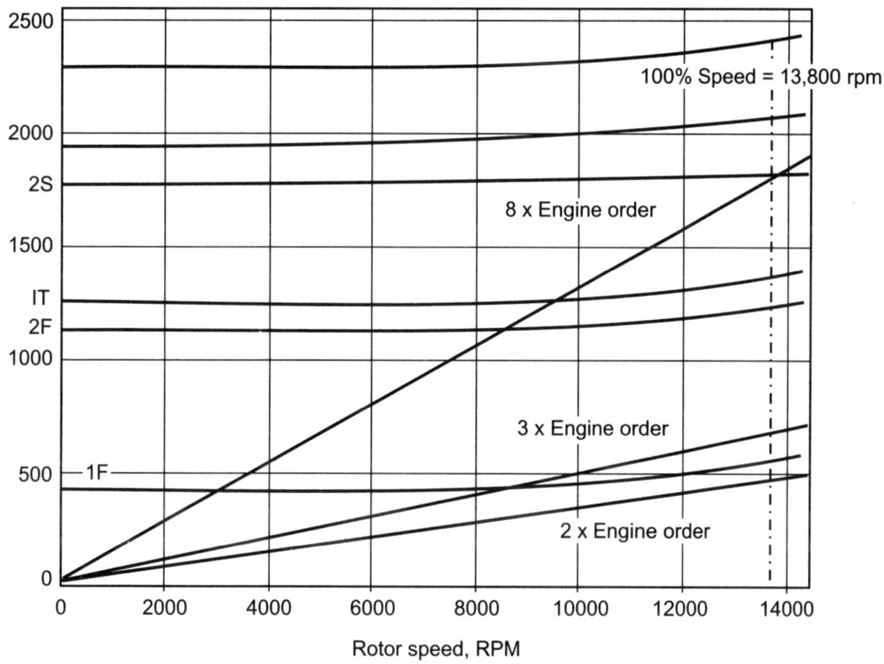

Fig. 13.1 Campbell Diagram of Fan Blade

This effect may be referred to as centrifugal stiffening of the airfoils. An associated effect of the enlarged centrifugal force may also be limited untwisting of the twisted geometry of the blades. It may also be argued that metal temperature in the turbine section increases as the engine speed goes up, which in turn may lead to some level of deterioration in the Young's modulus of the blade substrate. This effect will in turn reduce the stiffness, and hence the frequency of the airfoil, thus compensating the effects of centrifugal stiffening at least to some extent. In the case of compressor airfoils, air and metal temperatures do not soar to the same level and will have negligible compensating effects.

13.4 GAS PRESSURE LOAD ON AIRFOIL

Air pressure acts on both the suction and pressure surfaces of a compressor blade, as also pressure of the products of combustion on turbine blades. Figure 13.2 shows the profile of a fan blade at the tip. When the turbine engine is operating in the steady state mode, pressure on both surfaces remains at a reasonably uniform level.

Generally there is not a considerable difference in the magnitude of the pressure on the two sides. So it may be surmised gas pressure by itself does not create much stress in the blades, especially when compared to the stress produced from centrifugal effects in rotating blades. Centrifugal force being a function of the square of the speed, the stress at the root of the airfoil will be of greater significance at maximum rotor speed when compared with that due to gas pressure.

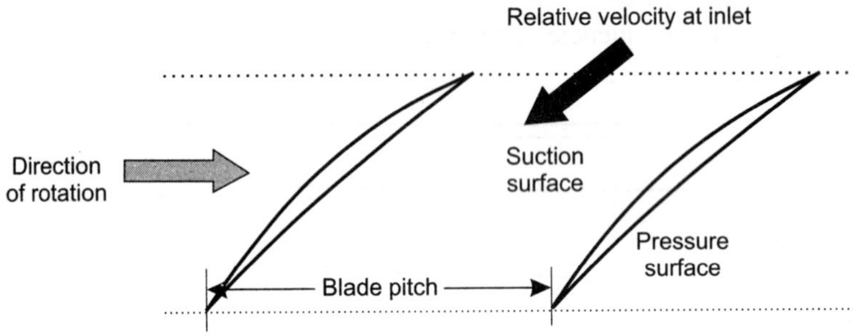

Fig. 13.2 Fan Airfoil Profile at Tip

13.5 DYNAMIC STRESS IN ROTATING BLADE DUE TO AERODYNAMIC INTERFERENCE

There is another aspect of gas pressure on the blades that arises as the rotating blades move in the flow stream past the stationary vanes. As the flow traverses along the surfaces of the vanes the streams are fairly uniform and free of turbulence. At the trailing edge, however, small eddy currents tend to form as the flow is shed and enters the dead space prior to entering the rotating blades. This causes some variation in the pressure in the stream, creating a finite region in its wake where the pressure will not be at the same consistent level. Figure 13.3 helps in understanding the pressure variation as the rotating blades cross the path past one stationary vane to the next.

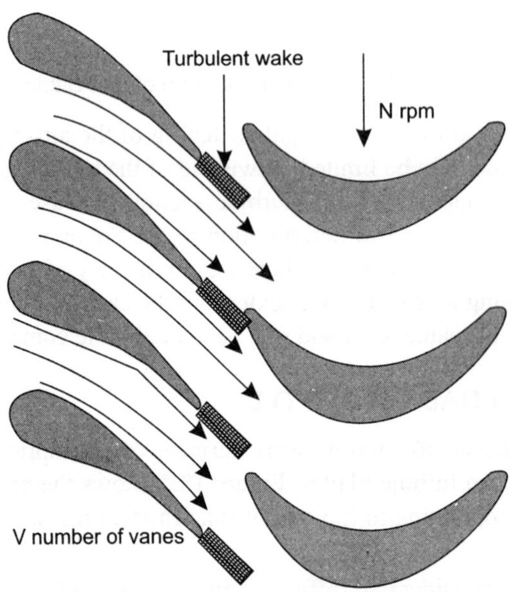

Fig. 13.3 Interaction between Rotating Blade and Wake Shed from Stationary Vane

If there are V stationary vanes, each rotating blade will run into V pockets of turbulence in one turn of the rotor. The variance in the pressure field of the stream will result in corresponding shifts in the stress developed in the blades. The magnitude of stress variation will depend on the difference in the gas pressure in the passage between the vanes and that in the wake shed by the vanes. So if the speed of the rotating blades is N rpm, each blade will then encounter $(V \times N)/60$ stress variations in one second. In the parlance of dynamic stress, the Vth excitation order is going to be of significance. Thus, on the Campbell diagram if the curves corresponding to a particular mode of blade vibration and the Vth excitation order intersect in the operating speed regime, the condition represents a potential source of failure for the rotating blades. The prospect of failure of the blades may be determined from the high-cycle fatigue characteristics of the blade material. As a rule, however, best practice calls for a blade design where the Vth excitation order and the blade vibration frequency curves do not meet in the operating speed regime.

It may be noted that the excitation order of the immediately preceding vane is not the only one that may affect a rotating row of blades. Excitation orders of any row of stationary vanes preceding or following a given rotating stage has been known to have the potential to create damage from high-cycle fatigue. But the damage due to the immediately preceding vanes is most likely to be greater than from other stages.

13.6 THERMAL STRESS IN TURBINE BLADE

Since the inlet temperature at the turbine has a large impact on the power produced in the unit, every effort is made by the turbine manufacturers to maximize it. Super alloys used to make the rotating blades are selected on the basis of most resistance to elevated temperatures. High operating temperatures in the turbine are possible because of the availability of super alloys. It may even be said that the development of super alloys has proceeded with the need to resist high temperatures in the turbine, which in turn permitted the turbines to operate at ultrahigh temperatures.

Thermal barrier coatings and cooling of the rotating airfoils from within has made operation of the turbine at temperatures close to the melting point of the blade substrate. Regardless of the multiple forms of protection, failure of the blades from thermal fatigue is a major problem faced by the turbine industry.

To understand the nature and development of thermal stresses, consider three relatively simple one-dimensional cases. In the first case, a steel bar of length L and Young's modulus E is clamped at both ends. Uniformly heating the bar from T_1 to T_2 will cause the bar to increase in length. Since the ends are clamped, this causes the bar to bend in the middle. From Hook's theory, $\sigma_{th}/\varepsilon_{th} = E$, where the suffix th indicates thermal stress and strain. Now thermal growth $= \alpha L \Delta T$, where α = coefficient of thermal expansion and $\Delta T = (T_2 - T_1)$. So $\varepsilon_{th} = \alpha L \Delta T / L = \alpha \Delta T$. Hence, thermal stress $\sigma_{th} = E\varepsilon_{th} = \alpha E \Delta T$. In the second case the same bar is lightly supported with a hinge at one end. Uniformly heating the bar from T_1 to T_2 will let the bar expand freely along the length, so there is no imposed load and thermal stress $\sigma_{th} = 0$. In the third case, the bar is lightly supported with a hinge at one end as in second case, but instead of uniform heating the bar is kept at one end at temperature T_1 and the other end at T_2. Even though the bar grows thermally without restriction, the temperature gradient along its length will produce thermal stress as in case one, $\alpha E \Delta T$.

The second and third case discussed above may be applied to a turbine blade since it is attached at one end to the disk with the dovetail. As the turbine engine is started, the blade is at temperature T_1.

Assuming uniform speed increase and warm up, the blade metal temperature will rise to T_2 when full speed is reached. During this acceleration period, the thermal stress developed will be as in the third case, $\sigma_{th} = \alpha E \Delta T$. During steady operation, gas and metal temperatures may be assumed to remain constant since the temperatures soak through the thickness and become uniform in the blade. So the thermal stress will reduce to zero, as in the second case. When the engine is to be shut down, the reverse of the acceleration will take place, and the thermal stresses will once again climb up but with one exception: during the heating period thermal stresses will be in tension as the metal fibers stretch and in compression during cool down.

13.7 INDIVIDUAL BLADE VIBRATION

Essentially a cantilever beam attached at its base through a dovetail to the rotating disk rim, compressor and turbine blades have natural frequencies of vibration and their own characteristic modes. The frequencies, mode shapes and related details about the dynamic behavior of the blades may be computed using finite element computer programs such as ANSYS. Several modes starting from the primary or the fundamental mode may be determined. But only the first few natural frequencies of vibration generally match those obtained from experimentally measured ones. The blades are excited with electro-magnets placed at various locations of the airfoil, with response measured with dynamic strain gauges. At higher frequencies, the calculated and measured frequencies tend to diverge. The associated mode shapes can also be identified to indicate torsional, flexural and stripe form of dynamic activity in different regions of the airfoil. Once again, higher modes tend to involve a combination of activities of the basic modes, and labeling of the mode shapes then becomes difficult.

A few notes pertaining to the dynamic analysis may be mentioned. The finite-element mesh created for stress analysis due to centrifugal effects will generally not be convenient for dynamic analysis. Since stresses tend to concentrate in regions where airfoil geometry is sharply changing, such as in the fillet radius at the root close to the trailing edge, a fine mesh is required in the area since the stresses rapidly change. In dynamic analysis, on the other hand, the fillet is short in length and does not contribute to the stiffness in the same region. So the fillet may be deleted from the solid model. Also, the element size does not need to be of fine size. A comparatively gross mesh should suffice, and this helps to reduce computation time since dynamic analysis is inherently more computer-intensive than stress analysis. Precautions are needed, however, to ensure the mesh is a good representation of the overall geometry of the airfoil.

Another point may be made about the effects of metal temperature on blade dynamics mentioned earlier. When thermal analysis is performed immediately following stress analysis, temperature distribution data throughout the airfoil is available. Through a process of iteration within the computer program, the temperature data may be calculated at the node points of the dynamic model from the data obtained earlier at the node points of the stress model. In addition, many of the finite-element programs include the feature of all of the material property data for the blade substrate at various temperatures. Thus, the material properties at a given node point will reflect values at its own temperature.

Likewise, the effects of centrifugal stiffening as the rotor speed increases may also be simulated when speed increment steps are included. At the start of the calculation sequence the speed is zero, then gradually increased in increments. At each step the centrifugal force is established throughout the airfoil, and from it the associated increase in stiffness.

13.8 TIP AND MID-SPAN SHROUDS TO REDUCE BLADE TWIST

The initial stages of the compressor and the later stages of the turbine are longer in length due to the lower fluid pressure flowing over the surface. The same is true of the steam turbine, where the last stage of the low-pressure module may be 48 in. long if made from a nickel-chrome alloy. Some steam turbine manufacturers also provide titanium blades. Titanium offers the advantage of lower density than steel but greater than that of aluminum. The strength of titanium is also between that of aluminum and steel, and the blades are usually 40 in. long.

The long and slender blades tend to lack torsional stiffness. Interlocking shrouds are used for turbine blades in aviation engines to provide additional resistance against twisting. One method calls for Z-shaped shroud at the outer tip with tongue and jibe arrangement that engages with adjacent blades on both sides. Essentially the blades lean against each other at the shroud to limit the twist in the airfoil. Clearances are provided to permit assembly. When the engine is running, centrifugal force causes the shrouds to form a nearly continuous ring.

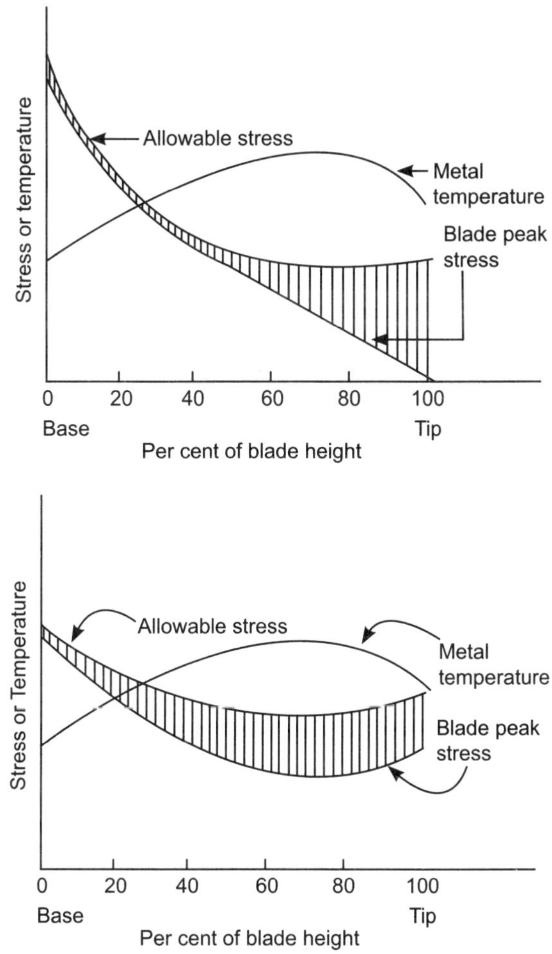

Fig. 13.4 Stress and Temperature Distribution in Free Standing (*upper*) and Shrouded (*lower*) Turbine Blades

The shrouds play the important of role of increasing the stiffness and limiting the angle of twist, and thus reduce the dynamic torsional stress in the airfoils. Figure 13.4 illustrates development of peak stress in a turbine blade standing freely and the same blade with shroud at the tip. As much as possible, turbine manufacturers avoid using shroud since it adds to weight, a definite demerit in aviation applications. Also, the shroud produces its own centrifugal force that has to be borne by the full length of the blade, and adds to the associated stress. At the same time, the shroud adds to the manufacturing cost. In spite of these negatives, it is necessary to use the shroud in order to enhance the integrity of the rotating blades from torsion-induced fatigue.

Steam turbine blades are even longer and slender. The last stage blades are the most productive, producing as much as 15 per cent of the total power, and much attention is devoted during the engineering and manufacturing of the blades. Besides the one at the tip, shrouds for the last stage LP section blades need to be provided at one-third span from root and at two-third span from the root. Various forms of shroud are in use. The one-third and two-third span shrouds may have a continuous 0.5 in. diameter tie-wire that goes through holes with boss in the airfoils. Other designs call for protrusions from the suction and pressure surfaces of the blades that lean against protrusions on adjacent blades to limit twisting and in plane and out of plane bending. At the tip the shroud may be made of metal strip in the form of 180° arc. One or two tenons are provided at the blade tip, which fit into similar shaped recesses machined in the strip.

Since steam turbines operate at much lower temperatures than their counterpart in gas turbines, their life expectancy is considerably higher. In the high-pressure section maximum temperatures of superheated steam is between 1050° and 1100° F. In the last stage of LP turbine the steam is close to saturation point. Droplets of condensate are formed that are flung towards the outer diameter of the blades. Millions of the droplets can cause erosion at the outer one-third at the leading edge. Protection from erosion is usually available with stellite strips attached at the edge. The stellite strips need to be periodically replaced to ensure high aerodynamic performance of the blades.

13.9 TANGENTIAL STRESS IN ROTATING DISK WITH ATTACHED BLADES

Turbine, compressor and impeller disks are subjected to loads arising from the centrifugal force of attached blades, rim lugs and bolts, and due to radial forces caused by the weight of the disk itself. In addition, the disk also provides a load path for attached shafts and spacers at the forward and aft ends. Steady and transient thermal gradients imposed on the disk create additional stresses in the disk. Disk material characteristics such as Young's modulus and thermal coefficient of expansion will play a major role in determining stresses throughout the geometry, as also the load carrying capability.

The circumferential direction load on the rim of the disk is given by the expression $P_c = [\rho\omega^2 R^2 A_{rim}/g]$ for a ring of radius R and cross-section A_{rim} spinning at ω angular speed (disk material density is ρ), and the corresponding hoop stress is given by $\sigma_t = [\rho\omega^2 R^2/g]$. Since the rim is also subjected to F_{rim} due to the mass of the blades, total tangential stress in the rim is given by: $\sigma_t = [\rho\omega^2 R^2/g] + F_{rim}/(2\pi A_{rim})$.

The hoop stress thus determined will be considerably higher than the ring's strength capacity. The ring's axial depth does not substantially affect the average tangential stress over the cross-section, since it mostly depends on the area A_{rim}. A disk is needed to reduce the stresses. Axial and shear stresses in the disk are negligible. Tangential and radial direction stresses will be axi-symmetric due to the symmetry in the radial loading. Radial stress at the disk's free bore will be zero if it is not shrunk on a shaft.

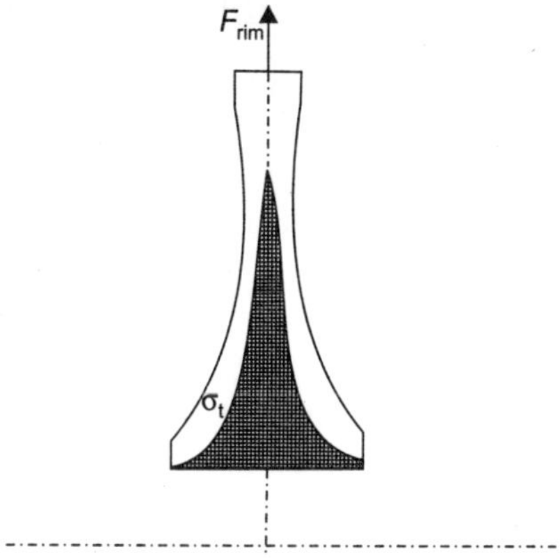

Fig. 13.5 Hoop Stress Distribution in Symmetric Rotating Disk with Center Bore

Many turbo-machinery applications call for a free bore to permit other components to pass through it, such as a center vent tube in aircraft engines. Where a disk is without a center hole both radial and tangential centers are non-existent at the center of the disk. With a free bore, maximum hoop direction stress in the disk's center bore, as shown in the symmetric disk in Figure 13.5. At steady operating conditions, the stress will be uniform since operating speed does not vary.

13.10 DISK BURST CAPABILITY DUE TO OVER-SPEED

An example of rotor increasing rapidly in speed is when a large amount of electrical load is abruptly shed due to a trip in a customer's large aluminum furnace. The electrical generator and the attached steam turbine would then go into an over-speed condition. To protect against the turbine running away in speed, speed control mechanisms are provided. The speed controller reduces the amount of steam entering the high-pressure turbine section, which in turn brings down the speed of the turbine and the attached generator.

But on occasion the speed control mechanism fails to perform its stated task. Failure may arise from corrosion weakening the links, springs and other functional elements of the speed controller.

Failure of the speed controller is the primary cause of turbines going into the over-speed regime, although malfunction in the inlet valve may be a contributing factor or may even add to the problem. As explained in the previous section, the limiting speed of the disks in the rotor assembly is established from the maximum hoop stress in the disk. When the hoop stress limit is reached in any of the several disks of the turbine unit, the disk will burst with catastrophic consequences. Incidents have occurred when the entire turbine installation is destroyed. The turbine unit will be destroyed totally. In addition, huge losses in life, property and related damages may be expected in the event of a rotating disk exceeding its burst capability due to failure to protect the turbine from going into over-speed condition.

Turbine manufacturers routinely carry out tests inside a pit to establish the burst speed. The final design of the disk is used, and the disk must have blades inserted for the tests. Since the turbine may use several disks, all of the disks must go through the test. The disk with the lowest burst speed will then establish the safe over-speed limit for the turbine.

13.11 ROTATING COMPONENT STRESS BUILDUP AND OPERATING LIFE

Many factors need to be closely evaluated in the design of blades and disks of turbo-machinery. Aside from efficiency, net power developed and exhaust gas emissions, the operating life of the rotating components top the list for aviation engines. The parts must be retired and replaced after a certain number of operating cycles or operating hours. Other aspects such as manufacturing cost and weight may then be of lesser significance.

A major factor is stress due to high cycle fatigue (HCF) in rotating blades. The primary role of the blades is to provide flow path and direct the fluid medium to maximize aerodynamic efficiency, but at the same time structurally be sound enough to withstand centrifugal, thermal and dynamic stresses. Manufacturing tolerances are needed on all dimensions. Sometimes referred to as mistuning, the dimensions may be within the tolerance limit. However, the natural frequencies of the airfoils will vary with the actual dimensions, while calculated vibration frequencies are calculated from nominal dimensions. The role of mistuning in HCF failures has been recognized and accepted throughout the industry.

Transient response of blades during startup and shutdown of the machines is another significant contributor. Failure due to resonance at lower modes such as the first bending and torsion may include fracture of an entire blade and its attachment to cause substantial secondary damage, and even greater if the fragments are not contained in the engine. In aviation engines, failure to contain the fragments and penetration of the casing is an important cause of concern since the fragments could hit and puncture the fuselage. At the higher second torsion or second and third bending modes mostly the outer portion of the blade is separated. At still higher modes only a portion of the tip maybe released.

13.12 CUMULATIVE DAMAGE ASSESSMENT IN COMPONENT LIFE PREDICTION

A number of cumulative damage theories have been proposed in making estimates of the life of a blade subjected to variable stress amplitude. A commonly used linear method in predicting fatigue life of a blade is based on work done by Palmgren and Miner. Dynamic stress around a critical speed may be split into a convenient number of steps on either side of peak stress. Operation at a stress level S_i gives a life of N_i cycles. If the blade is subjected to n_i cycles, it suffers a damage fraction of $D_i = n_i/N_i$. Failure is then predicted to occur when $\Sigma(n_i/N_i) \geq 1$. The assertion is that the damage fraction at any stress level is linearly proportional to the ratio of number of cycles of operation to the total number of cycles that would produce failure at that stress level. Since the blade is subjected to a mean stress, the $S - N$ plane is shifted to the location of the applied mean stress level on the fatigue failure surface.

A drawback to this theory is that it does not recognize the order of application of various stress levels, and damage is assumed to accumulate at the same rate at a given stress level without consideration of past history. Ample experimental evidence has been obtained to indicate non-linear accumulation of the fatigue damage, dependent on the alternating stress level.

13.13 OPERATING TIME AND CYCLES IN ESTABLISHING LIFE LIMIT

The operating life of rotating components generally must be determined both in terms of start and stop cycles as well as in number of operating hours. The need for this may be recognized from service conditions of the units. For instance, industrial and power generation gas turbines and steam turbines mostly operate over sustained periods of time continuously. Aviation engines, on the other hand, are made to run for far shorter duration.

Even in aviation applications, flight times vary considerably. Larger turbo-fan engines may run for several hours during transatlantic and trans-continental flights. Commuter flights equipped with turbo-prop and many turbo-fan engines may run for under 500 miles, covering the distance in less than one hour.

In the case of marine and offshore oil platforms, engines derived from aviation engine technology are used. Widespread disparity is observed in the matter of continuity of operation. Power generation and gas pipeline pumping applications call for the turbine engines to operate for periods running into thousands of hours. But naval vessels use the gas turbines sparingly for making rapid burst in speeds, relying more on diesel engines during regular operation.

A distinction based on the type of service thus needs to be made when the useful operating life of the components is to be established. Short start and stop cycles are bound to create greater damage to the components than those subjected to cycles of longer duration. Thermal fatigue in the hot path components is severe, with the severity taking a major toll on the component as the number of cycles increase. Variations in the centrifugal force in the compressor airfoils as the rotor starts and stops will also cause severe damage to the parts.

13.14 UNCONTAINED FAILURE FROM FRACTURE OF FAN HUB

In July of 1996, a McDonnell Douglas MD-88 experienced an engine failure during the take-off roll. Debris from the fan hub of the left engine penetrated the fuselage (Figure 13.6), and killed two passengers and seriously injured two others. The airplane was equipped with two Pratt & Whitney JT8D-219

Fig. 13.6 Damaged Airplane Area near Failed Engine (NTSB AAR-98/01)

turbofan engines. This engine has an axial flow fourteen stage split compressor, a nine can combustion chamber and a split four-stage reaction impulse turbine. The titanium fan hub had accumulated 16,542 hours and 13,835 operating cycles at the time of the accident. Service life of the fan hub is limited to 20,000 cycles.

The hub consists of a disk forging with dovetail slots for 34 fan blades. The aft end of the hub attaches to the stage 1.5 disk with 24 tie rods that pass through 0.5175 in. holes drilled in the rim just inside the dovetail slots. The 2.91 in. deep tie rod holes are located around the circumference of the hub bore that alternate with an equal number of smaller diameter stress distribution holes. The fan hub is forged from a titanium base alloy containing 6 per cent aluminum and 4 per cent vanadium. Figure 13.7 provides details of the fan hub.

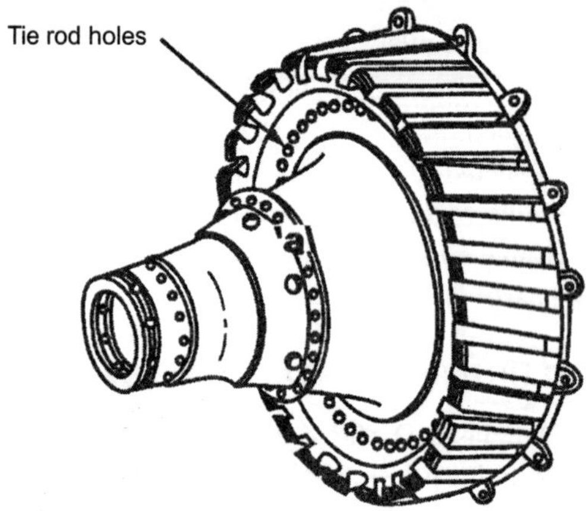

Fig. 13.7 Engine Fan Hub (NTSB AAR-98/01)

Engine debris was scattered along the airline's path. The nose bullet was found on the runway. The fan hub and blade assembly were separated from the engine, and the surrounding engine outer case and cowl were ruptured. The hub was separated at a 360° circumferential fracture located just forward of the of the stage 1.5 disk bore. The integral spacer had fractured into five pieces, while the fan hub fractured into three large pieces and a small fourth piece remaining in the number one bearing assembly. The largest piece, comprising nearly two-thirds of the hub rim and the adjoining conical section was located 700 feet to the left of runway centerline, shown in Figure 13.8. Another piece of the rim was found half mile away. The third piece was embedded in the right side fuselage.

The hub fractured through a tie rod hole and blade slot. Two fan blade roots were still in place on the small rim segment and thirteen roots on the larger rim segment. Three of the thirteen full-length blades were bent counterclockwise as viewed from aft looking forward. Three tangs on the hub rim near the dovetail slots were sheared. Twelve fan hub tie rods were recovered, appearing uniformly sheared near bolt head.

The hub rim's fracture surfaces were examined by a Safety Board metallurgist at the accident site and in the laboratory, and were found to have evidence of fracture cracking. The hub fractured radially in two locations. One radial fracture contained a fatigue crack that originated at two locations on the inboard side of a tie rod hole. Both initiating points were located within the hole for the tie rod about

Fig. 13.8 Fractured Hub Rim with Attached Conical Shaft (NTSB AAR-98/01)

0.4 in. from the aft edge of the hole. Fatigue fracture features extended a maximum of 1.5 in. hole radially towards the center of the engine. Outside the fatigue region the fracture features were consistent with an over stressed separation. Metallurgical examination of the surface of the wall of the hole revealed an area in which the surface finish was darker than the surrounding area at each fracture origin, displaying evidence of circumferential machining marks consistent with machining marks associated with boring operation performed during manufacture. There was no indication of honing in the dark areas. The remainder of the wall surface of the hole indicated a cross hatched pattern consistent with marks from a honing operation during manufacture. Magnified examination of the dark areas of the surface of the hole also showed a number of small parallel surface cracks aligned with the axis of the hole. A scanning electron microscope examination of the fracture face in the origin areas showed evidence of over-stress to a depth of 0.002 in., followed by an area about 0.006 in. deep that contained fracture features consistent with a fast-propagating fatigue crack.

According to the manufacturer's record, the imperfections in the tie rod holes were noted as chatter marks following the drilling operation, but were recorded as manufacturing marks after the subsequent boring and honing procedures. The marks were deemed to meet manufacturing standards, and were accepted. As a life-limited part, the airline operator must inspect the fan hub if the part was removed during engine overhaul using visual and fluorescent penetrant inspection procedures. The accident fan hub was removed from another engine following foreign object damage to the fan blades, and then installed in the accident engine. But the airline's inspector failed to detect a detectable crack.

13.15 REVIEW QUESTIONS

13.15.1 Discuss the differences in operation between power generation and shipboard gas turbines.

13.15.2 Does unbalance in a rotor affect static and dynamic stresses during steady operation?

13.15.3 How does wake shed from a stationary vane affect integrity of rotating blades?

13.15.4 What is centrifugal stiffening? What effect does it have on the natural frequencies of an airfoil?

13.15.5 Explain the merits and demerits of a shroud at the tip of a rotating blade.

13.15.6 Why must a disk be provided on a ring with attached blades?

13.15.7 What methods are used to determine the maximum speed capability of a turbine rotor?

13.15.8 What are the primary factors that cause a turbine to operate in a regime beyond its normal speed range?

13.15.9 What is mistuning, and what is its role in determining fatigue life of rotating airfoils?

13.15.10 Besides stress analysis calculations, what other factors are taken into consideration when establishing operating life of rotating components?

13.15.11 Why is estimation of component life prescribed in terms of operating hours not sufficient for helicopter engines?

13.16 EXAMPLE PROBLEMS

Example Problem # 13.16.1: Discuss the merits of various theories of elastic failure.

Solution: The 'maximum principal stress theory' may be applied to most brittle materials such as cast iron. The 'maximum principal strain theory', on the other hand, holds little significance. The 'maximum shear stress theory' is widely used for ductile materials, especially for a rotating shaft experiencing a combination of bending and torsion.

Experimental results on ductile materials tend to support the 'total strain energy theory', but are more in agreement with the 'Mises-Hencky criterion'. The latter finds extensive usage for the design of mechanical components.

Example Problem # 13.16.2: A steel alloy has a proportionality limit of 71×10^3 lb./in^2 in simple tension. In a two-dimensional stress system the principal stresses are 17.3×10^3 lb./in^2 tensile and 5.9×10^3 lb./in^2 compressive. Determine the factor of safety from the maximum shear stress theory.

Solution: The stresses are $\sigma_1 = 17.3$, $\sigma_2 = 0$ and $\sigma_3 = -5.9$

By the maximum shear stress theory the equivalent single tensile stress is:

$$s = \sigma_1 - \sigma_3 = 17.3 - (-5.9) = 23.2 \times 10^3 \text{ lbs/in}^2.$$

So the factor of safety is 71/23.2 = 3.06.

Example Problem # 13.16.3: Determine the factor of safety from the Mises-Hencky theory.

Solution: The Mises-Hencky theory for combined stresses is:

$$\begin{aligned} 2s^2 &= (\sigma_1 - \sigma_2)^2 + (\sigma_2 - \sigma_3)^2 + (\sigma_3 - \sigma_1)^2 \\ &= (17.3 + 0)^2 + (0 + 5.9)^2 + (-5.9 - 17.3)^2 \\ &= 872.34 \end{aligned}$$

Hence: $s = 20.88 \times 10^3$ lbs/in^2

The factor of safety $= 71/20.88 = 3.40$

Example Problem # 13.16.4: Discuss the greatest principal strain theory of St. Venant.

Solution: **Failure** refers to elastic breakdown and onset of permanent strain. The stress s at which this occurs in simple tension may be assumed to be the limit of elastic proportionality. Consider a three-dimensional complex stress system, where the principal stresses are σ_1, σ_2 and σ_3 in descending order, tensile being positive. The greatest principal strain theory of St. Venant considers the greatest strain as the relevant quantity. In this case the value is (using E for Young's modulus):

$$\varepsilon_1 = \sigma_1/E - \sigma_2\nu/E - \sigma_3\nu/E$$

In simple tension, the strain is s/E. By equating the strains, the expression for stress is:

$$s = \sigma_1 - \sigma_2\nu - \sigma_3\nu$$

Example Problem # 13.16.5: Discuss the Coulomb's maximum shear stress theory.

Solution: Coulomb's maximum shear stress theory is based on the maximum shear stress on an interface being half the difference of the corresponding principal stresses, or $(\sigma_1 - \sigma_3)/2$ for the complex stress system and $s/2$ when in simple tension. Hence, in the Example Problem # 13.16.4, using this theory, $\sigma_1 - \sigma_3 = s$.

13.17 EXERCISE PROBLEMS

Exercise Problem # 13.17.1: A certain steel has 0.2% yield strength of 123.5×10^3 lbs/in^2 in simple tension. If two-dimensional principal stresses are 21.85×10^3 lbs/in^2 tensile and 15.8×10^3 lbs/in^2 compressive, calculate the factor of safety from the maximum shear stress theory.

Exercise Problem # 13.17.2: Determine the factor of safety from the Mises-Hencky theory.

REFERENCES & BIBLIOGRAPHY

Corten, H. T., Dolan, T. J., "Cumulative Fatigue Damage", Proceedings, International Conference on Fatigue of Metals, ASME and IME, page 235, New York, 1956.

Hall, R. M., Armstrong, E. K., "The Vibration Characteristics of an Assembly of Interlock Shrouded Turbine Blades", Structural Dynamics Aspects of Bladed Disk Assemblies, Srinivasan, A. V., editor, ASME, New York, 1976.

Harris, C. M., Crede, C. E., "Shock and Vibration Handbook", Mc-Graw Hill Book Co., 4th edition, New York, 1995.

Henry, D. L., "Theory of Fatigue Damage, Accumulation in Steel", Transactions, ASME, volume 77, page 913, New York, 1955.

Marco, S. M., Starkey, W. L., "A Concept of Fatigue Damage", Transactions ASME, volume 76, page 627, New York, 1954.

Manson, S. S., Frecke J. C., Ensign, C. R., "Application of a Double Linear Damage Rule to Cumulative Fatigue", Fatigue Crack Propagation, STP-415, ASTM, Philadelphia, page 384, 1967.

Miner, M. A., "Cumulative Damage in Fatigue", ASME Transactions, Journal of Applied Mechanics, volume 67, page A159, 1945.

National Transportation Safety Board, NTSB Report # NTSB/AAR-98/01, "Aircraft Accident Report – Uncontained Engine Failure", National Transportation Safety Board, Washington, 1998.

Palmgren, A. "Die Lebensdauer von Kugellagern", ZV'DI, Volume 68, page 339, 1924.

Rao, J. S., Pathak, A., Chawla, A., "Blade Life: A Comparison by Cumulative Damage Theory", ASME paper # 99-GT-287, New York, 1999.

14

CHAPTER

STRUCTURAL FAILURE FROM MATERIAL FATIGUE

14.1 INTRODUCTION

Fatigue in the substrate of rotating components is a primary cause of failures in turbo-machinery. Repeated applications of cyclical loads resulting from loading and unloading cause the stresses to alternate. Even when the maximum cyclic stress is well below the ultimate tensile strength, and even below the yield strength, of the material, the part ruptures after a certain number of load/unload cycles. The expression 'metal fatigue' came into vogue because long ago it was thought of as a behavior of the metal getting tired.

The failure mechanism originates in areas where the geometry undergoes sharp changes and where the stress concentrates. Reducing the magnitude of the cyclic stress goes a long way in increasing the number of cycles before the part gives way and fractures.

The number of load cycles an engineered metal can accept before failing depends on many variables. The alloy, its heat treatment, hardness, presence of impurities, surface finish and stress due to static and dynamic loads are some factors. Operating conditions may also result from changes in environmental conditions such as operating temperature.

It is difficult to obtain fatigue life precisely since the endurance limit of a material is not constant nor is it repeatable. A number of identical samples machined from a bar when tested will have differing yield strength, ultimate strength and endurance limit. Each of the properties is obtained for the samples. Statistical procedures are required to account for the variability in the results of several identical tests carried out on a large number of samples of the metal. Test results are plotted in the form of stress and number of cycles, and are commonly referred to S/N diagram. A curve is then fitted through the statistical cluster to represent the behavior under fatigue of that material at the specific strength level.

14.2 FATIGUE INDUCING LOADS

Loads that create fatigue may take a variety of forms. In a zero to maximum and back to zero case, an unloaded part is subjected to a load, and then the load is removed after a while. Thus the part goes back to its initial unloaded condition. The blades attached to a disk have no load when the disk is stationary. When the disk is made to spin, centrifugal force is applied. When the disk stops turning, the load is removed.

Another fatigue type of load is a varying load imposed on a fixed load. For instance, the shaft carries the deadweight of the bladed disk. An electric motor is coupled to the shaft. At start the electric motor is standstill. Torque is applied gradually when the electric motor turns on. The torque load is removed when the motor is shutdown. The constant weight of the disk causes bending stress in the shaft.

In the maximum fatigue causing case, the load becomes fully reversing. In a single cycle the load causes buildup of tensile stress, then the load reverses to go to zero and continues to produce compressive stress of the same magnitude as the tensile stress before the load once again reduces to zero. The rotating shaft mentioned earlier is an example of the fully reversing load if it is assumed the shaft is of negligible weight.

14.3 MATERIAL STRENGTH AND FATIGUE CHARACTERISTICS

A battery of tests is required to establish the yield and ultimate strengths and the endurance limit for a material. The tests must be identical and carried out on the same metallurgy heat from which a large number of identical test samples are produced. The tests will provide varying values, and clusters of data are grouped through which the best curve is fitted.

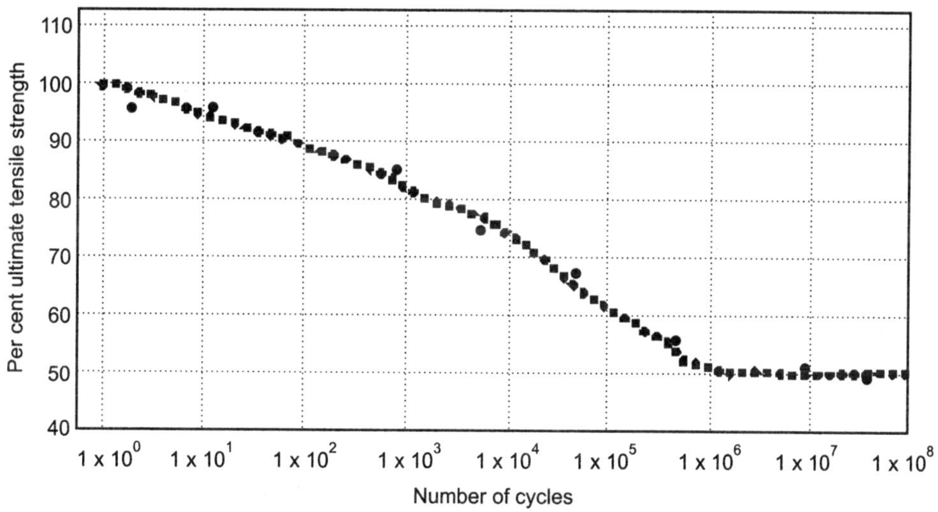

Fig. 14.1 Representative S/N Diagram of Engineered Metal

Figure 14.1 shows a sample S/N diagram of a material. Points marked in red indicate the cyclic stress level and the number of cycles when the test specimen gave way and fractured. Other points marked in blue indicate the cyclic stress level and the number of cycles when the test specimen did not fail.

Turbine manufacturers at the present time opt for the Marin method to estimate maximum fatigue loading of a part. The method calls for the test-determined endurance limit to be adjusted for factors such as surface finish, size, load, temperature and reliability as discussed in the next section. Other sundry effects arising from residual stress, corrosion, fretting, plating and metal spraying may also be included.

14.4 FACTORS AFFECTING FATIGUE LIFE

A number of factors affect the fatigue strength of a material:

- Surface quality factor, k_a, is used to reflect polished, ground, machined, forged or corroded surface.
- Size factor, k_b, is used to reflect differences in the dimensions of the part and those of the test specimen.
- Loading factor, k_c, is applied when differences are present in actual load types on the part and those on the test specimen.
- Temperature factor, k_d, is required when operating and test temperatures differ. Extreme high or low temperatures can decrease fatigue strength.
- Reliability factor, k_e, is used when substantial scatter is observed in test data. Thus, an 8 per cent standard deviation in test data requires $k_e = 0.868$ for 95 per cent reliability and $k_e = 0.753$ for 99.9 per cent reliability.

Other conditions may arise from residual stress, corrosion, fretting, plating and metal spraying. Grouping them under sundry factor k_f, these are:

- Cyclic stress state: Depending on the complexity of the geometry and the loading, one or more properties of the stress state need to be considered, such as stress amplitude, mean stress, bi-axial stress, in-phase or out-of-phase shear stress, and load sequence.
- Geometry: Notches and variation in cross section throughout a part lead to stress concentrations where fatigue cracks initiate.
- Material Type: Fatigue life, as well as the behavior during cyclic loading, varies widely for different materials. As an example, composites and polymers differ markedly from metals.
- Residual stresses: Welding, cutting, casting, and other manufacturing processes involving heat or deformation can produce high levels of tensile residual stress, which decreases the fatigue strength.
- Grain size: For most metals, smaller grains yield longer fatigue lives, but the presence of surface defects or scratches will have a greater influence than in a coarse grained alloy.
- Environment: Environmental conditions can cause erosion, corrosion or gas-phase embrittlement, which all affect fatigue life. Corrosion fatigue is a problem encountered in many aggressive environments.

All of the factors may be combined with the test-determined endurance limit to obtain the allowable cyclic stress using the equation:

Allowable cyclic stress = $[k_a \times k_b \times k_c \times k_d \times k_e \times k_f]$ × Test-determined endurance limit.

14.5 EFFECTS OF SURFACE FINISH

Surface roughness causes microscopic stress concentrations that lower the fatigue strength. Compressive residual stresses can be introduced in the surface if the material is shot-peened to increase fatigue life. Low plasticity burnishing, laser peening, and ultrasonic impact treatment can also produce this surface compressive stress and can increase the fatigue life of the component. This improvement is normally observed only for high-cycle fatigue.

The quality of the surface strongly varies with the manufacturing process and subsequent finishing of a turbine part. As a rule, the rougher the surface condition, the greater will be the concentration of stresses in the often times microscopic ridges. The intent behind carrying out processes such as turning, grinding and polishing is to eradicate the roughness on the surface.

In the forged and cast condition, irregularities are bound to appear even when superior forging or casting procedures are used that yield near net-shape for the parts. Many of the surface irregularities may be removed when the part is machined. Turning operation on the part causes many of the unevenness on the surface. Grinding further helps to eliminate some of the lines left after the turning process. Polishing of the surface further alleviates the unevenness on the surface and thus improves the load carrying capacity of the part.

Corrosion commonly takes place on the surface of components as the turbine unit accumulates operating hours. Harsh conditions prevail when the unit is required to operate in and close to marine areas, where the air is full of salt spray.

14.6 FRETTING BETWEEN PARTS

When the face of a part repeatedly rubs against the face of another of a pair of contacting parts under load, wear in the form of fretting occurs on both mating surfaces. Fretting may be defined as a special wear process occurring at the contact face between the parts subjected to dynamic loading that produces minute relative motion. Some examples are bolted flange, shafts coupled with splines and press-fitted gear and bearing on a shaft.

The mating components may be pressed together with a static load during initial assembly. Thus, in the case of a power generation gas turbine the compressor rotor and the turbine rotor are coupled with a bolted flange. In spite of the bolts being adequately pre-torqued, even a residual amount of unbalance in the assembled rotors will cause some vibrations in the mating faces of the two flanges. When the amplitude of motion is below 0.003 inch, fretting will set in.

Asperities in the contacting faces continually fuse and split under the action of the dynamic loads on the bolted flange assembly. The process leads to pitting of the two surfaces, with tiny metal particles splitting from the parent and exchanging between the faces. The trapped fragments may even oxidize in the presence of oxygen. The oxidized particles are harder than the base of most engineering metals, and create abrasive damage and scoring.

Certain features of fretting tend to aggravate the problem, such as increased contact load, slip amplitude and number of cycles. Providing lubrication may ensure better assembly, but at the risk of increasing slippage between the parts. Fretting induced corrosion is known to initiate fracture cracks at levels of stress well below the fatigue limit of a new part. Fatigue strength reduces rapidly when both fretting and cyclic stress are applied simultaneously.

14.7 STATISTICAL PROBABILITIES IN FATIGUE FAILURE

The scatter in the data points in Figure 14.1 is due to a certain level of randomness exhibited by the samples even though they are produced from the same material, subjected to the same stress level and test sequence. Adding to this uncertainty are the differences between laboratory-obtained endurance limit and observations from field failures.

The statistical feature of the endurance limit is based on optimized laboratory results. To enable the application of the laboratory-obtained endurance limits to the design of a given turbine component, it is preferable to reduce the laboratory endurance limit. The reduction will take into account differences between the observed and test values.

As coupons sampled from a homogeneous frame will manifest variation in their number of cycles to failure, the S/N curve should more properly be an S/N/P curve, capturing the probability of failure after a given number of cycles of a certain stress. Probability distributions that are common in data analysis and in design against fatigue include the lognormal distribution, extreme value distribution, Birnbaum–Saunders distribution, and Weibull distribution.

The intent is to take into consideration unpredictable factors as diverse as surface finish, actual load and temperature, machining tolerances, presence of impurities, alloy composition and its heat treatment.

14.8 DESIGN AGAINST FATIGUE FAILURE

Dependable design against failure from fatigue requires thorough expertise and supervised experience in structural engineering, mechanical engineering or materials science. There are three principal approaches to life assurance for mechanical parts that display increasing degrees of sophistication:

- The design must endeavor to keep stress below threshold of fatigue limit, or infinite lifetime concept.
- Use conservative design approach for a fixed life. The user must be instructed to replace the part with a new one. The method is also referred to as the 'lifed-part method', or the finite lifetime concept or 'safe-life' design practice.
- Instruct the user to inspect the part periodically for cracks and to replace the part once a crack exceeds a critical length. This method uses the non-destructive technology testing, and requires accurate prediction of the rate of crack-growth between inspections. This method is called damage-tolerant design or 'retirement-for-cause' procedure.

14.9 ARRESTING FATIGUE FAILURES

Drilling a hole in the path of a fatigue crack that has initiated and is beginning to propagate sometimes helps to arrest its further growth. This is not recommended as a general practice, because the hole represents a stress concentration factor that depends on the size of the hole and geometry, though the hole is typically less of a stress concentration than the removed tip of the crack. The possibility remains of a new crack starting in the side of the hole. It is always far better to replace the cracked part entirely.

As an interesting aside, during the American War of Independence the Liberty Bell developed a fatigue crack when it was rung to announce to the world America is a free nation. Starting from the base, the crack runs a length of about 15 inch and has an average width of 0.30 inch through the wall. The engineering expert of the time decided to drill a hole where the crack terminated, and inserted a rivet in the hole (Figure 14.2). The crack, hole and rivet are still in evidence on the nearly 5-foot diameter bell and about the same height at its home in Philadelphia. However, the citizens decided against ringing the bell again after the episode, so the question whether the engineering fix worked remains unanswered!

Change in the material used in the part can also improve fatigue life. For example, parts can be made from better fatigue rated metals. Complete replacement and redesign of parts can also reduce if not eliminate fatigue problems. Thus, helicopter rotor blades and propellers in metal are being replaced by composite equivalents. They are not only lighter, but also much more resistant to fatigue. They are more expensive, but the extra cost is amply repaid by their greater integrity, since loss of a rotor blade usually leads to total loss of the aircraft. A similar argument has been made for replacement of metal fuselages, wings and tails of aircraft.

Fig. 14.2 Liberty Bell

The durability and lifetime of dynamically loaded, welded steel structures are determined often by the welds, particular from the weld transitions. The method of selective treatment of weld transitions with 'High Frequency Impact Treatment' is considered to improve the durability of many designs significantly. This method is widely applied, requires only low-tech equipment and offers high reproducibility and control over quality.

14.10 SUPER ALLOYS FOR TURBO-MACHINES

Super alloys have made possible the development and operation of gas turbines at very high temperatures. The alloys used in the manufacture of jet engines and industrial and marine gas turbines represent the leading edge in materials technology, and in turn these engines have been the primary driving force for the development of the super alloys. Commercially available super alloys can be hardened from solid solution elements and in the grain boundaries. A combination of thermal and mechanical processing may be employed to strengthen the material by enhancing the density and by developing a substructure at dislocations. Advantage is taken of reinforcement with wires and directional solidification process in eutectics to upgrade the strength of some alloys and improve fatigue characteristics.

Over the years super alloys, have gone through many different refinements and improvements through the formation of new composition for the alloys and manufacturing processes. Military aircraft engines were the initial beneficiaries of the new technology, but application to other gas turbines did not lag too far behind.

Chromium is added to iron and nickel base to obtain resistance to oxidation, while smaller amounts of aluminum, titanium and columbium provide enhanced creep resistance characteristics. Iron gradually disappeared over a period of time as an alloy base, to be replaced by nickel and cobalt because of the more durable face-centered cubic structure. Although chromium may lead to some level of deterioration in strength, considerable reduction of this element can cause corrosion at high temperatures, leading to its more balanced use in IN-738. Excessive use of aluminum, titanium and columbium for forming γ' can also pose a variety of structural problems. Addition of molybdenum, initially in M-252, provided additional strengthening through solid solution and carbide effects. Other refractory elements, tungsten, tantalum and rhenium also later found their way to enhance the mechanical properties. Carbon remains as a complexing agent, with matrix carbides providing point strengthening in several solid-state reactions. Zirconium and boron improve on grain boundary effects, but are not needed in single crystal alloys since the boundaries are absent.

Nearly half the total weight of aircraft engines is comprised of parts made from nickel base alloys. The alloys are favored in elevated temperature regions of the power plant, in spite of the complex physical metallurgy. The tensile and creep rupture strength up to 5000 hours of these alloys in the 1200° F to 2000° F temperature zone make them the prime candidate for turbine blades. Industrial gas turbines, on the other hand, require creep rupture characteristics over longer periods, as also resistance to oxidation and corrosion at high temperatures.

Most of the commercially available cobalt alloys are melted in air or argon because of the absence of aluminum and titanium. These reactive elements necessitate the use of expensive vacuum melting procedures. Cobalt alloys with high temperature capabilities posses a complex chemical and crystallographic composition, as is true also of nickel and iron alloys. Consisting of an austenitic matrix and a variety of precipitated phases such as carbides and inter-metallic compounds in the geometrically and topologically closely packed family of structures, the alloys cannot be defined as stable at the operating temperatures encountered in gas turbines. This is because of the high levels of dynamic stress, time and interactions between the surface and the environment.

14.11 STRESS COMPONENTS LEADING TO LOW CYCLE FATIGUE

In many metals and alloys low cycle fatigue below 10^3 cycles and plasticity tend to occur in tandem. A strain-based parameter is required for fatigue life prediction in the metals. Testing is conducted with constant strain amplitudes between 0.01 and 5 Hz.

Practical aspects of turbine applications tend to result in a complex combination of loads. Oftentimes a random sequence of small and large loads needs to be weighed in order to assess the safe life of such a part. One possibility may be to reduce the complex loading to a series of simple cyclic loadings using a technique such as the rain-flow method of analysis. The intent is to create a histogram of cyclic stress to form a fatigue damage spectrum. For each stress level, the extent of cumulative damage incurred from the S/N curve is established. Individual contributions are then combined using an algorithm such as Miner's rule that had earlier been proposed by Palmgren. Sometimes referred to as the Palmgren-Miner linear damage hypothesis or simply Miner's rule, it states that where there are k different stress magnitudes in a spectrum, S_i $(1 \le i \le k)$, each contributing $n_i(S_i)$ cycles, then if $N_i(S_i)$ is the number of cycles to failure of a constant stress reversal S_i, failure occurs when:

$$\sum_{i=1}^{k} \frac{n_i}{N_i} = C$$

where C is experimentally found to be between 0.7 and 2.2. For design purposes, C is commonly assumed to be 1.

This can be thought of as assessing what proportion of life is consumed by stress reversal at each magnitude, then forming a linear combination of their aggregate. Though Miner's rule is a useful approximation in many circumstances, it has several major limitations. It fails to recognize the probabilistic nature of fatigue and there is no simple way to relate life predicted by the rule with the characteristics of a probability distribution. Industry analysts often use design curves, adjusted to account for scatter, to calculate $N_i(S_i)$.

There is sometimes an effect in the order in which the reversals occur. In some circumstances, cycles of low stress followed by high stress cause more damage than would be predicted by the rule. It does not consider the effect of overload or high stress that may result in a compressive residual stress. High stress followed by low stress may have less damage due to the presence of compressive residual stress.

Where the stress is high enough for plastic deformation to occur, the account in terms of stress is less useful and the strain in the material offers a simpler description.

The 'Coffin-Manson relation' usually characterizes low-cycle fatigue:

$$\frac{\Delta \epsilon_p}{2} = \epsilon'_f (2N)^c$$

where $\Delta\varepsilon_p/2$ is the plastic strain amplitude; ε_f' is the failure strain for a single reversal, an empirical constant known as the 'fatigue ductility coefficient', $2N$ is the number of reversals to failure (or, N cycles); c is an empirical constant known as the 'fatigue ductility exponent', and commonly ranges from −0.5 to −0.7 for metals in time independent fatigue.

14.12 COMPLEX AND RANDOM LOADINGS IN ROTATING BLADE

Most attention has historically focused on situations that require more than 10^4 cycles to failure where stress is low and deformation is primarily elastic. The S/N curve, also known as a Wöhler curve, provides interaction between magnitude of the cyclic stress (S) against the logarithmic scale of cycles to failure (N). The curves are derived from tests on coupon samples of the material to be characterized where a regular sinusoidal stress is applied with a testing machine, which also counts the number of cycles to failure. Each coupon test generates a point on the plot. Analysis of fatigue data requires techniques from statistics, especially survival analysis and linear regression.

In practice, a mechanical part is exposed to a complex, often random, sequence of large and small loads (Figure 14.3). In order to assess the safe life of such a part, usual practice is to reduce the complex loading to a series of simple cyclic loadings using a technique such as the 'rain-flow analysis'. The method calls for creating a histogram of cyclic stress from the rain-flow analysis to form a fatigue damage spectrum. For each stress level, the degree of cumulative damage is calculated from the S/N curve.

The individual contributions are combined using an algorithm such as Miner's rule, in spite of its several major limitations. Industry analysts often use design curves, adjusted to account for scatter, to calculate $N_i(S_i)$.

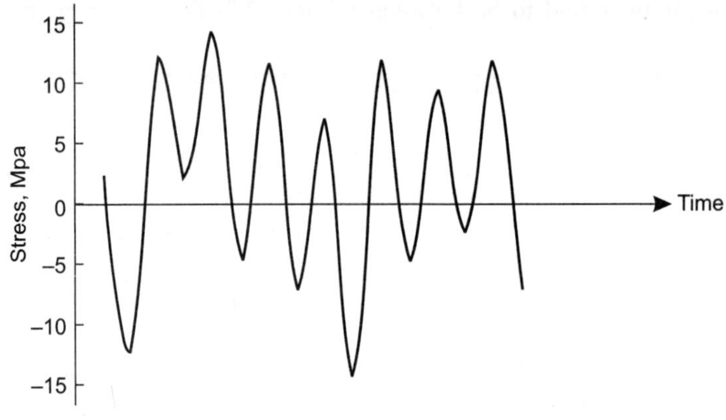

Fig. 14.3 Spectrum Loading

There is sometimes an effect in the order in which the reversals occur. In some circumstances, cycles of low stress followed by high stress cause more damage than would be predicted by the rule. It does not consider the effect of overload or high stress that may result in a compressive residual stress. High stress followed by low stress may have less damage due to the presence of compressive residual stress.

14.13 FATIGUE CRACK GROWTH

Mathematical relationships have been obtained based on fracture mechanics theory. Stage II crack growth with cycles N may be expressed in terms of the cyclical component ΔK of the Stress Intensity Factor K:

$$\frac{da}{dN} = C(\Delta K)^m$$

where a is the crack length and m is typically in the range 3 to 5 (for metals). This relationship may be modified to allow for the mean stress by introducing a factor depending on $(1 - R)$ where R = min stress/max stress, in the denominator.

In the presence of stresses superimposed on the cyclic loading, the Goodman relation can be used to estimate a failure condition. It plots stress amplitude against mean stress with the fatigue limit and the ultimate tensile strength of the material as the two extremes. Alternative failure criteria include Soderberg and Gerber.

The combination of steady and alternating stresses may be assessed with the aid of a modified Goodman diagram (see Figure 14.4). A well-defined criterion of fatigue or static failure may be obtained with this method.

Note that the Goodman diagram assumes cyclic loading of constant amplitude. Engine components will be experiencing different load spectra of varying frequencies and amplitudes. If it is assumed that damage from individual loads accumulates in a particular manner, failure may be predicted when underlying assumptions are satisfied.

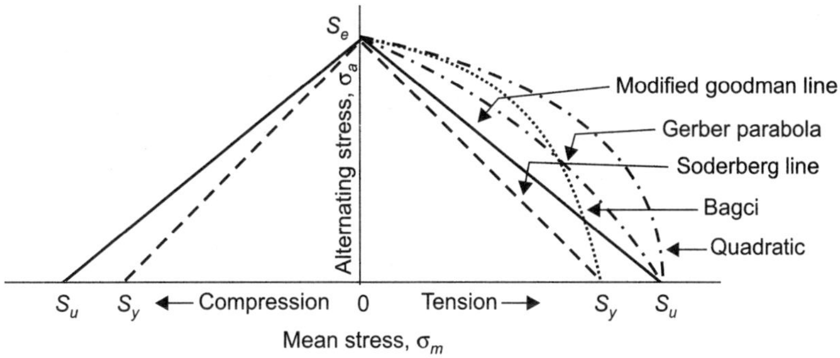

Fig. 14.4 Modified Goodman Diagram

For example, the Palmer-Minegren rule is based on a linear accumulation of damage, but the results do not agree with experimental results. Many different theories based on the method of damage accumulation have been proposed for the prediction of failure, but shortcomings are inherent. Damage accumulation based on comparative values between analytically predicted and experimental tests under similar conditions may provide a better estimate of the fatigue life.

14.14 LUBRICATION OIL PIPING FAILURE IN ROLLS ROYCE ENGINE

A Qantas Airlines passenger Flight 32 suffered an uncontained engine failure during takeoff from Singapore on 4th November of 2010. The failure was the first of its kind for the four-engine Airbus A380, the world's largest aircraft. The aircraft's # 2 engine on the port side near the fuselage was a Rolls Royce Trent 900 engine. A turbine disk was missing, and damage was sustained to the nacelle, wing, fuel system, landing gear, flight controls and controls for engine # 1.

The accident occurred at 10:01 AM Singapore time, and was caused by an uncontained failure of the port inboard engine # 2 while en route over Batam Island, Indonesia. Shrapnel from the exploding engine punctured part of the wing, damaged the fuel system and caused leaks. Also, one hydraulic system and the anti-lock brakes were disabled, and caused # 1 and # 4 engines to go into a degraded mode of operation, damaged landing flaps and controls for the outer left # 1 engine. Debris also landed on a school and houses, causing structural damage.

The cockpit pilots were alerted by 54 error messages generated by the aircraft's control system. The cockpit crew was able to control the airplane, took some time to assess its status, and decided to land back at Singapore's Changi airport. Consulting with the Control Tower, the plane landed at a speed 35 knots faster than normal, in the process blowing 4 tires. The crew could not shut down engine # 1, which had to be doused by emergency crew for 3 hours after landing until flameout. After the landing, fuel was leaking from the left wing and spilling on the hot brakes from maximum braking. No injuries were reported among the 440 passengers and the 29 crew members on board the plane.

Rolls Royce determined that the direct cause of the fire and resulting engine failure was a misaligned counter bore within a stub oil pipe, leading to a fatigue fracture. The Australian Transportation Safety Board (ATSB)'s preliminary investigation report confirmed Rolls Royce's determination.

Airbus, the plane's maker, determined that three fragments of the intermediate-pressure turbine (IPT) disk exited with high energy, resulting in systems damage. Segregated wiring routes were cut by two of the three pieces of disk debris. As a direct consequence, the engine could not be shut down after landing.

The ATSB's ongoing investigation found fatigue cracking in a stub pipe that feeds oil to the HP/IP bearing housing. Bearing lubrication oil leaked from that crack, causing oil fire in the engine. The fatigue fracture was a result of misalignment of the stub pipe during counter-boring. The inaccuracy resulted in one side of the stub pipe becoming too thin to resist fracture from fatigue. The consequence is increased risk of crack initiation and growth, leading to leakage of oil and potential catastrophic engine failure from the fire.

The fire led to the release of IPT disk. At the time of writing, ATSB, Rolls Royce and Airbus have not addressed the question of failure mechanism of the IPT disk. Singapore lies close to the equator at sea level, and in the month of November usually experiences high air temperature and humidity around midday. When the aircraft carries a full load combination of passengers and cargo and has a full gas tank (since the aircraft was headed towards Sydney, a few thousand miles away), the engines will need to produce maximum thrust to enable the aircraft to get off the ground and climb to altitude. A possible explanation of the failure of the disk is that the failed bearing is required to balance axial thrust, causing the disk to slide axially and rub against stationary components. The disk then disintegrated from the combined action of running at full speed while rubbing. The maximum operating speed will result in the disk and attached blade fragments having substantial kinetic energy, and cause them to penetrate the engine casing. News reports also indicated one of the disk fragments was embedded into an aluminum spar running along the length of the wing. It is also worth pointing out that without the benefit of recovering all of the fragments of the failed disk, the investigators would not be able to fully explain its failure.

14.15 REVIEW QUESTIONS

14.15.1 How does a varying load imposed on a fixed load differ from the maximum fatigue-causing load?

14.15.2 Explain the Marin method to estimate maximum fatigue loading of a part.

14.15.3 Discuss some factors that affect the fatigue strength of a material.

14.15.4 How does surface finish affect the life of a part? Explain this in the case of a forged and machined-turned finish.

14.15.5 Explain the process of fretting. What is its role in failure of turbine parts?

14.15.6 How does an S/N/P curve differ from an S/N curve? Why is the S/N/P curve preferred?

14.15.7 What are the primary requirements to ensure a component does not fail against fatigue?

14.15.8 Mention some benefits of super alloys that make them suitable for use in gas turbines.

14.15.9 Explain the technique of rain-flow analysis in forming a fatigue damage spectrum.

14.15.10 How do industry professionals account for probabilistic nature of fatigue and scatter in measured test data when using Miner's rule for determining fatigue life of a component?

14.15.11 Name some methods to assess the integrity of a component when subjected to a combination of steady and alternating stresses.

14.16 EXAMPLE PROBLEMS

Example Problem # 14.16.1: A 0.875 in. diameter and 3.90 in. long test coupon is subjected to a torque load of 2250 in-lbs. Calculate the shear stress induced from the load.

Solution: From the theory of strength of materials,

$$(T/I_p) = (\tau_s/R) = (G\theta/L)$$

where T = torque, I_p = cross-section polar inertia, τ_s = shear stress, R = radius, G = shear modulus, θ = angle of twist and L = length of specimen.

$$\text{Shear stress } \tau_s = (TR/I_p) = (2250 \times 0.875/2)/[(\pi/64) \times 0.875^4]$$
$$= 34227.78 \text{ lbs/in}^2$$

Example Problem # 14.16.2: In *Example Problem # 14.16.1* calculate the angle of twist if material shear modulus is 11.5×10^6 lbs/in².

Solution: $$(T/I_p) = (\tau_s/R) = (G\theta/L)$$

Then angle of twist: $$\theta = (TL/GI_p) = (2250 \times 3.90)/[11.5 \times 10^6 \times (\pi/32) \times 0.875^4]$$
$$= 0.0133 \text{ radian}$$
$$= 0.76 \text{ degrees}$$

Example Problem # 14.16.3: A 6.75 in. long cantilever beam of cross-section 1.25 in. depth and 0.425 in. width is subjected at its free end with a bending moment of 577 in-lbs. Calculate the maximum bending stress in the beam.

Solution: From the theory of strength of materials,

$$(M/I_{xx}) = (\sigma_b/y) = (E/R)$$

where M = bending moment, I_{xx} = area moment of inertia, σ_b = bending stress, R = radius of curvature of deflected beam, E = Young's modulus, y = deflection in beam.

$$\text{Bending stress } \sigma_b = (My/I_{xx}) = (577 \times 1.25/2)/(0.425 \times 1.25^3/12)$$
$$= 5213.36 \text{ lbs/in}^2.$$

Example Problem # 14.16.4: In *Example Problem # 14.16.3* calculate the radius of curvature of deflected beam if material Young's modulus is 28.5×10^6 lbs/in².

Solution: $$(M/I_{xx}) = (\sigma_b/y) = (E/R)$$

Then radius of curvature of deflected beam: $R = (EI_{xx}/M)$

$$= 28.5 \times 10^6 \times (0.425 \times 1.25^3/12)/577$$
$$= 3416.70 \text{ ins.}$$

Example Problem # 14.16.5: In *Example Problems # 14.16.3* and *14.16.4* calculate the maximum deflection in the beam.

Solution: $$(M/I_{xx}) = (\sigma_b/y) = (E/R)$$

Then maximum deflection: $$y = (\sigma_b I_{xx}/M) = 5213.36 \times (0.425 \times 1.25^3/12)/577$$
$$= 0.625 \text{ in.}$$

14.17 EXERCISE PROBLEMS

Exercise Problem # 14.17.1: A 1.25 in. diameter and 4.550 in. long test coupon is subjected to a torque load. Calculate the load if the shear stress induced is 24525 lb./in².

Exercise Problem # 14.17.2: In *Exercise Problem # 14.17.1* calculate the angle of twist if material shear modulus is 11.5×10^6 lb./in^2.

Exercise Problem # 14.17.3: A 8.725 in. long cantilever beam of cross-section 2.5 in. depth and 0.565 in. width is subjected at its free end with a force of 822 lb. Calculate the maximum bending stress in the beam.

Exercise Problem # 14.17.4: In *Exercise Problem # 14.17.3* calculate the radius of curvature of deflected beam if material Young's modulus is 29.25×10^6 lb./in^2.

REFERENCES & BIBLIOGRAPHY

Airbus, "Qantas Flight QF32 Engine Failure/In Flight Turn Back to Singapore".

Anderson, Gomez and Paris, J., "Fracture Mechanics", McGraw-Hill, New York, 1994.

ASM International, "Failure Analysis and Prevention Handbook", V-11, 1998.

Australian Transport Safety Bureau, "Aviation Safety Investigation Report 089 – Qantas Airbus A380 – In Flight Engine Shut Down, Indonesia", Retrieved 5 November, 2010.

Australian Transport Safety Bureau, "In Flight Uncontained Engine Failure – Over Head Batam Island, Indonesia", ISBN 978-1-74251-118-4, Retrieved 3 December, 2010.

Baumeister, H., et al, "Mark's Handbook for Mechanical Engineers", McGraw-Hill, New York, 1978.

Buckingham, E., "Analytical Mechanics of Gears", McGraw-Hill, New York, 1988.

Eschmann, F., "Ball and Roller Bearings: Theory, Design, Application", McGraw-Hill, New York, 1985.

National Transportation Safety Board – Aircraft Accident Report, NTSB/AAR-90/06, Washington, DC, 1990.

Oberg, J, Jones, P., Horton, H., Ryffel, F., "Machinery's Handbook", McGraw-Hill, New York, 1988.

Pilkey, W. D., "Peterson's Stress Concentration Factors", Prentice-Hall, Clifton, NJ, 1997.

Rakhit, A. K., "Heat Treatment of Gears", McGraw-Hill, New York, 2000.

Rolls Royce, "Trent 900 Update", 12 November, 2010, Retrieved 23 November, 2010.

Shigley, N., Mischke, G., Budynas, M. F., "Mechanical Engineering Design", McGraw-Hill, New York, 2004.

15

CHAPTER

THERMAL DISTORTION IN HOT PATH

15.1 INTRODUCTION

Steep temperature gradients and species concentration are the hallmark of the combustion process. The burned gases are far higher in volume and temperature then fresh air and fuel mixture. In the detonation part of the combustion process a shock wave connected with and supported by the chemical reaction zone propagates rapidly. Instead of the combustion wave spreading through a static gas mixture, the flame stabilizes to a steady condition by supplying it with a continuous flow of combustible mixture.

Physical and chemical reactions are embraced during combustion. A primary objective of a good combustion system is to achieve satisfactory mixing within the liner and a stable flow pattern throughout, with no parasitic losses and minimal pressure loss. Inside the liners containing the primary combustion zone, considerable flow re-circulation is essential to stabilize the flame and to obtain maximum benefit from the cooling air along the walls.

Mechanical stresses in the combustor liner, also called flame tube, are relatively low. But the liner is required to withstand high temperatures and considerable temperature gradients. Materials for the liner are selected because of their ability to withstand elevated temperatures. The maximum temperature is restricted to 1100 K, beyond which mechanical strength of the nickel and cobalt based alloys rapidly deteriorates. The liners are lightly supported in the axial, radial and tangential directions to avoid buildup of stresses at the attachment points.

Among the many schemes employed to extract heat and cool the liner, conduction and convection methods are used. Cooler air extracted from the compressor is used, with the coolant entering through a large number of holes machined in the liner. Film-cooling calls for annular slots along the length, through which the coolant is axially injected along the inner wall of the liner. Other devices in the form of wiggle strips and rings of various profiles are used to suit the conditions of pressure and temperature and of manufacturing ease.

15.2 TEMPERATURE GRADIENT AND FATIGUE FAILURE

Temperature gradients existing through the walls of stationary vanes and rotating blades are the primary cause of buildup of thermal stresses. Combined with frequent start and shutdown of the engine then pave the way for failures associated with metal fatigue. Mechanical stresses compound the problem in rotating blades due to centrifugal effects.

During steady state operation of the turbine the temperatures soak through the thickness of the airfoils, and practically all of the gradients dissipate. Theoretically, then, thermal stresses cease to be a problem. However, conditions during startup and shutdown of the engine are altogether different. Gas and surface metal temperatures climb up when the engine is fired at startup. Heat gradually traverses through the thickness of the walls of the components and sets up thermal gradients. Thermal stress is a function of coefficient of thermal expansion and of the elastic modulus of the material of the component, and also of the temperature differential between the hot gases and the metal. A similar situation occurs when the engine is shutdown, except that the heat flow reverses in direction from the hot metal to the cooler air. Also, the stresses will also change from tensile to compressive between the two cases.

The complete sequence of engine startup and shutdown constitutes one full load cycle. Aside from thermal stresses, mechanical stresses primarily from centrifugal effects of the rotating blades also develop, with the tensile stress depending on the square of rotor speed. At a cross-section close to the tip of the airfoil the centrifugal effect is negligible, varies continuously along the length and reaches peak value at a section close to the root. Note, however, that the thermal and mechanical stresses develop at different time points during the full cycle, and overlap of the two stress types may occur only for short durations. The overlap will mostly occur when the engine is close to reaching maximum speed during startup and also when the shut down process is in its initial stages.

Both in the compressor and in the turbine rotating blades, stresses concentrate near the trailing edge close to the root where the airfoil meets the dovetail, mostly because of rapid change of geometry in the region. In the case of cooled turbine blades, it is difficult to internally cool the blades, although some relief may be expected from film cooling as it separates the edge from the main gas stream. Field failures of turbine blades tend to corroborate these observations.

Life expectancy of the rotating blades, more so in the turbine but also in the compressor, is based on the number of operating hours and start/stop cycles using the low cycle fatigue criteria. The metal substrate's yield, ultimate and endurance strength in the operating temperature range play a major role. Maximum steady and alternating stresses developed in critical portions of the part aid in establishing the safe number of operating cycles.

15.3 COMBUSTOR LINER CONSTRUCTION

The liner of a combustor functions to contain the combustion process and to permit circulation of cooling air in various zones in set amounts. Besides the pressure differential on either side, the liner must have thermal resistance to withstand continuous and cyclic operation at elevated temperatures. Materials such as Nimonic 75, Hastelloy X and HS 188 are used for the liner, with maximum temperature restricted to 1100 K, beyond which mechanical strength of the nickel and cobalt based alloys rapidly deteriorates.

Engine specifications and efficient use of available space influence the type and layout of the combustion chamber. Larger aviation engines generally call for the air to flow nearly parallel to the axis of the combustor, but in smaller engines the flow reverses direction to provide compactness and closer connection between the compressor and the turbine. Reduced pressure losses because of the relatively straight flow path of the gases and compactness in design make the annular combustor suitable for aviation engines.

Annular combustors have the liner inside an annular casing to give a compact design and a clean aerodynamic flow path with little loss in pressure. But the absence of radial load carrying members between the inner and outer segments generates buckling problems in the outer liner.

A tubular can form of construction calls for a cylindrical liner mounted concentrically inside a cylindrical case. The larger length and weight of the resultant assembly restricts their usage to industrial turbines, where relative simplicity and accessibility are of prime significance. Can combustor designs call for the liner placed concentrically inside a cylindrical housing. From 6 to 16 cans are placed equally spaced around a central core casing in many early jet engines. The construction offers extra rigidity, but the extra length adds to engine weight. Industrial turbines extensively use this combustor type because of accessibility and for ease of maintenance.

15.4 COOLING OF LINER

Innovative means of removing the heat from the liner walls are provided. A sizable part of the heat transferred from the hot gases contained within the liner to the liner walls is by radiation. An effective barrier between the hot gas and the liner wall is created when the cooling air is injected in regions that are primarily heated through internal radiation. Cooler air is generally bled from the compressor. Internal convection is difficult to estimate, because the hot gases are undergoing rapid physical and chemical changes.

The arrangement of cooling and dilution holes distinguishes the design of combustor cans. The number, diameter and location of the holes is optimized to promote circulation of cooler and hotter gases in order to lower the temperatures and thus reduce formation of NO_x. An annular vortex generator is mounted in the combustor cap around the fuel nozzle. A flexible seal at the downstream end of the combustor liner fits into the transition piece.

Crossfire tubes at the upstream end are used to equalize pressure between the chambers and to propagate the flame upon ignition. By placing the mixing and dilution holes closer to the fuel nozzle, air is redistributed from the downstream end towards the upstream side to produce a more fully aerated diffusion flame. The technique achieves a reduced turbulent flame length while decreasing the amount of time spent in the stoichiometric flame zone, thus lowering NO_x emissions.

15.5 COOLING METHODS FOR HIGH PRESSURE TURBINE AIRFOILS

Increasing the inlet temperature of a gas turbine results in greater net power available, and is also the underlying factor for achieving improved thermal efficiency of combined cycle power plants. But the chief obstacle to indefinite increments in the firing temperature is the ability of blade metal to withstand the thermal assault. Material properties degrade rapidly after the metal temperature reaches a certain level. Beyond this limit, it is impractical to impose a greater thermal burden on the components.

The key to this problem is to cool the airfoils, thus reducing the metal temperature to a certain extent. Most gas turbines use air as a coolant, but alternatives such as water and steam have been identified.

Steam at intermediate pressure from the re-heater or exhaust from the high-pressure steam turbine may also be available to cool the gas turbine's vanes. The design of the vane has similarities with the air-cooled version, employing the impingement technique and an inserted core (Figure 15.1). Steam enters the tip area to the insert core at the leading edge and collected in the root area. The steam then flows to the insert core of the mid-chord region and recovered at the tip. The trailing edge is cooled with air by using film cooling and blow-off holes in the area.

In one power generation turbine the rotating blades gas temperature is estimated at 1266° C and gas pressure at 1.11 MPa. Available cooling air temperature is 410° C. Figure 15.2 shows a cross-section of the blades. The coolant is supplied in the root region through three independent channels located at

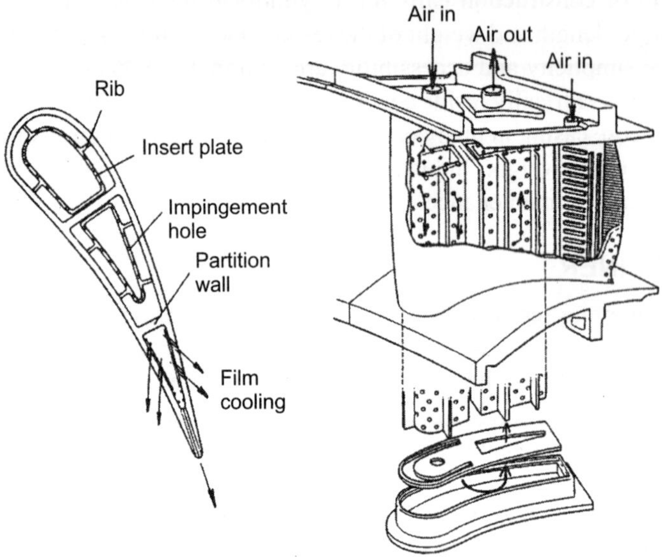

Fig. 15.1 Intermediate Pressure Steam Cooled Vane (Nomoto, 1996)

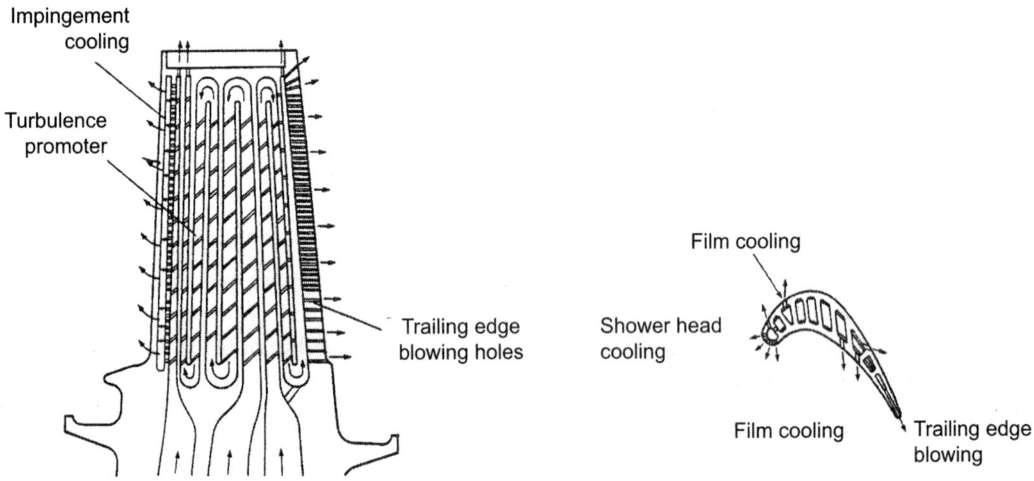

Fig. 15.2 Air Cooled Blade Cross-section (Nomoto, 1996)

the leading edge, mid-chord and the trailing edge. Cooling air impinges on the inner surface of the leading edge, flowing out in the form of a showerhead through four staggered rows of exit holes. A five-pass serpentine passage is used at the mid-chord channel to enhance heat transfer by convection. Two rows of film holes are placed on the pressure surface and three rows on the suction surface. Turbulence is promoted in the channel with ridges leaning in the direction of airflow to improve cooling effectiveness. The trailing edge channel is a three-pass serpentine, with air flowing out of ejection holes into the gas path.

15.6 ADDITIONAL PROTECTION USING THERMAL BARRIER COATING

The hostile environment of elevated temperatures and corrosive products of combustion in which the combustor liner and turbine airfoils operate call for multiple levels of protection. Creative cooling methods are required to deliver and circulate the coolant in the rotating blades. In addition, thermal barrier coatings (TBC) are applied to the surfaces of the heated parts to separate the hot gases in the turbine from the metal substrate of the components. The extra protection is required both for the turbine inlet vanes and some of the initial rotating blade stages.

Air plasma spraying (APS) and electron beam physical vapor deposition (EB-PVD) processes are well established for the deposition of TBC. In the EB-PVD method the coating is made from yttria-stabilized zirconia (YSZ) that has a columnar structure. This coating material has a long operating life even under severe cycling conditions, when compared with the porous APS coating. Both processes are of the 'line-of-sight' type, implying that the application of the coating is limited to simple shaped components, unless complex robotic motions are used. Deposition on shadowed regions, for instance on nozzles, poses considerable difficulties. Also, the equipment for EB-PVD is capital-intensive, which has hindered the spread of this technology for industrial gas turbines. Chemical vapor deposition (CVD) is another procedure for deposition of YSZ layers.

An important goal of the coating process is to obtain uniform thickness of the coat even in the most complexly contoured regions, such as the trailing edge close to the platform of an airfoil. Also, there must be good bonding between the coating and the substrate. If chipping of the coating occurs chiefly from handling, means are available for chemically stripping the coat and applying a new layer.

15.7 LIQUID AND GAS FUELS

Aviation engine fuels are restricted to liquid fuels, but industrial and power generation turbines are capable of running on both liquid and gas fuels. Of the latter turbine types, turbine manufacturers provide dual fuel delivery and injection systems for both fuel categories. The intent is to permit turbine users to switch from one to the other in case of exigencies arising from availability or costs. Many turbine users are even successful in shifting from one fuel to the other without fully shedding the driven load. In multi-fuel engines, however, the possibility remains of lower heat content fuels using up more of the combustor mass flow and combustion zone volume to cause a mismatch between the compressor and the turbine.

Liquid fuels for gas turbine may be split into two main groups. Lighter weight high performance aircraft engines require higher grades, while heavy duty industrial engines operate on a wide range of heavier and more easily obtainable fuels. Natural gas is now the preferred fuel for power plant and many other industrial gas turbines. Derivative aircraft engines for marine and industrial applications use modified combustion systems to enable them to burn commercial distillate fuels.

JP-4 and JP-5 are used widely by the US military for turbo-jet operation. JP-4 is a naptha fuel with vapor pressure of about 2.5 lb./in^2 and aromatic content under 25%. JP-5 is a blended kerosene fuel, has a flash point of 140° F and freezing point of –51° F. Airplanes operating at supersonic speed up to 3.5 Mach require thermally stable fuels. JP-7 and JP-8 provide varying degrees of flash and freezing points, thermal stability, aromatic content and flame luminosity. Commercial airlines rely on ASTM jet aviation turbine fuel. This is also a kerosene-based fuel with a flash point of 110° F and freezing point of –36° F.

Atomization of liquid fuels into droplets with extensive surface area is necessary for ignition and combustion, since they are not sufficiently volatile to produce vapors. The rate of evaporation is enhanced if the size of the droplets is reduced. Forcing the fuel under pressure through an orifice aids in atomization. In a simplex atomizer a swirl chamber is placed before the orifice, producing a sheet of spray with a cone angle of about 90º.

Pressure-swirl injectors, or atomizers, for liquid fuels can sustain combustion at weak fuel and air mixture strength, but may get plugged with suspended contaminants in the fuel, and have a tendency to form soot at higher pressures. Another concept calls for the fuel to flow at lower pressure over a lip in a high velocity air stream, causing atomization by the air as it enters the combustion zone. When the liquid sheet at the atomizing stream is exposed to the high velocity air on both sides, drop size is minimized to provide maximum contact between the air and the fuel. In this form of air-blast atomizers the airflow pattern controls the distribution of the fuel to reduce soot and exhaust smoke formation. However, the system has limited stability limits, and at startup lack of air velocity leads to poor atomization. Combining the features of a pilot pressure-swirl atomizer to obtain easy light-up and improved stability limits may control the problem.

Gaseous fuels such as natural gas are comparatively easier to burn when the calorific value is higher. Gas fuel injection may be accomplished by using orifices, swirlers, slots and venturi nozzles.

15.8 CONTROLLING EXHAUST GAS POLLUTANTS

Formation of unacceptable products of combustion of a fuel has been the bane of human life ever since pre-historic ages. Burning wood, for instance, provides fire for warmth, cooking meals and other necessities. Besides producing heat, however, the fire also produces smoke, soot and sometimes carbon monoxide (CO).

Inherent to the process of combustion, the higher the firing temperature the lesser will be the production of CO and unburned hydrocarbons (UHC). Unfortunately, the elevated temperature increases production of oxides of nitrogen (NO_x, most notably nitrogen oxide and nitrogen dioxide). And therein lies the crux of the problem of controlling exhaust gas pollutants.

Considerable success has been achieved in controlling all three of the offending components using truly innovative methods, but the methods tend to have their own merits and demerits. So most turbine manufacturers rely on a combination of methods in achieving the ever-stringent standards imposed by government agencies.

With conventional combustors of gas turbines, any modifications to alleviate generation of smoke and NO_x invariably results in increased levels of emission of CO and UHC, with the reverse also being true. Regulating the amount of air entering the primary combustion zone through a variable geometry mechanism has been successful to a considerable extent in overcoming this problem. Larger quantities of air are admitted at higher pressures to obtain more complete combustion and to minimize the incidence of soot and NO_x formation. When the pressure is low the primary airflow is restricted to raise the air-fuel ratio and to reduce the velocity, thereby improving combustion efficiency and lowering emission of CO and UHC. This feature also is helpful in initiating the combustion process during light up. The variable inlet geometry arrangement for controlling the flow of air calls for complex control and feedback mechanism, adding to the cost and weight while raising reliability questions, but may be justifiable for larger industrial gas turbines.

The concept of staged combustion attempts to achieve the same objectives by using two or more separate zones, each designed specifically to optimize certain features of combustion. In the lightly loaded primary first zone, high combustion efficiency and minimized CO and unburned hydrocarbon production are achieved by operating at a relatively high equivalence ratio between the actual fuel ratio and the stoichiometric fuel ratio of around 0.8. The primary zone, placed at the upstream end along the central axis of the chamber provides the temperature increase needed at low power settings up to idle speed. At higher power requirement, it acts as a pilot heat source for the main combustion region located downstream in the axial staging version, or at different radial distance from the pilot in the radial staging style. The second and subsequent zones are supplied with the fuel-air mixture of suitably adjusted equivalence ratio to all zones at varying levels of power setting. Staged combustion is extensively used in industrial turbines using gas fuels, achieving acceptable levels of pollutant emission. Ports are provided in the combustor liner to admit fresh air at various locations, which increases its complexity and cost of production.

Lean pre-mixed combustion is the preferred method for controlling the pollutants in natural gas fired turbines, but as the system operates close to the edge of combustion stability to reduce emissions of NO_x, modest upsets in operating conditions or variations in fuel composition tend to have repercussions in the overall power generating capability of the unit. Constituent concentration for natural gas available in United States tends to vary substantially. Such large swings pose a considerable challenge for the combustor to maintain optimal performance. As a consequence, delineation of the key phenomena is necessary to render the combustor insensitive to the variations.

Current emission regulations call for combustion efficiency in excess of 99 per cent. Failure to achieve this level is objectionable partly because fuel is wasted and increases production costs, but more so because it manifests in the form of excessive pollutants in the environment. For aircraft engine combustors, a design hallmark calls for it to be large enough to ensure an adequate level of combustion efficiency during engine restart at the highest altitude at which re-light is required. After a flameout during flight, combustion efficiency in the 75 to 80 per cent range is justified, because with the engine wind milling the temperature and pressure of ambient air are low enough to affect the stability of the flame. The engine control system attempts to compensate by supplying more fuel to the combustor, preventing the flame to stabilize due to the overly rich air-fuel mix. As a consequence, the combustion efficiency deteriorates.

Proper mixing in the primary zone enhances the rate of burning, while also minimizing soot and nitric oxide formation. A satisfactory temperature distribution in the exiting gases relies on the degree of mixing of the air and the products of combustion in the dilution zone.

15.9 STABILITY OF COMBUSTOR FLAME

Starting an aviation gas turbine engine has an additional requirement when an in-flight shutdown occurs at altitude and the light in the combustor is lost. The mixture of air and fuel must be capable of re-igniting as the engine continues to windmill. Ignition of fuel at start is sensitive to pressure inside the chamber.

The difficulty of igniting increases as the pressure reduces. Re-lighting of the engine at the aircraft's peak altitude is considered a benchmark. As a precaution to avoid flameout when an aircraft is taking off or coming in for landing during adverse weather when substantial amounts of water may be ingested into the engine, turning on the ignition system is usually recommended. Engine operation during ascend is at maximum speed, and mostly at flight idle during landing.

Combustion may not even initiate when the mass flow is too large. The stability curve for a typical combustor is shown in Figure 15.3, where limiting airflow and air/fuel ratio define the regions of stable operation. In addition, further difficulties are encountered during startup and shutdown periods. For instance, the mixture of air and fuel may be too rich as the engine accelerates and the fuel throttle is activated, since the flow of air takes time to build up to the necessary level. This leads to an unacceptably low ratio of air and fuel. One possibility for rectifying the situation is to measure variation in fuel flow rate and control it with the engine's microprocessor to avoid loss of flame during the transient periods, and thus also prevent rapid temperature changes in the turbine. The defined stable flame region in the diagram is also affected by pressure in the chamber. Chemical reactions slow down as the pressure decreases, which tend to restrict the flame's stability region. Aircraft engines also have to contend with reduced pressure of air exiting the compressor when operating at peak altitudes.

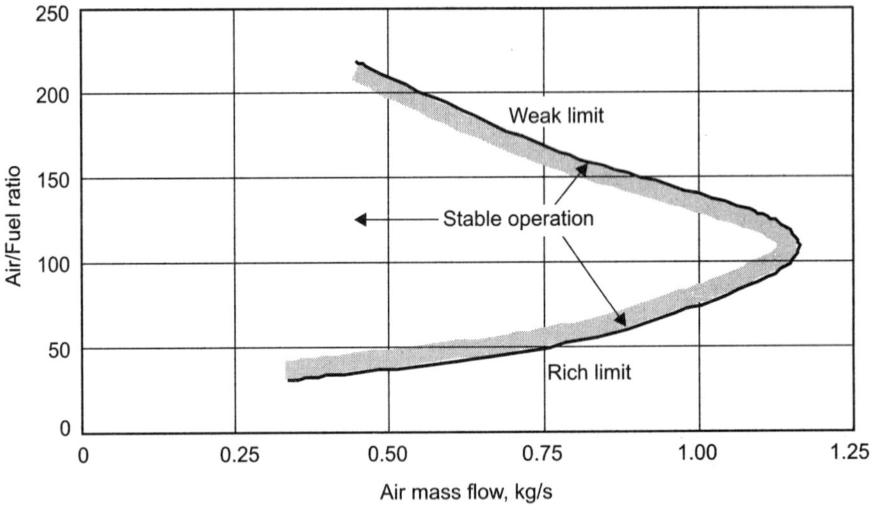

Fig. 15.3 Combustor Flame Stability Limits

Stable ignition at start may be represented in the form of a loop lying within the flame stability loop shown in the figure. Thus, for a given air mass flow, the range of air/fuel ratio in which the mixture can be ignited is smaller than that for which continuous ignition occurs. High-tension spark plugs similar to those used in piston engines may suffice for starting at sea level, but a higher energy spark is needed for ignition at lower chamber pressures. Periodic replacement of the igniter is essential since the sparks cause the electrode material to erode.

15.10 ACOUSTIC PRESSURE OSCILLATIONS IN COMBUSTOR

Combustion of the air and fuel mixture is accompanied by noise directly as a consequence of the process and indirectly due to the flow of burned gases through the turbine and exhaust nozzle. The sudden expansion of cool air of certain volume to a significantly larger volume after combustion is accompanied by noise.

Combustion noise can become detrimental when instabilities arising in the burning process couple with acoustic modes inside the chamber. The natural frequencies of the combustor can be excited by resonant pressure waves in the main gas flow along the axial and radial directions, as also by lateral modes in the tangential direction Sustained oscillating phenomena due to a higher level of mixing of the fuel and air prior to combustion lead to engine noise and associated vibration problems. Some gas turbine makers set the limit of 2.5 lb./in^2 in the dynamic pressure oscillations, pointing to the significance of the problem. The limit is based on the fact that the combustor liner is exposed to very high temperatures and is lightly supported. Imposition of higher dynamic pressures will lead to cracks and disintegration of the liner, and possible spread of combustor fire to the full engine.

Premixed combustion in gas turbines helps produce low levels of NO_x emissions, but practical application of this concept is limited due to self-excited combustion oscillations. When operation in a lean, premix combustor is close to the flammability limit, slight changes in operating conditions can lead to sudden flame extinction or to excessive CO emissions. In addition to static stability, lean premix combustors must achieve dynamic stability, meaning the combustion must not oscillate. Oscillation must be eliminated in a combustor design because the related pressure oscillations tend to have life shortening consequences. Figure 15.4 shows cracks developed in a transition piece of a power generation turbine due to excessive acoustic oscillations.

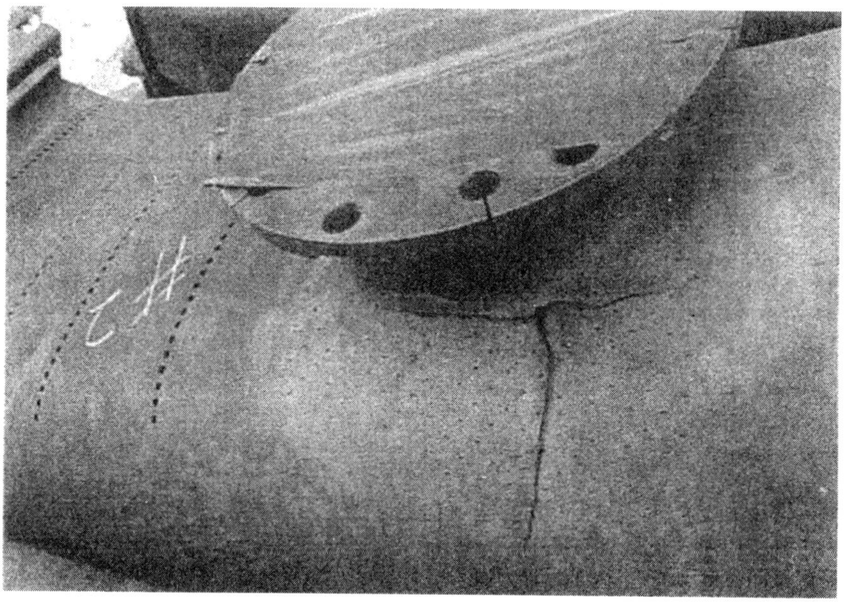

Fig. 15.4 Acoustic Oscillations Damaged Transition Piece (Lieuwen, 2002)

The task of studying and eliminating combustion oscillations in a gas turbine is complicated by the specific acoustic response of a combustor's design. The combustion process interacts with the acoustic field, leading to instabilities. Rapid changes in air and fuel supply and aerodynamic disturbances may lead to the instability because of a sequence of extinction and re-ignition of the flame in parts of the combustor. If the heat release rate does not take place uniformly and periodic spikes occur, acoustic waves of the same frequency are produced in the combustion zone. Reflection from the liner causes pressure waves to be returned to the combustion zone after a time delay, and the waves are reinforced when the heat release and pressure wave peaks coincide. As defined by Lord Rayleigh's criterion, oscillations set in when changes in heat release are in phase with acoustic pressure disturbances. Conversely, oscillations are dampened when heat release fluctuations are out of phase with pressure fluctuations. This criterion serves as the cornerstone for the development of combustion oscillation analysis. Variation in heat release results from changes in flame structure produced by acoustic pressure disturbances. Time delay between pressure disturbance and heat release variation determines the phase and, consequently, the stability of the system.

Based on these observations, lean premix combustors can be characterized by a simple time lag approach. Figure 15.5 shows for a specific case a schematic of important processes, where a sinusoidal pressure disturbance produces a sinusoidal variation in airflow 180° out of phase with the pressure. Time lag τ is estimated from the distance between the point of fuel injection and the flame front divided by average axial velocity, or:

$$\tau = (L + L')/U_{avg} \tag{15.1}$$

where L is the distance from fuel injection point to nozzle tip, L' is distance of nozzle tip to flame front and U_{avg} is average velocity of air-fuel mixture in the nozzle.

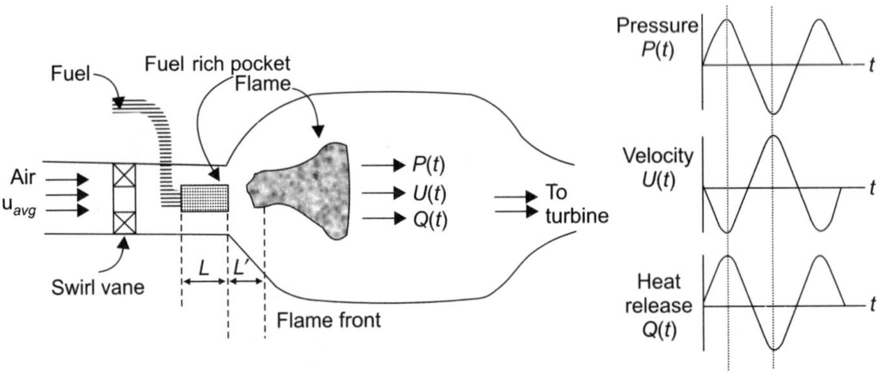

Fig. 15.5 Flow Characteristics during Acoustic Oscillation

15.11 NEED FOR UNIFORM TEMPERATURE DISTRIBUTION AT COMBUSTOR EXIT

Combustion of fuel in the presence of air discharging from the compressor diffuser produces extremely hot and pressurized gases. The gases then proceed towards the turbine inlet at the combustor exit. The thermal energy drives the turbine rotor, and develops enough power not only to drive the compressor but also the driven load. The driven load may be in the form of thrust to propel an aircraft, electrical signal in a power generator, compression of natural gases or other process applications.

It is imperative to have uniform temperature distribution in the complete cross-section of the flow area as the gases depart the combustor and prior to entering the turbine. If there are variations in the temperature profile, appreciable risks of experiencing hot spots in the turbine inlet vanes and in the subsequent rotating blades can be expected. As a consequence, thermal gradients are produced in the zones where the uneven temperature gases flow over the surfaces of the turbine components.

Unacceptably high thermal stresses will develop in the affected zones, and rapid thermal fatigue will ensue in the localized areas.

Precautionary features are designed in the combustor to obtain maximum level of mixing of cooler air and hot products of combustion, and thus produce an even temperature distribution at its discharge. Swirl cups at the entrance to the combustor are mostly intended to produce thorough mixing of the fuel and air to facilitate combustion of all of the fuel, but a side benefit is also to produce more uniform temperatures following combustion. The staged method of admitting fresh air at multiple locations in the combustor also helps towards the same objective. Other devices are also built in to enhance mixing of the cooler air and heated products.

Industrial and power generation turbine combustors have the benefit of axially longer combustors so that the two flow components are more thoroughly mixed, thus producing more uniform temperatures. In addition, after exiting the circular liners the gases go through a transition piece before entering the turbine. The transition piece basically provides flow path for the burnt gases from the circular cross-section at one end to the circular arc segment at the turbine inlet end. But weight is not of special significance in stationary turbines that a longer combustor path entails. Aviation engines, on the other hand, cannot enjoy this benefit due to stringent weight restrictions. The preferred method then is to induce maximized mixing of the hot and cold gases.

15.12 TURBINE INLET VANE THERMAL DEFORMATION

Direct exposure of the hot gases proceeding from the combustor poses critical problems in the turbine inlet vanes in the form of thermally induced deformation. Engine components at startup are in the cold condition. As the combustor fires and gases gradually heat up, metal temperature of the inlet vanes lying directly in the line of flow will also climb. Thermal growth is produced in the vanes.

The thermal growth, however, is not uniformly occurring in all regions of the vanes. The airfoils are attached at the outer and inner annulus walls with specially designed mount attachments to ensure proper fixity. Thus, at the two extremities the vanes have substantial rigidity to obtain restricted deformation from the thermal effects. In the airfoil section there is limited availability of the metal, making them far more flexible and exposed to thermal induced deformation. Also, the leading edge of the airfoil is wider with more metal then at the thinner trailing edge. This causes the trailing edge to warp from its cold condition state.

As the hot and pressurized gases leave the fixed vanes, they enter the rotating blades. A prime requirement of the stationary vanes is to direct the flow and feed the gases at the right angle of incidence. Variations in the angle of attack will lead to degraded power production. So the warping at the trailing edges of the inlet vanes has substantial ramifications for power developing ability of the machine.

In the design stage the thermal impact is taken care of using finite element methods of computation to calculate the differential thermal growth in all parts of the vane geometry. Cold dimensions are then determined to achieve best performance at multiple levels of power output. But as the turbine engine accumulates a number of start and stop cycles, thermal fatigue in the metal takes its toll and permanent, or plastic, deformations set in. The thermal growth does not go back to the original shape of the vane, as may be expected when the stresses and deformations are in the elastic range.

15.13 CREEP DEFORMATION FROM THERMAL RATCHETING

Creep is a material behavior characteristic that is a highly non-linear phenomenon. Creep behavior in a metal is dependent on four parameters: load (hence stress) on a component, temperature, time and creep-related material characteristics. Larsen-Miller curves are commonly used to depict creep-related characteristics, taking into account stress developed in a part, exposure to temperature and time to which it is subjected.

A consequence of creep is the metal part develops a combination of elastic and plastic (or permanent) deformations even when the part is subjected to stresses well within its elastic range. Thus, after a turbine part is exposed to a certain load and temperature for a period of time and following its removal, the elastic portion of the strain reduces to zero but the plastic deformation remains.

Rotating parts of a turbine are susceptible to creep-related deformation, but more so in the turbine section with its harsh operational environment than in the compressor. Centrifugal force from the spinning effects of the rotor chiefly causes tensile stresses throughout the airfoil. In the first turbine stage the hot gases flowing over the airfoils leads to elevated metal temperature. Subjecting the blades to the conditions for a period of time will lead to permanent growth at the tips when the engine is shutdown. Turbine inlet vanes mentioned in the previous section are another example of creep deformation.

The permanent portion of the strain tends to rise rapidly in the initial stages of application of the load. As the machine runs for a period of time, the plastic strain levels out. So if the unit runs for extended periods of time, as in power generation turbines that may run over several months continuously, development of plastic strain does not become an issue due to its limited magnitude.

Aviation engines, on the contrary, are required to run for far shorter periods of time, running into hours. But the engines have to go through a combination of several long and short flights. As the engine is fired up at the start and shut off at the end of the flight, the turbine blades will be exposed to what is known as thermal ratcheting. Blade metal temperatures will rise and fall during each flight. The plastic strains will accumulate at a higher pace than in the longer running power generation turbines. In particular, growth at the blade tips may get severe enough to cause rubbing against the casing and possibly lead to chipping and breaking of the blades.

15.14 EFFECTS OF FUEL CONTAMINANTS ON TURBINE BLADE SUBSTRATE

Suspended contaminants in natural gas tend to pose some problems in turbo-machinery because of their potential to produce erosion of the parts in the flow path from scratching. Natural gas invariably carries with it oily and gritty forms of sand and iron oxide, which may also stick to the surfaces and reduce the cross-sectional flow area. The fuel may even be restricted from flowing through small holes at the end of the fuel delivery tubes in the combustor.

Dissolved and suspended contaminants in liquid fuels are more difficult to resolve. Sodium, potassium, calcium and phosphorous compounds are present practically in all liquid fuels. Sulfur is even more insidious in its effects when the fuel is burned in a gas turbine's combustor. Following combustion the sulfur combines with the metal substrate of the airfoils in the elevated temperature zone to produce chromium and nickel sulfates. The products corrode the surface and create wavy patterns that obstruct the smooth flow of gases past the airfoils.

The effects of erosion and corrosion thus necessitate removal of the offending elements. In the case of natural gas and other gaseous fuels the cyclonic type of particle separators provide a high degree of reliability. One type calls for fixed vanes inside a vertically oriented cylindrical vessel that provide a whirl to the entering gases. The action causes heavier sand to fly outwards from centrifugal force of the whirling flow. The sand particles may stick to the walls or fall down to the base of the vessel. The inner core of the flow then goes through a set of dense filters that hold back tiny particles and let dust-free gas to be admitted to the combustor.

Liquid fuels depend on a set of filters of varying mesh size to hold back solid particles. Sodium, potassium, calcium and phosphorus elements are not difficult to chemically combine with agents to produce granules that are then washed off. Sulfur, on the other hand, is more difficult to treat chemically, and requires multiple steps for its removal.

15.15 DIFFERENTIAL GROWTH OF ROTOR AND STATOR AND TIP CLEARANCE

Operation of a gas turbine with excess clearance between the rotating blade tips and the stationary casing holds prospects of deteriorated performance, while potentially carrying serious risk of malfunctions. The effects will be felt both on the compressor and on the turbine sides.

Metal temperatures on the compressor side are cooler then on the turbine side. However, thermal mass of the rotating and stationary modules also plays a role in the clearances. The thermal mass takes into account overall weight, material density and thermal conductivity of the parts in the two assemblies. Though metal temperatures are lower in the compressor, relative thermal growth in the range of few thousandths of inch can affect its performance.

Air flowing through the annulus of an axial compressor performs the more aerodynamically difficult task of going against an adverse pressure gradient. As the compression ratio increases, the danger of stall troubles rises correspondingly, which is all too prevalent in axial compressors.

In individual airfoils, stalling is encountered when the difference between the flow direction and the blade's angle of incidence becomes excessively large. The radial clearance between the blade tips and the outer casing in the compressor is also a major culprit since it directly affects the stall margin and ability of operation. The stability of the flow is jeopardized due to the formation of random eddy currents that prevail in the dead space. The thickness of the turbulent flow regime varies with the available clearance at the blade tips. A thickened turbulent layer can gradually spill over to the flow space between adjacent airfoils to act as a blockage. Flow reversals can then occur and lead to a stall.

Physical effects of a surge in an aircraft engine's compressor can be grave. A sudden stoppage in the airflow and accompanied by a loud noise emanating from the engine is common during such an event. Damage to the engine structure may be expected when repeated a number of times, particularly in the compressor and fan areas.

Centrifugal and thermal loads are responsible for maximum radial variations in tip clearance in the turbine section. When a disk of mass m is mounted on a flexible shaft of length l between two bearings and spins at speed ω, the center of gravity of the disk is displaced at a radial distance e away from the centerline of the shaft. Then a rotating centrifugal force $m\omega^2e$ acts on the disk. The rotating force further causes the disk to execute vibratory motions, in the process increasing the clearance at the tip of the blades.

Differences in the temperature of the casing and rotor also will cause increments in radial clearance at the blade tips. A direct consequence is spilling over of the hot gases past the blade tips without doing useful work. The deterioration in performance is further aggravated under the simultaneous action of rotor unbalance.

The diminished power output will be detected through the torque sensor, sending the signal to the microprocessor. The processor will then order more fuel into the combustor to make up for the reduced power output. But the extra fuel will cause the metal temperatures to rise and even worsen the situation. The problem is further compounded as the turbine exhaust gas temperature also increases and consumes more of its valuable margin.

15.16 PASSIVE COOLING OF STATOR CASING

Cooling the turbine casing to shrink its size is one method of controlling the radial tip clearances at the rotating blades in aviation engines. Cooler air extracted from one of the compressor's intermediate stages is then sprayed over the full 360º of the casing in the plane of the rotating blades. Tubular rings with holes to spray the cool air are used for this purpose. The temperature differential between the hot metal surface and the coolant effectively reduces the tip clearances, with the intent to minimize spillage of the hot and pressurized gases past the blade tips. The net result is to increase power output of the turbine and to improve efficiency.

Advantage is taken of the thin walls of the turbine casing of aviation engines to readily reduce its diameter through this procedure. Land-based stationary engines have heavier walls, and are not easily vulnerable to shrinking with the cooler air.

Continuous cooling of the casing provides simplicity in design and fabrication of the system to ensure achieving its benefits. Maximum benefits are gained when the turbine engine is producing maximum thrust during the takeoff phase of the aircraft, since the casing temperature will be at peak. Operation at cruise conditions will cause somewhat lesser increase in metal temperatures, but will still offer benefits. Operation at flight and ground idle conditions can accrue even lesser gains.

Diverting the flow from the compressor to cool the compressor detracts from producing more hot gases, since the reduced availability of oxygen means lesser fuel can be burned in the combustor. So when the cooler air keeps flowing over the turbine casing without stopping while the engine is running at any speed, consequent losses will also be appreciable. Thus, some of the benefits gained are offset.

15.17 ACTIVE CLEARANCE CONTROL METHOD

Partial recovery of the losses attributed to the continuous flow of coolant in the passive cooling method can be achieved if the flow volume can be controlled during the various operating conditions of the engine. This serves as the basis in the active clearance control method of cooling the turbine casing. Dependent on the operational phase, the appropriate amount of air is made available for shrinking the casing diameter.

Elaborate sensors and control devices are required in the active clearance control method that work in conjunction with the turbine engine's microprocessor. Pyrometers that measure the temperature of the metal are placed around the circumference in each of the planes of the rotating turbine stages. The detected temperature readings are forwarded to the microprocessor, with its logic controller deciding the amount of air to be sprayed for individual stages. The command is then executed with valves located in the coolant supply line. Proportional control valves then open and release the required flow for each stage.

Substantial improvements in turbine performance are enjoyed using the active clearance control method to cool the turbine casing and to reduce the clearance at the blade tips. The advantages over the passive method cannot be overstated. The control method helps in providing the correct amount of coolant flow, and reduces losses pertaining to excess flow when it is not required.

But the method requires careful design and installation of all the control mechanisms. Efficient operation of the rather cumbersome system is necessary if the full benefits are to accrue. In the event of failure of any of the temperature detectors or the controlling valves, portions of the cooling system would not be functioning. The malfunctions, however, should not interfere with the regular operation of the turbine engine.

15.18 FUEL COMBUSTION EFFICIENCY

Complete combustion of fuel is the prime factor in maximizing the operating efficiency of a gas turbine engine. Even small amounts of fuel left unburned causes drop in the efficiency, while also representing financial losses for the engine operator.

Special features are built in the design of the combustor to procure close to 100 per cent efficiency during combustion. An important aspect is the mixing of fresh air with the fuel. Gaseous fuels generally mix with the air without much difficulty when the incoming airflow is given a swirl. A cup with curved vanes placed inside deflects the air and causes it to flow into the combustion zone in a swirling motion. The consequent turbulence reduces the flow velocity to enhance better mixing of the components. The incidence of particulate formation is then reduced substantially. Reduction in the axial flow velocity component also helps towards the same objective, since air exiting the compressor diffuser flows at a high velocity.

Liquid fuels need to be admitted in the form of a pressurized spray. Fuel injectors, or atomizers, have holes at the tip that deliver the spray in a conical sheet. The injectors are carefully designed so that the fuel splits into tiny spherical elements. The diameter of the spherical fuel droplets is controlled. The intent is to have maximum surface area of the droplets exposed to the oxygen, and to combine more easily during the combustion. The swirling motion of the airflow also helps in the process. In the design of the fuel injectors, the diameter of the droplets and the necessary air pressure are monitored to achieve the best combination.

Routine maintenance tasks performed during scheduled outages of the turbine engine also hold significance. Unburned hydrocarbon particles invariably tend to stick to the component surfaces in the proximity of the combustion zone. As the deposits accumulate with the hours of operation, total combustion of the fuel is inhibited. Chipping of the thermal barrier coat and cracks forming in the liners must also be repaired to achieve balanced and efficient operation of the combustor. Leakage of gases past the cracks in the liner will affect functionality of the combustor, while also affecting the engine generally.

15.19 EXHAUST GAS EMISSIONS AND COMBUSTION TEMPERATURE

The temperature in the firing zone plays a crucial role in the formation of exhaust gas emissions. Aside from inevitable products of combustion such as carbon dioxide and moisture (or water), certain objectionable elements are formed when the fuel is burned in the combustor of a gas turbine or the furnace of a boiler in a steam turbine installation. Poisonous gases such as carbon monoxide (CO) and unburned hydrocarbon (UHC) particles and oxides of nitrogen (NO_x) that are generally detrimental to human physiology are formed.

The amounts of the pollutants vary with the temperature of combustion. When the temperatures are low, say in the range of 800° F to 1500° F, the formation of CO and UHC runs at high levels. The ratio of air and fuel will also play a role. As the temperature is increased to above 2000° F and higher, CO and UHC levels reduce and diminish rapidly above this level. But the formation of NO_x follows the exact opposite trend, with NO_x levels rising rapidly in the higher temperature range.

In this regard, note that the higher gas temperatures also help to boost the level of power output produced in the turbine. So the trend in the combustion temperature to control CO and UHC is in the right direction. The problem, however, arises due to the oxides of nitrogen. The obvious solution is then to aim for combustion temperatures somewhere in the middling zone so all three objectionable elements can be controlled.

Turbine manufacturers thus focus their attention on achieving the firing temperatures in the middle range using a combination of methods such as staging, dry-low NO_x and selective catalytic reduction. The emphasis is more on reducing the formation of NO_x than to chemically reduce it to its elements, oxygen and nitrogen.

15.20 REVIEW QUESTIONS

15.20.1 How does exposure to high gas temperatures affect parts in a gas turbine?

15.20.2 Why is thermal barrier coating of hot parts required? What are its requirements to provide protection?

15.20.3 What is meant by the expression flame stability? How is a stable flame achieved?

15.20.4 What are the characteristics of a good temperature profile of the gases entering the turbine?

15.20.5 What is thermal ratcheting? What is its role in the operation of a turbine?

15.20.6 How are oscillating pressure waves caused? How do they affect the liner of a combustor?

15.20.7 How does sulfur in the fuel affect performance of a gas turbine?

15.20.8 Why is excess clearance at the blade tip objectionable in a compressor?

15.20.9 Under what circumstances are the passive and active methods of controlling tip clearance preferable in a turbine engine?

15.20.10 Name some methods to achieve the best results when burning a liquid fuel.

15.20.11 What is the one factor chiefly responsible for formation of NO_x during combustion of a fuel?

15.21 EXAMPLE PROBLEMS

Example Problem # 15.21.1: A 0.895 in. diameter and 11.25 in. long bar is pinned at one end, hanging freely. The temperature of the bar is uniformly raised from 128° F to 1150° F. If the bar material's coefficient of thermal expansion $\alpha = 6.285 \times 10^{-6}$ in/in/° F and Young's modulus is 28.72×10^6 lbs/in², calculate the thermal growth in the bar lengthwise and associated thermal strain.

Solution: Thermal growth $= \alpha L \Delta T = 6.285 \times 10^{-6} \times 11.25 \times (1150 - 128)$

$= 0.07226$ in.

and thermal strain $\varepsilon_{th} = \alpha L \Delta T / L = \alpha \Delta T$

$= 6.285 \times 10^{-6} \times (1150 - 128)$

$= 0.006423$ in./in. length

Example Problem # 15.21.2: A 1.275 in. diameter and 8.375 in. long bar is clamped at both ends. The bar is uniformly heated from 75° F to 290° F. If the bar has coefficient of thermal expansion $\alpha = 7.737 \times 10^{-6}$ in/in/°F and Young's modulus is 11.63×10^6 lbs/in^2 calculate the thermal stress in the bar, ignoring lateral growth in the bar.

Solution: Thermal growth = $\alpha L \Delta T$ and thermal strain $\varepsilon_{th} = \alpha L \Delta T / L = \alpha \Delta T$

Then thermal stress $\sigma_{th} = \alpha E \Delta T = 7.737 \times 10^{-6} \times 11.63 \times 10^6 \times (290 - 75)$

$$= 19345.98 \text{ lbs/in}^2$$

Example Problem # 15.21.3: A 2.135 in. deep, 1.735 in wide and 7.375 in. long rectangular bar is pinned at one end and hangs freely. The temperature of the bar uniformly varies from 145° F at the top end to 642° F at the lower end. If the bar material's coefficient of thermal expansion $\alpha = 8.7375 \times 10^{-6}$ in/in/° F and Young's modulus is 26.53×10^6 lbs/in^2 calculate the thermal stress.

Solution: Thermal stress in bar with uniformly varying temperature along its length:

$$\sigma_{th} = \alpha E \Delta T = 8.7375 \times 10^{-6} \times 26.53 \times 10^6 \times (642 - 145)$$
$$= 115207.52 \text{ lbs/in}^2$$

Example Problem # 15.21.4: A rotating stage in a steam turbine rotor has a design cold clearance at the blade tip of 0.0073 in. Static weight deflection is 0.0018 in. and calculated dynamic deflection from unbalance is 0.0041 in. Determine the maximum and minimum clearances on that stage.

Solution: Assume cold design clearance is uniform around the circumference. Deadweight deflection will be maximum at the base. When rotor is spinning, dynamic deflection will be uniform around the circumference.

Then maximum clearance at top of rotor = 0.0073 + 0.0018 – 0.0041

= 0.0050 in.

And minimum clearance at base of rotor = 0.0073 – 0.0018 – 0.0041

= 0.0014 in.

Example Problem # 15.21.5: In *Example Problem # 15.21.4* assume dynamic deflection varies linearly with unbalance at this stage. What percentage of unbalance increase will cause the tips to rub at the base?

Solution: The given unbalance leads to 0.0041 in. deflection. Additional unbalance will cause 0.0014 in. deflection, so unbalance has to increase by:

$$= (0.0014/0.0041) \times 100$$
$$= 34.15 \text{ per cent.}$$

15.22 EXERCISE PROBLEMS

Exercise Problem # 15.22.1: A long and uniform bar is pinned at one end, hanging freely. The temperature of the bar is uniformly raised from 79° F to 550° F. If the bar material's coefficient of thermal expansion $\alpha = 7.723 \times 10^{-6}$ in/in/° F, calculate the length of the bar if it grows lengthwise 0.0118 in.

Exercise Problem # 15.22.2: A freely hanging bar is pinned at one end. The temperature of the bar is uniformly raised from 129° F to 1273° F. Calculate the bar material's coefficient of thermal expansion if the length of the bar is 13.25 in. and thermal strain lengthwise is 0.0083 in./in

Exercise Problem # 15.22.3: Calculate the thermal stress in a 9.25 in. long bar clamped at both ends if the bar is uniformly heated from 1235° F to 1490° F, the bar has coefficient of thermal expansion

$\alpha = 6.277 \times 10^{-6}$ in/in/° F and Young's modulus is 27.42×10^6 lbs/in^2 the bar, ignoring lateral growth in the bar.

Exercise Problem # 15.22.4: A gas turbine stage has static weight deflection of 0.0029 in. and calculated dynamic deflection from unbalance is 0.0052 in. Determine the design clearance at the blade tip if minimum clearance on that stage is limited to 0.0025 in.

Exercise Problem # 15.22.5: A dynamic unbalance in a turbine rotor of 5 oz.-in. produces 0.00245 in. deflection in a rotor at a stage. Design clearance is set at 0.0075 in. If running clearance is required to be 0.00425 in., calculate the maximum unbalance acceptable.

REFERENCES & BIBLIOGRAPHY

Anderson, D. N., "Effects of Equivalence Ratio and Dwell Time on Exhaust Emissions from an Experimental Premixing Pre-Vaporizing Burner", ASME paper # 75-GT-69, New York, 1975.

Bahr, D. W., "Technology for the Design of High Temperature Rise Combustors", Journal of Propulsion and Power, volume 3, number 2, page 179 – 186, 1987.

Buecker, B., "Emissions: SCR Design", Power Engineering, page 24 – 28, August, 2002.

Candel, S. M., "Combustion Instability Coupled by Pressure Waves and Their Active Control", 25th International Symposium On Combustion, Sydney, Australia, 1992.

Darling, D., Radhakrishnan, K., Oyediran, A., Cowan, E., "Combustion-Acoustic Stability Analysis for Premixed Gas Turbine Combustors", NASA TM 107024, 1995.

Feitelburg, A. S., Tangirala, V. E., Elliott, R. A., Pavri, R. E., Schiefer, R. B., "Reduced NO_x Diffusion Flame Combustors for Industrial Gas Turbines", ASME paper # 00-GT-085, New York, 2000.

Gupta, A. K., Lewis, M. J., Daurer, M., "Swirl Effects on Combustion Characteristics of Pre-Mixed Flames", ASME Journal of Engineering Gas Turbines & Power, volume 123, page 619 – 626, New York, 2000.

Kaya, H., "Catalytic Combustion Technologies", Journal of Gas Turbine Society of Japan, volume 25, number 98, page 48 – 51, Tokyo, 1997.

Kuznetsov, V. R., "Influence of Turbulence on High Non-Equilibrium Concentrations of Atoms and Free Radicals in Diffusion Flames", Fluid Dynamics, volume 17, page 815 – 820, 1982.

Lefebvre, A. H., "Gas Turbine Combustion", Taylor and Francis, Philadelphia, 1999.

Leong, M. Y., Smugeresky, C. S., McDonell, V. G., Samuelsen, G. S., "Rapid Liquid Fuel Mixing for Lean Burning Combustors: Low Power Performance", ASME paper # 00-GT-116, New York, 2000.

McManus, K. R., Poinsot, T., Candel, S. M., "A Review of Active Control of Combustion Instabilities", Prog. Energy Combustion Science, volume 19, page 1 – 29, 1993.

Nealy, D. A., "Combustor Cooling – Old Problems and New Approaches", Gas Turbine Combustor Design Problems, A H. Lefebvre editor, page 151 – 185, Hemisphere, Washington, DC, 1980.

Nicol, D. G., Steele, R. C., Marinov, N. M., Malte, P. C., "The Importance of Nitrous Oxide Pathway to NO in Lean Pre-Mixed Combustion", ASME Journal of Gas Turbines & Power, volume 117, page 100 – 111, New York, 1995.

Ohtsuka, M., Yoshida, S., Inage, S., Kobayashi, N., "Combustion Oscillation Analysis of Premixed Flames at Elevated Pressures", ASME paper # 98-GT-581, New York, 1998.

Rangwala, A. S., "Turbo-machinery Dynamic: Design and Operation", McGraw-Hill, New York, 2004.

Richards, G. A., Janus, M. C., "Characterization of Oscillations During Premix Gas Turbine Combustion", ASME paper # 97-GT-244, New York, 1997.

Rink, K. K., Lefebvre, A. H., "Influence of Fuel Drop Size and Combustor Operating Conditions on Pollutant Emissions", International Journal of Turbo and Jet Engines, volume 6, number 2, page 113 – 122, 1989.

Sato, H., Mori, M., Nakamura, T., "Development of a Dry Low NO_x Double Swirler Stage Gas Turbine Combustor", ASME paper # 96-GT-134, New York, 1996.

Sawyer, T., "Gas Turbines – Volumes I, II and III", International Gas Turbine Institute, Atlanta, ASME, 1982.

Schlatter J. C., Dalla Betta, R. A., Nickolas, S. G., Cutrone, M. B., Beebe, K. W., "Single Digit Emissions in a Full Scale Catalytic Combustor", ASME paper # 97-GT-57, New York, 1997.

Segalman, I., McKinney, R. G., Sturgess, G. J. Huang, L. M., "Reduction of NO_x by Fuel Staging in Gas Turbine Engines – A Commitment to the Future", AGARD Conference Proceedings 536, page 29/1 – 17, 1993.

Seume, J. R., Vortmeyer, N., Krause, W., Hermann, J., Hantschk, C. C., Zangl, P., Vortmeyer, D., Orthmann A., "Application of Active Combustion Instability Control to a Heavy Duty Gas Turbine", ASME paper # 97-GT-119, New York, 1997.

Snyder T. S., Rosfjord, T. J., McVey, J. B., Chiappetta, L. M., "Comparison of Liquid Fuel/Air Mixing and NO_x Emissions for a Tangential Entry Nozzle", ASME paper # 94-GT-283, New York, 1994.

Touchton, G. L., "Influence of Gas Turbine Combustor Design and Operating Parameters on Effectiveness of NO_x Suppression by Injected Steam or Water", ASME paper # 84-GT-IPGC-3, New York, 1984.

White, D. J., Batakis, A., Le Cren, R. T., Yacabucci, H. G., "Low NO_x Combustion Systems for Burning Heavy Residual Fuels and High Fuel-Bound Nitrogen Fuels", ASME Journal of Engineering Gas Turbines & Power, volume 104, page 377 – 385, New York, 1982.

16

CHAPTER

PITFALLS IN FAN AND COMPRESSOR OPERATION

16.1 INTRODUCTION

Aerodynamically speaking, the compressor is far more difficult to design than the turbine. The problem arises from the fact that the flow of air takes place against the pressure gradient in a compressor, while it takes place with the gradient as the gases expand in the turbine. Similar to a man climbing up a hill against gravity and thus more tiring, likewise the compression process may be considered to be climbing up in pressure against the gradient. Once he reaches the summit, gravity makes it much easier to go down the slope, much like expanding gases in the turbine flowing with the pressure gradient. Further extending the analogy, stumbling of the climber due to a difficult obstacle would be comparable to a pressure surge in the compressor.

The role of the rotor is to propel or push the air and impart kinetic, or velocity, energy to the air. This energy is then converted into pressure or potential energy through the fairly simple mechanism of expanding the cross-sectional flow area. The result is to cause the flow velocity to reduce and convert into pressure energy according to the dictates of principle of conservation of energy.

The impeller in a centrifugal compressor imparts mostly velocity to the air as it enters the eye, or the inducer, and turns 90° to exit at the exducer nearly in the radial direction. The flow passages between the blades attached to the disk are configured to convert little kinetic energy into potential energy. Almost all of the conversion takes place in the stationary diffuser section of the centrifugal stage, where the gradually expanding space aids in the conversion process.

The high velocity in the rotating stage, as also in the dead space between the blade tips and the diffuser's leading edge, will imply higher friction losses since they vary with the square of flow velocity. Also, the reduced access area to the flow at the eye limits the amount of air admitted.

Axial flow compressors have a much larger flow area at the inlet to the compressor. Nearly 50 per cent of the energy conversion takes place in the rotating blade passages and the remainder in the stationary vane passages. The intent here is to meet the twin objectives of maintaining cross-sectional flow area and flow velocity within limits and the energy losses associated with friction. At the same time, however, the flow has to go up the pressure gradient in both the stationary and rotating stages, thereby limiting the amount of compression in one stage.

A centrifugal stage has the capacity to provide compression to the extent of 5:1 in one single stage. Further increase beyond 7:1 will result in substantial drop in efficiency that cannot be ignored. One axial flow stage can at the most provide pressurization running between 25 to 30 per cent due to the more difficult flow regime in the expanding section.

16.2 FAN TANGENTIAL TIP SPEED LIMITATION

A large portion of the thrust produced with an aircraft's high bypass turbo-fan engine is developed in the fan, with the gases exiting from the turbine's exhaust delivering a smaller part. The arrangement offers the benefits of improved specific fuel consumption and reduced levels of noise.

The level of thrust produced is dependent on the diameter of the fan and its spin speed. The available thrust increases in proportion to the area of the fan blades as the air flows through the annulus region between the fan hub and the fan casing. The thrust generated also increases with the speed of the rotor. But the tangential speed at the blade tips limits the amount of thrust produced. When the tip speed exceeds the speed of sound, shock waves are set up that can potentially jeopardize the integrity of the blades. And since both the span of the blades and the rotor speed dictate the speed at the fan's periphery, the two parameters need to cope with the restriction.

Both the velocity of the flowing air over the surface of the rotating blades and of the peripheral tip speed must not go past the acoustic speed to prevent the blades from being affected with flutter. In industrial and power generation engines, as also in aviation engines, there are instances where the initial stages of the compressor operate beyond Mach 1.0. The outer sections of the blades are referred to as operating past the sonic range, with the inner portions within the acoustic limit, giving rise to what is known as transonic compressor blades.

16.3 VARIABLE INLET VANES IN POWER GENERATION TURBINE

Gas turbines dedicated to electrical power generation operate at a steady 3600 rpm for 60 Hz signal or at 3000 rpm if the power is to be produced at the frequency of 50 Hz. However, at startup time the unit needs to go from standstill to the maximum speed. Inlet guide vanes are provided to direct the flow of outside air through the intake manifold, bell-mouth and into the compressor's rotating first stage.

Problems arise at certain ambient conditions of pressure and temperature during the startup roll when excessive amounts of air flows in than can be handled, and set up choking conditions at entry to the compressor. Backup in the flow and consequent flow reversal can cause the engine to stall when the engine has not attained full speed. The immediate impact is to cause the engine system to vibrate, emit noise and generally interrupt the startup of the unit. During shutdown the problem is not of concern since the engine speed is coming down.

Variable inlet vanes provide the solution for the problem. Individual inlet vanes are actuated to permit them to swivel about their own axis to restrict the flow area available for the outside air to enter the compressor's first stage. All of the vanes are made to turn simultaneously through a definite schedule. An actuating ring placed around the casing and activated with a hydraulic power cylinder causes individual mechanical linkages to move, thus forcing the vanes to open or close.

During the startup process the vanes are in position to substantially restrict the flow. As the engine picks up speed, the vanes are required to gradually open and admit more outside air. At full speed the vanes are in their regular position and permit the turbine to perform.

16.4 BYPASSING AIR THROUGH BLOCKER DOORS

Many aviation engines have variable blocker doors built into the fan frame's structure, just prior to the air entering the high-pressure compressor module. The doors permit some of the air to bypass the regular flow coming from the fan and the low-pressure compressor during certain operational phases. Especially during the startup mode and when the engine is running at idle the flow of air needs to be restricted, since a larger mass flow tends to choke the initial lower-pressure stages. As the speed increases into its regular operating regime, the doors gradually close and allow all of the air to flow into the high-pressure compressor.

The doors are located on the inner fan frame hub at discrete circumferential places. The automatically actuated doors scoop some of the air from the flow channel and direct it into the fan outlet duct. The consequent reduced mass flow then allows the engine to go through the paces during the start, without fear of being overwhelmed by excess supply of air.

The variable blocker doors are somewhat akin to the bleeding of air from the higher-pressure stages. At and near full thrust settings, the latter higher-pressure stages may experience choking, meaning the blades are not tall enough to let all of the air coming through. Some of the air is bled through holes machined in the casing just before the air flows into the succeeding stator stage. The air may be bled from more than one stage when the compressor is built in several stages.

The bled air is used for carrying out multiple tasks, such as cooling the combustor liner and high-pressure turbine blades and disks. A customer bleed port with control sensors and valves for cabin pressurization may also be available.

16.5 VARIABLE STATOR VANES IN AIRCRAFT ENGINE COMPRESSOR

Military and commercial service aircrafts are required to fly over periods of time ranging from few hours for short trips to several hours for longer flights between continents. Within the flight envelope the aircraft goes from ground idle to flight idle, and takeoff during the start, followed by cruise, flight idle, ground idle and final shutdown at the end of the mission.

A direct implication is that both the low and high-pressure modules of a transport aircraft engine operate over a large speed range. A case in point is General Electric Company's CF6-80C engine for Boeing 747 airplane and other applications. The low-pressure module operates between 800 and 3500 rpm range, while the high-pressure modules turns between 6600 rpm and 10600 rpm.

The first six stages of the high-pressure compressor are provided with variable stator vanes. A compressor stage is susceptible to stall when the angle of attack of the air entering the rotating stage blades is not within its required upper and lower limits. When the incidence angle falls outside the two bounds, turbulence in the form of eddies form around the leading edge of the moving row of blades. The effect is to partially block the flow of air, and in severe conditions may even cause the flow to reverse itself and lead to a full-blown stall.

Stator vanes that can be variably set for different running speed regimes provide the answer to this problem. As the running speed changes, the stator vanes turn to a pre-determined angle and direct the flow into the rotating blades smoothly to minimize buildup of eddy currents around their leading edge. Manipulation of the vanes in all six stages is through rings with mechanical links connected to individual vanes. A single hydraulic actuator connects with a mechanical member that in turn connects with the rings. The variable vanes also aid in controlling the flow area, so that at low speeds the flow volume may be restricted. The combined effect is to closely control the operation of the compressor and substantially reduce the risk of the engine going into a stall.

16.6 FRICTION LOSSES FROM FLOW OVER AIRFOIL

Flow of air over the surfaces of the airfoils inevitably leads to losses arising from friction. A minute layer of air sticks to the surfaces, with progressive layers of air away from the areas gradually assuming the velocity of the free stream of air. Shear forces are produced in the streams of differing velocity around the airfoils that result in friction.

Surface finish of the airfoils is the controlling factor in the generation of friction losses. When the surface has a high degree of finish from the manufacturing processes, the air flows smoothly without any obstructions. If tiny ridges are left over after machining, the ridges will retard the flow since air particles tend to cling between them. And when there are greater deficiencies in manufacturing, surface roughness will result in even greater obstruction and cause the efficiency to drop.

Operational aspects also are responsible for high friction losses. Power generation turbines ingest dust and biological matter through the intake and get past the air filters in the intake and into the compressor inlet. Particles also stick to the annulus walls and other parts, gradually restricting the passages. The larger flow velocity increases the friction and brings down the efficiency of the compressor.

16.7 PRESSURE SURGE AND STALL IN AXIAL COMPRESSOR

The chief culprit behind an engine going into a stall is incorrect angle of attack into the rotating blades and the consequent creation of pockets of turbulence at their leading edge. The surging pressure behind the turbulent blockage of the rotating blades causes the flow of air to backup. The natural tendency of air is to flow from a point at higher pressure to a point at a lower pressure. At the same time, the rotor is compelling the air to move forward, thereby setting up a struggle.

Another significant factor causing the pressures to surge is the radial clearance between the stationary casing and the rotating blade tips of the compressor. Above a certain magnitude, the air currents move in a mostly random whirling pattern within the dead space.

The radial clearance is set during initial manufacture and assembly as a proportion of the chord length at the tip of the blades. Minimum clearances are required to assure the rotor does not rub against the casing due to unbalance, misalignment and other assorted factors. But the clearances must not be excessive if the whirling airflow in the region is to be avoided.

The risk of the turbulent regime spilling over into the defined channels between adjacent airfoils is inherent with excessive clearance. Aircraft engine manufacturers go through the routine of measuring outer diameters around the periphery at all stages of the rotor and also the inside casing diameters. Computer programs are used to match a given rotor with the selected casing from the group in order to obtain the best combination of clearances at the outer tips.

16.8 STALL MARGIN DETERMINATION

Aircraft engine manufacturers carry out extensive performance tests to accurately establish the stall margin and ensure operation of the engine does not stray into the prohibited region. Problems arise from the fact that the compressor can only deliver a restricted level of pressurization for a stated amount of air flow mass. When the pressure exceeds its upper bound for any given flow, reversal in the direction of flow and its associated ramifications will be felt throughout the engine.

A typical compressor performance map displays the stall curve distinctly, with the horizontal axis representing mass flow rate and the vertical axis indicating compression ratio or pressure increase. The two parameters are appropriately non-dimensionalized to permit taking into account related parameters such as speed, pressure and work done. A number of curves are shown on the map for various parameters such as corrected speed and power required.

The observed stall margin data obtained during engine tests is valuable information for the programming of the engine's microprocessor-based controller. Strict protocols must be set up to prevent the engine from running in the restricted zone of the performance map. Repeatability of the acquired data is not guaranteed, so the tests must be carried out over prolonged periods of engine run. And to avoid variations from one unit to another, the performance tests must be carried out over a number of engines of the same type.

Analytical prediction of the surge curve is not without risks. Aside from the geometric parameters defining all the stages, the task is complicated due to the many different operating variables. At best, attempts may be made to use observed stall margin of an operating engine and use geometric similarities to interpolate data for calculating the stall margin. Extrapolation of the data is even more hazardous if the compressor is larger than that from the test data.

16.9 EFFECTS OF STALL

Variable speed turbine engines for aircraft are more susceptible to the phenomenon than the essentially fixed speed machines for power generation and other industrial uses. A complete reversal in flow and subsequent stall will cause the engine and the aircraft structure to vibrate heavily, accompanied by loud noise. Pressure and temperature at the compressor discharge will spike, together with the exhaust gas temperature at the turbine exit. Excessive vibrations will also lead to severe rubs between the stator and rotor in the compressor and the turbine, with consequent opening up of clearances and metal chips dislodging from the blade tips.

The measured signals from the pressure, temperature and vibration transducers will cause the engine's microprocessor system to order reduction in fuel injected into the combustor. Power output and rotor speed will then fall and prevent the stall from further spreading. The sequence of events taking place during the occurrence is so rapid it is not possible for the pilot to take timely corrective action. The engine's control system is generally powerful and sensitive enough to take remedial action on its own.

After a stall incident, ground personnel need to bore-scope the compressor and turbine sections to evaluate the damage at the blade tips. When the engine is restarted, the engine's vibration signals will generally be higher than before the incident. Also, there will be deterioration in the turbine's exhaust gas temperature margin since the clearances will have opened up.

16.10 RADIAL COMPRESSOR FOR CHEMICAL PROCESSING

Radial flow centrifugal compressors are ideal where the mass flow requirement is limited. Higher compression than is available from a single stage is possible if two or more stages are placed in series with each other on the same shaft. The discharge from one stage is then directed into the impeller eye of the following stage through suitably machined ports in the casing. The net result is to provide high levels of pressurization at acceptable levels of efficiency.

The small size of the impeller and diffuser combination results in a compact size, making them suitable for many process chemical applications. The centrifugal concept also finds usage in helicopter engines and in auxiliary power units for larger transport aircrafts.

The vanes on the rotating disk direct the flow towards the outer radius while imparting velocity to the flow. The pressure at the disk periphery remains essentially the same as at the eye of the impeller. For chemical processing, after clearing the gap between the impeller and the diffuser, the gaseous chemical enters the radially expanding space in the diffuser, where conversion of the velocity into pressure occurs. The size and weight of the impeller is rarely of concern in process applications, so substantial pressurization can be obtained with just one centrifugal stage. Mass flow can also be increased with suitable diameters at the inducer and at the exducer. For greater compression, more stages may be added on the same shaft, with flow proceeding in series from one stage to the next. Thus, if a single stage results in compression ratio of 5 to 1, four stages connected in series will produce 5^4 to 1, or 625 to 1, pressure ratio.

16.11 SURGE PROBLEM IN CENTRIFUGAL COMPRESSOR

The nearly absent conversion of velocity energy into pressure energy in the rotating stage means there is little opportunity for the flow to reverse directions, with the rotor forcing the gas to move outward. In the diffuser section the role is reversed. The flow rate is reduced as the velocity decreases and pressure increases. The flow is now going through the uphill task of increased pressure gradient, and possibly set up conditions for its reversal.

But the task of traversing from a lower-pressure region of the diffuser into a higher-pressure region is easier when compared with that of the axial flow compressor, where turbulence may be encountered at the leading edges of the rotating blades and its consequent blockage. Excess radial clearance at the impeller tip with the diffuser casing may create whirling eddy currents in the dead space. Fixed stator vanes in the diffuser with rounded edges at the tip may help facilitate the flow.

Overall, it may be surmised there is lesser risk of surging pressure in a radial compressor, and improved reliability may be expected.

16.12 FLOW VELOCITY AND EFFICIENCY IN COMPRESSORS

Maximized flow velocity and efficiency in a compressor are desirable features for developing greater compression. But the two parameters are at loggerheads. Higher flow velocity in an annulus permits greater cramming of the air. At the same time, friction losses increase with the square of velocity and adversely affect efficiency.

Centrifugal compressors attempt to develop peak velocity in the rotating stage as the air departs at the tip and proceeds into the stationary diffuser. Rotor speed is controlled to obtain compression in the range of 5:1 to 7:1 at the exit from the diffuser. Any further attempts to increase the pressure ratio through higher rotor speed will impede the efficiency of operation, since friction losses increase exponentially with velocity. Even when flow surfaces are in optimum condition and friction coefficient is low, efficiency dramatically falls to unacceptable levels at higher velocity of radial flow.

Axial flow compressors are restricted from achieving compressibility above 1.3:1 in a single stage due to the surge factor. Three-dimensional wider chord compressor blades have been designed in a number of aviation gas turbines that can deliver remarkably greater pressurization in one stage, coupled with satisfactory stall margin. Both stationary and rotating blade profiles are developed to obtain partial conversion of velocity to pressure energy. The objective is to get the most mass flow and encounter least resistance due to friction in the stages.

The rotating blades are tasked to impart velocity to the flow while also participating in the conversion from kinetic to potential energy. What is referred to as the 50 per cent design calls for half of the pressurization in the stationary vanes and the remainder in the rotating blades. Satisfactory compressibility and surge margin are the main beneficiaries with the arrangement.

16.13 COMPRESSOR FOULING AND WASH

A power generation facility located in industrial areas generally has a high incidence of suspended particles of foreign material that may get ingested through the turbine's intake. Even if air filters in the intake are in good condition, some of the particles will get ingested in the compressor inlet. Sand and iron oxide particles are particularly insidious since they tend to scratch the airfoils.

A large number of oily and greasy particles can also stick to the airfoils and gradually reduce the area of flow passages between the airfoils. As a consequence, the velocity of flow increases, as also friction since the losses depend on the square of velocity. Larger foreign debris may also cause chips of metal to dislodge from the airfoils.

Washing the compressor is effective in removing the adherents from the airfoils. A spray of water injected into the compressor inlet with the engine running is many times beneficial in raising the performance of the compressor. Other operators prefer using small amounts of chemical detergent mixed with water, with the engine on turning gear.

16.14 INLET AIR COOLING AT HIGH AMBIENT TEMPERATURE

Daytime temperatures in the summer in many parts of the world routinely climb up to 115° F or more. So people in the area must resort to turning on their air conditioners to be able to carry on their regular activities. But the power turbine generators cannot cope up with limitless demands. Since the ambient temperature at the intake to the compressor is high, the consequent reduced density implies the mass flow of air is also reduced.

One option for turbine units operating in a region where the relative humidity is low is to cool the incoming air through water injection at the inlet. The Mojave desert in Southwestern United States and the Saudi Arabian desert are just two examples that can benefit from cooling the air during peak summer days. The combined effect of the dry and hot air and the cool water is to shrink the volume of air, thereby permitting a higher mass flow into the compressor. Thus, greater fuel may be injected in the combustor to raise the power available at the coupling with the generator rotor. The method has been successfully used at many locations to provide valuable additional power.

Precautions must be taken to ensure the water injected is free of dissolved and suspended contaminants. Compounds of sodium, potassium and other elements must be chemically removed before admitting the water into the unit. Also, large amounts of water will be required, which can create problems in many desert regions. Still another problem is corrosion of steel parts throughout the turbine as the water is sprayed at the intake.

16.15 OPERATION AT SUB-ZERO AMBIENT TEMPERATURE

Frigid sub-zero air temperatures are all too prevalent in regions both north and south of Arctic in the northern hemisphere, as also around the Antarctic in the southern hemisphere. The low temperatures cause embrittlement of many grades of carbon steel. The properties of the material degrade to the extent that when load is applied to a machine part, the component fails due to the metal becoming brittle and causing it to fail, essentially because the part is exposed to very low air temperature.

The compressor spool of a power generation gas turbine is one such example. Since the part weight may run into several thousand pounds, turbine manufacturers avoid using superior grade metals that can withstand low temperatures in order to keep costs under control. When ambient temperatures dip to –30° F, starting the turbine unit from the cold condition carries substantial risks for the integrity of the compressor rotor. The rush of incoming cold air will also have further chilling effect on the metal.

Heating the incoming air with air heaters provides a reasonable solution to the problem. Circulating hot gases or steam through a bank of tubes may be placed at the inlet to the compressor. The air heater system may be rolled in place only when environmental conditions dictate.

16.16 REVIEW QUESTIONS

16.16.1 Why is it essential to limit the speed at the tip of fan blades?

16.16.2 Why is stalling a bigger problem in aviation gas turbines than in power generation gas turbines?

16.16.3 Why is improved efficiency not achievable in a centrifugal compressor stage with a higher compression ratio?

16.16.4 What is the role of variable stator vanes? How are they actuated?

16.16.5 Explain what transonic operation of low-pressure compressor blades implies.

16.16.6 Explain the merits and demerits of a centrifugal compressor.

16.16.7 How is the stall curve established on the performance map of a compressor?

16.16.8 How does a stall manifest itself in a compressor?

16.16.9 Why is surging pressure in an axial flow compressor a larger problem than in its radial counterpart?

16.16.10 Why is a radial flow centrifugal compressor not suitable for a turbo-fan engine of transport aircrafts?

16.16.11 How is operation of a gas turbine affected when outside air temperature is very low? Also when it is very high?

16.17 EXAMPLE PROBLEMS

Example Problem # 16.17.1: The barometric pressure of air outside an aircraft is 29.24 in. of mercury, pressure at the nozzle exhaust is 80 in. of mercury and area of the exhaust nozzle is 275 in^2. Determine the thrust developed in the engine due to the pressure differential.

Solution: At sea level 29.829 in of Hg = 14.69 lb./in^2 pressure.

Hence, 29.24 in of Hg = $14.69 \times 29.24/29.829 = 14.40$ lb./in^2 pressure.

And 80 in of Hg = $14.69 \times 80.0/29.829 = 39.398$ lb./in^2 pressure.

Then thrust $f_n = (p_j - p_{atm})a_j$.

$f_n = (39.398 - 14.4) \times 275$

$= 6874.44$ lb.

Example Problem # 16.17.2: Fuel flows into a combustor at speed of 1040 ft/s, and consumes fuel at the rate of 5275 lb./hour. Determine the related thrust component.

Solution: Fuel flows in at $v_f = 1040$ ft/s.

Thrust $f_n = w_f v_f / g$

$= (5275/3600) \times 1040/32.2$

$= 47.33$ lb.

Example Problem # 16.17.3: A jet engine develops thrust of 7250 lbs. Determine the velocity of the gas as it exits the nozzle if it handles 225 lb. of air per second.

Solution: Velocity of gas leaving the exhaust nozzle $v_g = \text{Thrust} \times g/w_a$

Hence: $v_g = 7250 \times 32.2/225$

$= 1037.56$ ft/sec.

Example Problem # 16.17.4: Calculate the work done by gravity when an object weighing 17.5 lbs falls through a distance of 127.8 ft.

Solution: The product of force and distance gives work done.

Hence: Work $= 17.5 \times 127.8 = 2236.5$ ft-lb.

Example Problem # 16.17.5: Determine the rate of doing work when an object weighing 23.7 lbs travels through a distance of 144.8 ft in 18.7 seconds in horsepower.

Solution: Rate of work done = work/time

$= 23.7 \times 144.8/18.7 = 183.52$ ft-lb./sec.

Corresponding horsepower $= 183.52/550 = 0.3337$

16.18 EXERCISE PROBLEMS

Exercise Problem # 16.18.1: An object weighing 287.5 lb. is subjected to acceleration of 3.23 ft/sec^2. Calculate the force acting on it.

Exercise Problem # 16.18.2: Calculate the work done when a weight of 43.35 lb. is subjected to acceleration of 5.47 ft/sec^2, and travels a distance of 47.5 in.

Exercise Problem # 16.18.3: A jet engine develops thrust of 9887 lb. while allowing gas to flow at the exit at the nozzle at 2089 ft/sec. Calculate the flow entering its inlet.

Exercise Problem # 16.18.4: The barometric pressure of air outside an aircraft is 17.28 ins. of mercury, and pressure at the nozzle exhaust is 83.5 ins. of mercury. Calculate area of the exhaust nozzle if thrust developed due to the pressure differential is 9344 lbs.

Exercise Problem # 16.18.5: Calculate the fuel flow velocity into a combustor if it consumes fuel at the rate of 3786 lb./hour and produces related thrust of 58.46 lb.

REFERENCES & BIBLIOGRAPHY

Benvenuti, E., "Design and Test of a New Axial Compressor for the Nuovo Pignone Heavy Duty Gas Turbines", ASME paper # 96-GT-145, New York, 1996.

Benvenuti, E., Bianchi, D., Gusso R., Sabella D., "The PGT10 Heavy Duty Gas Turbine", ASME paper # 88-GT-319, New York, 1988.

Camp, T. R., "A Study of Acoustic Resonance in a Low Speed Multi-Stage Compressor", ASME paper # 97-GT-259, New York, 1997.

Camp, T. R., Day, I. J., "A Study of Spike and Modal Stall Phenomena in a Low Speed Axial Compressor", ASME Journal of Turbo-machinery, volume 120, page 293–401, New York, 1997*a*.

Cumpsty, N. A., Whitehead, D. S., "The Excitation of Acoustic Resonance by Wake Shedding", Journal of Sound and Vibration, volume 18, No. 3, page 353–369, 1971.

Day, I. J., "Active Suppression of Rotating Stall and Surge in Axial Compressors", ASME paper # 91-GT-87, New York, 1991.

Day, I. J., "Stall Inception in Axial Flow Compressors", ASME Journal of Turbo-machinery, volume 115, page 1 – 9, New York, 1993.

Day, I. J., Breur, T., Escuret, J., Cherret, M., Wilson, A., "Stall Inception and the Prospects for Active Control in Four High Speed Compressors", ASME paper # 97-GT-281, New York, 1997.

Ehrich, F. F., Spakovszky, Z. S., Sanchez, M. M., Song, S. J., Wisler, D. C., Storace, A. F., Shin, H. W., Beacher, B. F., "Unsteady Flow and Whirl Inducing Forces in Axial Flow Compressors: Part II - Analysis", ASME paper # 2000-GT-566, New York, 2000.

Ehrich, F. F., "Rotor Whirl Forces Induced by the Tip Clearance Effect in Axial Flow Compressors", ASME Journal of Vibration and Acoustics, volume 115, page 509 – 515, New York, 1993.

Epstein, A., Williams, J., Greitzer, E., "Active Suppression of Aerodynamic Instabilities in Turbo-machines", Journal of Propulsion, volume 5, no. 2, page 204–211, 1989.

Freeman, C., Wilson, A., Day, I. J., Swinbanks, M. A., "Experiments in Active Control of Stall on an Aero-engine Gas Turbine", ASME Journal of Turbo-machinery, volume 120, No. 4, New York, 1998.

Garnier, V. H., Epstein, A. H., Greitzer, E. M., "Rotating Stall Anticipation and Initiation in Axial Compressors", ASME paper # 90-GT-156, New York, 1990.

Haynes, J. M., Hendricks, G. J., Epstein, A. H., "Active Stabilization of Rotating Stall in a Three Stage Axial Compressor", ASME Journal of Turbo-machinery, volume 116, page 226–239, New York, 1994.

Hoying, D. A., Tan, C. S., Vo, D., H., Greitzer, E. M., "Role of Blade Passage Flow Structures in Axial Compressor Rotating Stall Inception", ASME paper # 98-GT-588, New York, 1998.

Khalid, S. A., Khalsa, A. S., Waitz, I. A., Tan, C. S., Greitzer, E. M., Cumpsty, N. A., Adamczyk, J. J., Marble, F. E., "End Wall Blockage in Axial Compressors", ASME paper # 98-GT-188, New York, 1998.

Mailach, R., Lehman, I., Vogler, K., "Rotating Instabilities in an Axial Compressor Originating from the Fluctuating Blade Tip Vortex", ASME paper # 2000-GT-506, New York, 2000.

Muir, D. E., Saravananamuttoo, H. I. H., Marshall, D. J., "Health Monitoring of Variable Geometry Gas Turbines for the Canadian Navy", ASME Journal of Engineering for Gas Turbines and Power, volume 111, page 244 – 250, New York, 1989.

Paduano, J., Epstein, A. H., Valavani, L., Longley, J. P., Greitzer, E. M., Guenette, G. R., "Active Control of Rotating Stall in a Low Speed Axial Compressor", ASME paper # 91-GT-88, New York, 1991.

Parker, R., Stoneman, S. A. T., "An Experimental Investigation of the Generation and Consequences of Acoustic Waves in an Axial Flow Compressor: Large Axial Spacing Between Blade Rows", Journal of Sound and Vibration, volume 99, number 2, page 169 – 182, 1985.

Rangwala, A. S., "Dynamic Analysis of Turbo-machinery", McGraw-Hill, New York, 2004.

Rao, J. S., "Turbo-machine Blade Vibration", Wiley Eastern Limited, John Wiley & Sons, New Delhi, 1991.

Saravananamuttoo, H. I. H., Rogers, G. F. C., Cohen, H., "Gas Turbine Theory", Prentice Hall, Harlow, England, 2001.

Spakovszky, Z. S., Paduano, J. D., Larsonneur, R., Traxler, A., Bright, M. M., "Tip Clearance Actuation with Magnetic Bearings for High-Speed Compressor Stall Control", ASME paper # 2000-GT-528, New York, 2000.

Tryfonidis, M., Etchevers, O., Paduano, J. D., Hendricks, G. F., Epstein, A. H., "Pre-Stall Behavior of Several High Speed Compressors", ASME Journal of Turbo-machinery, Volume 117, page 62 – 80, New York, 1995.

Tsalavoutas, A., Mathioudakis, K., Stamatis, A., Smith, M. "Identifying Faults in a Variable Geometry System of Gas Turbine Compressor", ASME paper # 2000-GT-33, New York, 2000.

Tyler, J. M., Sofrin, T. G., "Axial Flow Compressor Noise Studies", Transactions, Society of Automotive Engineers, volume 70, page 309 – 332, 1962.

Wadia, A. R., Wolf, D. P., Haaser, F. G., "Aerodynamic Design and Testing of an Axial Flow Compressor with Pressure Ratio of 23.3:1 for the LM2500+ Gas Turbine", ASME paper # 1999-GT-210, New York, 1999.

Weigl, H., Paduano, J., Frechette, L., Epstein, A., Greitzer, E., Bright, M., Strazisar, A., "Active Stabilization of Rotating Stall and Surge in a Transonic Single Stage Axial Compressor", ASME Journal of Turbo-machinery, volume 120, page 625 – 636, New York, 1998.

Wisler, D. C., "Aerodynamic Effects of Tip Clearance, Shrouds, Leakage Flow, Casing Treatment and Trenching in Compressor Design", Von Karman Institute Lecture Series 1985-05 on Tip Clearance Effects in Axial Turbo-machinery, 1985.

17

CHAPTER

PROBLEMS ENCOUNTERED IN COMBUSTION OF FUEL

17.1 INTRODUCTION

High efficiency gas turbines with their continuous flow system have demonstrated the capability to use a large variety of fuels. Both gaseous and liquid fuels may be further split by their calorific value, chemical composition and volatility. Traditional and non-traditional fuels have emerged in the search for sustainable alternative energy channels. Any sensible usage of alternative fuels, including process streams from industrial and plants such as refineries, petro-chemical plants, iron ore refineries and steel plants must meet generation of economic and environmental benefits.

Blast furnace gas, a by-product fuel gas produced during the combustion of coke in blast furnaces, is an ultra-low calorific value gas. Recovery to a higher grade is possible when mixed with other by-products, such as coke oven gas and converter gas, as fuel for power and steam generation. Flexible fuel handling capabilities are required to obtain reliable solutions.

Market conditions oftentimes require operators to switch from one fuel type and grade to another in order to maintain operations. A degree of complexity sets in with the variation in the fuel when the content of inert gases and hydrocarbons changes. A control system is required to measure and adjust to changes in fuel composition and link directly to the operability boundaries of combustion dynamics, emissions, and blowout.

Chemical, physical and combustion reactions are inherent during combustion of fuel in air. The rapid rate at which fuel is burned in a gas turbine makes it mandatory to ensure complete combustion of fuel, minimum production of objectionable exhaust components and stability of flame are achieved. Added to the list is the need for minimal pressure loss, even temperature profile at the turbine inlet and maximum thermal protection to maintain integrity of the combustor liners. Each of the objectives require careful attention, making successful design and operation of the combustor of a gas turbine a challenge.

17.2 GAS FUELS FOR POWER GENERATION

The most commonly used natural gas for industrial and power generation gas turbines is mainly methane, with some small amounts of volatile hydrocarbons and inert gases. In regions where the natural gas supply is limited, secondary gases and liquids are used to meet power generation needs. To supplement the available supply, there has also been an increased emphasis on the development of liquefied natural gas.

Among other high calorific value gases are the more volatile hydrocarbons with minor fractions of inert gases, which are usually very clean and perform well in gas turbines. Refinery by-products typically are propane, butane or a mixture of the two. They often contain some amounts of hydrogen.

Medium calorific value gases tend to be either weak natural gases that are made of methane with high fraction of inert gases, process gases, or gasified coal. Process gas is a broad classification of by-product gases, with wide range of composition, mainly containing methane, hydrogen and carbon monoxide. Sometimes called 'syngas' is mostly derived directly from abundant fossil carbon, and may be derived from refinery residuals, coal, lignite, tar sands and shale oil. Representing some potential for the carbon-constrained economy, high hydrogen gaseous fuels are a subset of these syngas fuels, with hydrogen making up more than 65% of the total gas volume. These fuels are made either by methane reforming or when oxygen is blown in during gasification of mineral coal.

Low calorific, or BTU, gases contain carbon monoxide and hydrogen diluted with a large fraction of inert nitrogen and carbon dioxide components. Derived from the chemical, oil and gas, or steel industries, many of these fuels cannot be transported or stored. Their appeal arises from reducing reliance of the fuel on market conditions and the need to burn the fuel in any way before releasing into the environment.

17.3 LIQUID FUELS

Three groups may be distinguished under the liquid classification. True distillate fuels are refined products that are mostly free of components causing ash. Naphtha, kerosene and diesel fuels can normally be used without further processing, except for minor cleanup. On the other hand, crude oils and other refined residuals need to be heated to acceptable levels to enable the needed viscosity for gas turbine combustion.

Of great interest are the bio-liquid fuels due to their overall neutral carbon balance and concern about greenhouse gas emissions. One approach is to use fuels that do not add any additional carbon to the environment. One such solution is bio-fuels, which essentially recycle carbon already in the environment. There are many diverse bio-fuels and feed stocks for bio-fuel under consideration. The feed stocks include corn, soy, palm, rapeseed, and jatropha. Multiple chemical processes take these raw plant-based elements and convert them into alcohol-based fuels, such as methanol and ethanol, or petroleum like fuels, such as bio-diesel. Some popular bio-liquid fuels are vegetable oils, alcohols (methanol or ethanol) and bio-diesel.

Bio-diesel has been successfully tested as a gas turbine fuel, demonstrating the performance over a range of operational conditions. Power and efficiency levels were very close to the ones achieved with diesel oil. When compared with diesel oil, emissions were the same, SO_x reduced to about 1 ppm, no visible plume of smoke, CO and volatile organic compounds comparable, lower NO_x emission and NO_x abatement effect of water injection normal and similar. Particulate matter, polycyclic aromatic hydrocarbons and aldehyde emissions were below the detection limits.

17.4 FUEL MANAGEMENT AND CONTROL SYSTEM

Specialized system hardware is unnecessary when changing the fuel type or grade, in addition to existing control sensors to account for shifts in humidity, fuel manifold pressure and related parameters. The control system employs physics-based models of gas turbine operability boundaries to stabilize exhaust emissions and dynamics of the combustion process. The turbine unit's microprocessor executes in real time to continuously estimate current boundary levels. Mathematical simulations combined with field test data are used to validate the system.

Closed-loop simulations model the gas turbine and control system. The control computer hardware and software are coupled to data from the field in real-time. Results from the simulations are needed to demonstrate impact from a rapid change in fuel composition on maintaining NO_x levels and impact on combustion.

The effects of transition such as from natural gas to liquefied natural gas with wide fuel heating value variation may then be evaluated. Given the wide range of fuel sources, the turbine units need the flexibility to operate in a variety of application spaces. Process off-gas from industrial steel mill blast furnaces, coke ovens and converters are also used for co-generation alternatives to steam boilers and integrated gasification combined cycle systems.

17.5 EFFECTS OF IMPURITIES IN FUEL ON TURBINE PERFORMANCE

The challenge of preventing harmful effects of ash-bearing fuels on the turbine metal and causing corrosion, erosion and fouling must be overcome. Thus, sodium and potassium salts are removed by washing the fuel with water, the corrosive vanadium is inhibited by using magnesium based additives, and frequent offline water washes of turbines need to be performed to remove ash deposits. Washing the turbine and compressor aids in removing unwanted deposits on turbine parts.

An alternate procedure to enhance turbine performance when a heavy fuel oil is used in the combustor is to modify the gas turbine controls with the intent to diminish heat rate degradation due to fouling. As the ash deposits build up on the hot gas path airfoils, the turbine efficiency drops due to increasing surface roughness and decreasing nozzle throat areas. With traditional schedule-based gas turbine controls, this loss of turbine efficiency results in a gradual drop in firing temperature, effectively magnifying the performance loss due to fouling. Using an on-board simulation model directly calculates the firing temperature from sensor readings, thus enabling the control system to compensate for the loss in turbine efficiency, maintain firing temperature and recover lost performance.

17.6 SUSPENDED CONTAMINANTS IN GAS FUEL

Natural gas arriving from the wellhead invariably carries certain levels of sand, iron oxide and other foreign particles. The size and abrasiveness of the particles are of concern as they flow over metal surfaces in the combustor and in the turbine sections. The greasy particles generally do not burn and melt away in the hot flames, and have to be extracted from the fuel gas prior to its admission in the combustion zone.

Millions of particles flowing over the polished metal surfaces of the airfoils will scratch them and affect the surface finish, raising friction losses. A more insidious effect from the erosion is corrosion setting in within the crevices. As a consequence, fatigue induced damage is enhanced to increase the risk of premature failure in the heavily stressed rotating blades.

Because of their mass, the particles may be separated if the gas stream is given a strong whirl. Centrifugal forces will cause the heavier granules to flow away from the central core of the stream. Effectively the larger particles will fly out and the fuel gas will proceed further to be cleansed of the tinier dust form of contaminants. The latter may then be removed with specialized filtering elements.

17.7 MIXING OF AIR AND FUEL

Excessively rich mixture of fuel and air even in isolated parts of the combustor can lead to some of the fuel not being burned, thus lowering the efficiency of combustion. Introducing turbulence in the air as it enters the combustor gives the fuel a better chance to mix with the air. The turbulent stream also retards velocity of the flow, and gives the fuel more time to burn. But some loss in pressure will be inevitable due to higher friction against the walls.

The mixing process may be accelerated using fixed vanes placed inside a cup that deflect the flow of air to produce a swirling form of motion in the stream. Gas fuels have an advantage since the air can mix with the fuel more easily. Liquid fuels need to be split into tiny drops, since a continuous stream of the fuel cannot participate to mix with the air.

Special fuel injectors are designed to force the fuel out of its tips using compressed air. The fuel then sprays out of holes in the injector nozzle in the form of a conical sheet of droplets of controlled dimension. The droplets then mix with the turbulent air stream to improve combustion. If even small particles were to be present in the fuel, the risk of the holes at the injector tip being blocked is raised, necessitating the use of proper fuel filtering elements.

17.8 EXCESS PRESSURE OSCILLATIONS IN COMBUSTOR

The liner of a combustor is required to enclose the combustion zone and thus prevent the outer engine casing from exposure to the elevated temperatures. Liner dimensions will increase when the engine is started and goes to full speed, and metal temperatures rise from the cold to operating condition. Thermal stresses are avoided if the supports for the liner permit free growth of the metal in all principal directions.

Static pressure of air within the liner will produce some stresses, but it is necessary to avoid imposition of fluctuating pressure since rapid fatigue will set in at the extremely high metal temperatures. Cracks may then develop in the liner, and cannot prevent smoke from escaping. If the cracks were to grow and pieces of the liner were to separate, the risk of fire spreading to other parts of the engine and to the aircraft increases exponentially.

One source that can cause pressures to oscillate is the fuel delivery system. The fuel does not arrive into the combustor in a continuous stream, but instead comes in as a pulsating flow. This is because liquid fuels for gas turbines require a fuel pump. Fuel pumps for gas turbines may be reciprocating single or double-acting type and rotating vane or screw type. The running speed of the pump controls the frequency of pulsations. A single-acting reciprocating pump produces same pulsations as the running speed, and double-acting pumps develop two times speed. Rotary pumps produce pulsations of frequency dependent on running speed and number of vanes or lobes on the screw. The pulsating fuel flow develops pulsations in heat waves as the fuel burns, which in turn sets up pressure waves that travel both circumferentially and axially inside the combustor.

The significance of these oscillating pressures may be recognized from the limit of 2.5 lb./in^2 double amplitude set by turbine manufacturers. One solution to this is to adjust the position of the tube delivering the fuel away from the face of the flame. Minute adjustments in its location permit the heat and pressure waves to go out of phase with each other and thus reduce the amplitude of the pressure waves.

17.9 COMBUSTION NOISE

A humming noise is sometimes emitted from a combustor as it burns the fuel. The noise signal may be strong enough to be distinguished from other prevailing sources. Turbulence is the primary noise production mechanism, since laminar flow passes over a surface with no hindrance and little noise. The cause of the hum may be attributed to reasons as diverse as engine running speed coinciding with natural frequency of a part or assembly, frequency of pulsating fluid flow or due to an ancillary component.

When a turbine engine is in operation, a number of sources contribute to the overall noise level. Among them, the leading source is noise that accompanies combustion of the fuel. The mixture of air and fuel suddenly goes through expansion as the fuel burns and increases in both pressure and temperature. The abrupt process sets up pressure waves, and associated noise waves propagate all around the engine. Since the combustion takes place continuously, noise production is also continuous.

Blade passing frequency noise is also prominent. In aviation engines the noise due to fan blades passing fixed vanes is emitted from the fan inlet and outlet ducts, as also from the turbine exhaust nozzle. Some noise may be suppressed from emitting through the fan case with acoustic panels placed inside the duct.

Power generation units have the flexibility of noise suppression arrangement in tandem with intake air filters. Silencers with baffle plates and acoustic liners may also be provided on the exhaust plenum. The sides of the turbine are protected from noise with an enclosure to trap the noise. Local government rules are mandatory concerning noise production, especially when the unit is in the proximity of residential areas.

17.10 STAGED COMBUSTION OF FUEL

The need for reducing production and emission of dangerous contaminants in exhaust gases from turbine engines cannot be overemphasized. A number of innovative solutions have appeared in the recent past for meeting the goals, although the methods also tend to carry with them unwanted effects. The dry-low NO_x method and injection of water or steam into the diffuser or directly into the combustion zone focus on preventing formation of effluents, while catalytic reduction is dedicated to chemically reducing NO_x into its elements after its formation.

Staged combustion of fuel attempts to lower the firing temperature to maintain production of carbon monoxide, unburned hydrocarbons and oxides of nitrogen within limits. The objective is to have flame temperatures in the region where all three components are produced within set limits. Towards that goal, the fuel is made to burn in multiple regions of the combustor instead of only in the area surrounding the fuel injector. One possibility is to have multiple fuel injectors so that only a portion of the total requirement is burned around each one.

Another option in staged combustion is to restrict the amount of air around the main fuel injector. Only a portion of the fuel then gets burned since the mixture of fuel and air is too rich. As the flow proceeds downstream, more air from the diffuser is introduced through a number of ports machined in the liner. The new air also helps to set up greater circulating flow to induce further combustion of the fuel left unburned in the primary zone. The overall effect of burning the fuel in several areas is to bring down the combustion temperature and thus reduce production of NO_x. At the same time, within the zones of combustion the temperature is high enough to prevent formation of carbon monoxide and unburned hydrocarbons.

The complexity arising from the provision of a multitude of ports for the air to enter into the combustor adds to the cost of its fabrication. But the benefits available from reduced emissions may be surmised to outweigh the extra cost.

17.11 SELECTIVE CATALYTIC REDUCTION OF NO_x

It is possible to reduce the level of NO_x emitted into the environment from the turbine exhaust through chemical reduction into its elements nitrogen and oxygen. The methods call for a catalytic agent that facilitates the reaction. Ammonia gas is required in the selective catalytic reduction (SCR) process to break down the offending component into nitrogen and moisture, or steam. The catalytic converter may be placed before the products of combustion enter the turbine, or may be located after flowing through the turbine.

A drawback of the SCR method is its reliance on ammonia. The gas itself is objectionable since it can cause problems in the form of skin irritant and even burns. The gas needs to be stored in tanks at the site and sent to the catalytic converter through a delivery system. If a leak were to spring in the system, the gas would affect the population nearby.

A number of catalytic agents and chemicals have been experimented with, and some of them have produced remarkable results. Many catalysts operate in a specific temperature range. Some agents are more efficient around 750° F and others at cooler temperature of about 400° F. The selection of the catalyst and placement of the converter in the turbine layout must then proceed accordingly. Platinum-based and other elements have been employed, some of which get exhausted after a number of hours of operation, and have to be replenished at a significant cost. Chemicals such as urea have also been tried in an attempt to secure more efficient and inexpensive method for tackling the problem.

17.12 STABILITY OF FLAME AND IN-FLIGHT SHUTDOWN

The ratio of air and fuel in the mixture participating in the combustion process is of special interest where stability of the flame is concerned. When the mass of air is high for a given amount of fuel, a lean combination is the result, while a rich mixture is the result when the amount of air is not sufficient. A direct effect of either too lean or too rich mixture is felt on the flame. The flame may be lost in the event of the ratio of air and fuel going beyond the upper and lower bounds.

Loss of flame in the combustor is a strenuous experience in aviation turbines since it leads to the engine shutting down during flight. Cockpit crew will attempt to restart the engine as it windmills, but may not be successful. The loss of thrust from the engine will cause the plane to lose some altitude. Outside air pressure will change if the descent is too rapid. The air pressure plays a role in conjunction with the ratio of air and fuel, so re-ignition attempt may fail. The microprocessor based control system of the engine may not be able to respond fast enough due to inherent time lag in the system to order the amount of fuel to be provided to the combustor. The requisite adjustments to the air and fuel ratio may then not be made in time. As a result, the flame may be lost.

Fighter jet engines are particularly vulnerable to the engine shutting down when the pilot makes the plane go through unusual exercises such as the cobra and J-maneuvers. The rapid changes in speed, direction and altitude hold an increased risk of the flame being lost during the exercises.

Land-based power generation engines are provided with cross firing tubes between the combustor cans. If the flame were to be lost in one combustor, the cross tubes aid in getting the flame from a neighboring can.

Another effect of an in-flight shutdown will be felt in the turbine. The thermal mass of the rotor and stator are different and cause the casing to shrink faster than the rotor. For several turns in the rotor, the available clearance at the blade tips is essentially lost. The blades rub against the casing for sometime before the rotor and stator reach a state of thermal equilibrium. The resultant opening of the clearances will let some of the gases to flow over the blade tips without doing useful work and bring down the efficiency of the turbine. Temperature of gases at the turbine exhaust will also deteriorate as a direct consequence of an in-flight shutdown of an engine.

17.13 PRESSURE LOSS IN COMBUSTOR

The turbine section of a gas turbine is reliant on the highest pressure and temperature of the gases received from the combustor for developing maximum amount of power. Any deterioration in either of the two parameters directly affects the level of power. Hence, it is imperative that the pressurized air entering from the compressor and flowing through the combustor be maintained without suffering excessive losses. In this regard, the diffuser section between the compressor exit and entry into the combustor plays a useful role of converting the kinetic energy of the flowing air into potential energy, and thus lower the flow velocity and associated friction losses.

But pressure losses are inevitable as the burned gases rub against the walls of the combustor liner. The problem is aggravated due to the necessity of inducing turbulence as the flow goes through. The turbulent flow regime causes better mixing of the air and the fuel, thus promoting the combustion process. At the same time, however, the swirling flow also leads to greater rubbing against the walls and consequent increase in friction losses.

Abrupt changes in the geometry of the combustor liner can also lead to obstructions to the flow. The obstructions may arise due to ports that admit fresh air in other parts of the combustor, cooling channels for delivering coolant or the need to accommodate other engine parts outside the liner. An overall loss of pressure of around 2 per cent is generally assumed in the combustor.

17.14 REVIEW QUESTIONS

17.14.1 Why is it easier to burn a gaseous fuel compared with its liquid counterpart?

17.14.2 Prepare a list of objectives to be met when fuel is burned in the combustor of a gas turbine.

17.14.3 How do solid particles in the fuel interfere in the running of a gas turbine?

17.14.4 Name some sources of noise when a turbine unit is in operation.

17.14.5 What is staged combustion, and how does it prevent NO_x emission?

17.14.6 How does the diffuser section of a gas turbine help in lowering loss of pressure in the combustor?

17.14.7 What is a major source of fluctuating pressure inside a combustor liner?

17.14.8 What problems need to be faced in the selective catalytic reduction method of controlling NO_x formation?

17.14.9 What is needed to control formation of dangerous exhaust gas components in a turbine?

17.14.10 Why is it sometimes difficult to re-light an engine after it shuts down during flight?

17.14.11 Name some reasons how loss of light in the combustor affects overall operation of a gas turbine.

17.15 EXAMPLE PROBLEMS

Example Problem # 17.15.1: Determine the potential energy and kinetic energy of a 34.3 lb object falling from a height of 27.5 ft at rest (*i*) at the start and (*ii*) when it falls back to the ground.

Solution: Potential energy = wh and kinetic energy = $mv^2/2$

(*i*) At start the potential energy = 34.3 × 27.5 = 943.25 ft-lbs

Object is starting from rest with no velocity, so kinetic energy = 0

(*ii*) At ground height h is zero, so potential energy = 0

Velocity at ground: $v^2 = u^2 + 2gx = 0 + 2 \times 32.2 \times 27.5 = 1771$

Or: $v = 42.08$ ft/sec

So kinetic energy = $(34.3/32.2) \times 42.08^2/2 = 943.25$ ft-lbs.

Example Problem # 17.15.2: A ball weighing 3.45 lbs is shot vertically up at a speed of 29.4 ft/sec. Calculate the time required for it to return back, and velocity as it reaches ground.

Solution: At maximum height velocity is zero.

As it goes up, $v_1 = u_1 + gt_1 = 29.4 - 32.2 \times t_1 = 0$

So $t_1 = 0.913$ sec.

Also, distance traveled $x_1 = u_1t_1 + gt_1^2/2 = 29.4 \times 0.913 + 32.2 \times 0.913^2/2$

$= 40.265$ ft

On return initial velocity is zero. So $u_2 = 0$

Also, distance traveled going up and down is same. So $x_1 = x_2$

$$v_2^2 = u_2^2 + 2gx_2 = 0 + 2 \times 32.2 \times 40.265 = 2593, \text{ or } v_2 = 50.92 \text{ ft/sec.}$$

$$v_2 = u_2 + gt_2 = 0 + 32.2 \times t_2, \text{ so } t_2 = 50.92/32.2 = 1.58 \text{ sec.}$$

$$\text{Total time} = t_1 + t_2 = 2.49 \text{ sec.}$$

Example Problem # 17.15.3: An aircraft engine has airflow rate entering inlet of 248 lb./sec, velocity of air at inlet is 0, velocity of gas at exhaust is 1629 ft/sec, exhaust nozzle area is 372 in^2, pressure at exhaust nozzle is 6.12 lb./in^2 and ambient pressure is 0 lbs/in^2. Neglecting the effects of fuel, calculate the engine's thrust.

Solution: Total thrust is sum of thrust from exhaust and thrust from fuel burned.

Thrust
$$f_n = w_a \times (v_2 - v_1)/g + (p_j \times p_{atm}) \times a_j$$
$$= 248 \times (1629 - 0)/32.2 + (6.12 - 0) \times 372$$
$$= 14822.98 \text{ lbs.}$$

Example Problem # 17.15.4: A jet engine develops thrust of 27320 lb. Determine the velocity of the gas as it exits the nozzle if it handles 755 lb. of air per second.

Solution: Assume flow velocity reduces to zero after exiting the nozzle.

Thrust $f = w_a \times v = (755/32.2) \times v = 27320$ lb.

Hence: velocity of gas leaving exhaust nozzle $v = f/w_a$

$$= 27320 \times 32.2/755 = 1165.17 \text{ ft/sec.}$$

Example Problem # 17.15.5: The velocity of gas exiting the turbine exhaust in an aircraft engine is 982 ft/sec. If the engine handles 689 lbs of air per second, the airplane travels at a speed of 377 miles/hour and weight of fuel consumed is 4391 lbs/hour, determine the net thrust.

Solution: Aircraft speed v_1 = 377 mph = 377 × 5280/3600 = 552.93 ft/sec.

If velocity of fuel coming into the burner relative to the aircraft is negligible,

Velocity of gas exiting the turbine exhaust v_2 = 982 ft/sec.

Total thrust is sum of thrust from exhaust and thrust from fuel burned.

So total thrust $f_n = w_a \times (v_2 - v_1)/g + w_f v_f/g$

$$f_n = 689 \times (982 - 552.93)/32.2 + (4391/3600) \times 982/32.2$$
$$= 9218.16 \text{ lbs.}$$

17.16 EXERCISE PROBLEMS

Exercise Problem # 17.16.1: Calculate the exhaust nozzle area of an aircraft engine when airflow rate entering the inlet is 271 lbs/sec, velocity of air at inlet is 0, velocity of gas at exhaust is 1874 ft/sec, pressure at exhaust nozzle is 4.97 lbs/in^2 and ambient pressure is 0 lbs/in^2 and engine thrust is 16850 lbs.

Exercise Problem # 17.16.2: Determine the thrust in an engine if velocity of the gas as it exits the turbine nozzle is 1712 ft/sec and it handles 893 lbs of air per second.

Exercise Problem # 17.16.3: Calculate the kinetic energy of a 401.44 lb object falling from a height of 81.45 ft at rest when it falls to the ground.

Exercise Problem # 17.16.4: Determine the potential energy of an object weighing 28.75 lbs placed 107.85 ft above the ground.

Exercise Problem # 17.16.5: Calculate the maximum kinetic energy and potential energy of a 37.43 lb object falling from a height of 432.47 in. at rest when it falls to the ground.

REFERENCES & BIBLIOGRAPHY

Bahr, D. W., "Technology for the Design of High Temperature Rise Combustors", Journal of Propulsion and Power, volume 3, number 2, pages 179–186, 1987.

Barooah, P., Anderson, T. J., Cohen, J. M., "Active Combustion Instability Control with Spinning Valve Actuator", Proceedings, ASME Turbo Expo, paper # GT-2002-30042, Amsterdam, June, 2002.

Buecker, B., "Emissions: SCR Design", Power Engineering, pages 24–28, August, 2002.

Candel, S. M., "Combustion Instability Coupled by Pressure Waves and Their Active Control", 25th International Symposium on Combustion, Sydney, Australia, 1992.

Claeys, J. P., Edward, K. M., Mick, W. J., Symonds, R. A., "Combustion System Performance and Field Test Results of MS7001 F Gas Turbine", ASME Journal of Engineering Gas Turbines & Power, volume 115, pages 537–546, New York, 1993.

Cohen, J. M., Rey, N. M., Jacobson, C. A., Anderson, T. J., "Active Control of Combustion Instability in a Liquid Fueled Low NO_x Combustor", ASME paper # 98-GT-267, New York, 1998.

Dalla Betta, R. A., Schlatter, J. C., Nickolas, S. G., Cutrone, M. B., Beebe, K. W., Furuse, Y., Tsuchiya,

T., "Development of a Catalytic Combustor for a Heavy Duty Utility Gas Turbine", ASME paper # 96-GT-485, New York, 1996.

Dodds, W. J., Ekstedt, E. E., "Broad Specification Fuel Combustion Technology Program", Phase II, Final Report, 1989.

Fishenden, M., Saunders O., "An Introduction to Heat Transfer", Oxford University Press, New York, 1950.

Gupta, A. K., Lewis, M. J., Daurer, M., "Swirl Effects on Combustion Characteristics of Pre-Mixed Flames", ASME Journal of Engineering Gas Turbines & Power, volume 123, pages 619–626, New York, 2000.

Koff, B. L., "Aircraft Gas Turbine Emissions Challenge", ASME paper # 93-GT-422, New York, 1993.

Lefebvre, A. H., "Gas Turbine Combustion", Taylor and Francis, Philadelphia, 1999.

Lefebvre, L., Norster, P., "Technical Advances in Gas Turbine Design", Institution of Mechanical Engineers Symposium, 1969.

Lieuwen, T., McManus, K., "That Elusive Hum", Mechanical Engineering, pages 53–55, June, 2002.

Marshall, A. W., Ph. D. Thesis, University of Maryland, College Park, MD, 1996.

Maughan, J. R., Lutts A., Bautista, P. J., "A Dry Low NO_x Combustor for the MS3002 Regenerative Gas Turbine", ASME paper # 94-GT-252, New York, 1994.

McCarty, J. G., "Kinetics of PdO Combustion Catalysis", Proceedings, International Workshop on Catalytic Combustion, H. Arai, editor, page 108, 1994.

McManus, K. R., Poinsot, T., Candel, S. M., "A Review of Active Control of Combustion Instabilities", Prog. Energy Combustion Science, volume 19, pages 1–29, 1993.

Nealy, D. A., "Combustor Cooling – Old Problems and New Approaches", Gas Turbine Combustor Design Problems, A H. Lefebvre editor, pages 151–185, Hemisphere, Washington, DC, 1980.

Rangwala, A. S., "Turbo-machinery Dynamics: Design and Operation", McGraw-Hill, New York, 2005.

Richards, G. A., Janus, M. C., "Characterization of Oscillations During Premix Gas Turbine Combustion", ASME paper # 97-GT-244, New York, 1997.

Schlatter J. C., Dalla Betta, R. A., Nickolas, S. G., Cutrone, M. B., Beebe, K. W., "Single Digit Emissions in a Full Scale Catalytic Combustor", ASME paper # 97-GT-57, New York, 1997.

Segalman, I., McKinney, R. G., Sturgess, G. J. Huang, L. M., "Reduction of NO_x by Fuel Staging in Gas Turbine Engines – A Commitment to the Future", AGARD Conference Proceedings 536, page 29/1 – 17, 1993.

Seume, J. R., Vortmeyer, N., Krause, W., Hermann, J., Hantschk, C. C., Zangl, P., Vortmeyer, D., Orthmann A., "Application of Active Combustion Instability Control to a Heavy Duty Gas Turbine", ASME paper # 97-GT-119, New York, 1997.

Snyder T. S., Rosfjord, T. J., McVey, J. B., Chiappetta, L. M., "Comparison of Liquid Fuel/Air Mixing and NO_x Emissions for a Tangential Entry Nozzle", ASME paper # 94-GT-283, New York, 1994.

Sotheran, A., "The Rolls-Royce Annular Vaporizer Combustor", ASME paper 83-GT-49, New York, 1983.

Touchton, G. L., "Influence of Gas Turbine Combustor Design and Operating Parameters on Effectiveness of NO_x Suppression by Injected Steam or Water", ASME paper # 84-GT-IPGC-3, New York, 1984.

White, D. J., Batakis, A., Le Cren, R. T., Yacabucci, H. G., "Low NO_x Combustion Systems for Burning Heavy Residual Fuels and High Fuel-Bound Nitrogen Fuels", ASME Journal of Engineering Gas Turbines & Power, volume 104, pages 377–385, New York, 1982.

18
CHAPTER

DIFFICULTIES IN HOT GAS EXPANSION IN TURBINE

18.1 INTRODUCTION

The ease with which expansion of hot gases proceeding from the combustor into the turbine section has been discussed in the introduction of chapter # 16, mostly because of the flow taking place with the pressure gradient. Unless an obstruction to the flow is encountered, the generally expanding gases smoothly enter from the stator vanes into the rotating stage while producing power to drive both the compressor and the driven load. The obstruction may be in the form of geometric irregularity or defect from missing metal on the stationary vanes or the blades, causing the direction of flow to deviate from its assigned path.

The expanding flow lets the gases occupy the increasing flow path without running into the problem of separation of boundary layer from the annulus walls. In contrast, the compression process is hindered when the flow against the pressure gradient diffuses and velocity converts into pressure energy. Note that the gases also perform work while expanding, a feature that is the exact reverse in the compressor where work is done to pressurize the air. The factors highlight the differences in the aerodynamic design of the compressor and turbine sections.

But the elevated temperature and pressurized gases in the expander section have their own litany of problems to contend with. Temperatures in the proximity of melt point of the metal substrate cause the material properties to rapidly deteriorate and the endurance strength to fall. Fatigue problems have to be reckoned with, both due to thermal and mechanical loads. In the rotating blades, for instance, centrifugal forces and dynamic stresses arising from aerodynamic interference have to be withstood. As the rotating blades traverse across the stationary vanes, variations in the pressure field cause dynamic stresses. Thermal gradients during the startup and shutdown of the engine also produce thermal stresses.

Since power developed in the turbine is dependent on the operating temperature, cooling of the components is necessary if the turbine airfoils and disks are to produce maximum power while withstanding the onslaught of excessively hot gases. Creative techniques are required to admit the coolant inside the rotating blades. The hollow interior of the airfoils requires efficient cooling in multiple ways. At the same time, a thermal barrier coating of the initial turbine stages also helps in shielding the metal substrate from the hostile flow of gases.

18.2 ELEVATED TEMPERATURE RESISTANCE

Development of super alloys stands out at the forefront among the many different technologies that have made possible advances in gas turbines. The alloys used in the manufacture of jet engines and industrial and marine gas turbines represent the leading edge in materials technology, permitting them to operate at very high temperatures. In turn, the gas turbine engines may also be said to be the primary driving force for the development of the super alloys, with the twin technologies proceeding hand in hand.

All heat engines rely on high operational temperatures for efficiency in performing useful work. Reciprocating engines are still in use for propulsion of small aircraft, their prime advantage coming from relatively low exhaust temperatures, which boosts their thermodynamic efficiency. But they suffer from the twin disadvantages of poor thrust/weight ratio and excessive dynamic loads due to the mass of the reciprocating pistons and connecting rods. For large transport aircrafts, the piston engines would become too heavy, making them impractical. The dynamic inertia loads would need to be borne by the engine and the aircraft structure. In contrast, airfoils made from super alloys are much lighter while producing large amounts of power.

Resistance to oxidation and enhanced creep characteristics are prime requirements for turbine components, and super alloys are eminently qualified for providing the two features. Addition of chromium to the alloy gives good protection from corrosion at high temperatures, a more balanced use of the element producing IN-738. Varying amounts of aluminum, titanium, columbium and molybdenum providing additional strengthening through solid solution and carbide effects. Tungsten, tantalum and rhenium also enhance mechanical properties. Carbon remains a mainstay, with matrix carbides providing point strengthening in several solid-state reactions. Thermal and mechanical processing is required to strengthen through densification and development of substructures at dislocations.

18.3 DIRECTIONAL SOLIDIFICATION TECHNOLOGY

Conventional casting methods do not produce grain boundaries that can be oriented normally to the stress axis. Rotating turbine blades have a large component of stress acting along its length due to centrifugal effects close to the root. Application of directional solidification, or columnar grain, form of alloys provides substantial strength while also improving ductility. The grain boundaries are deleted, where failures may be expected to initiate. Single crystal alloys go a step further and totally eliminate all grain boundaries. Considerable improvement in the resistance to thermal fatigue is also available as a consequence of the solidification process in a specific direction, a feature for turbine blades that cannot be overemphasized.

Since the array of grain boundaries run parallel to the principal axis of stress in the direction of solidification, the stresses at the weaker grain boundaries are minimized. The initiation of failure is then prolonged, and creep rupture life is improved.

Orientation of the primary and secondary axes in specific directions is possible by using seed crystals. Columnar grains may also be obtained in the solid state from the growth of secondary grains. The process calls for the transition from the liquid to the solid phase according to a controlled thermal gradient to produce elongation of the grains in the direction parallel to the temperature variation. Thus, the thermal gradients are aligned parallel to it to obtain elongated grains along the principal stress axis.

The single crystal casting requires a helical constriction between the mold and the starter block, allowing only a single grain to pass through it. Solidification takes place in the form of dendrites growing in all three directions, so the continuously changing direction of the helix then restricts the growth of the dendrite to a single grain releasing at the top of the helix. The selected grain then expands to fill the cavity in a manner similar to the growth in columnar grain castings. Depending on the design of the shell, the process can produce hollow single crystal and columnar grain airfoils with cooling passages. For single crystal castings the seed crystal is assumed to be of the same alloy with the same or higher melting temperature, and must be oriented to allow its duplication in the alloy material in the mold.

18.4 COOLING TURBINE AIRFOIL

The primary role of air entering the engine inlet is to provide maximum mass flow to burn the fuel in the combustor, and thus generate maximum power. If some of the pressurized air from certain stages of the compressor is diverted to cool the hot components, then that represents loss of power. But since operating temperature must also be very high for the most power to be generated, bleeding a percentage of mass flow to cool the turbine components is justified. In some aviation engines, as much as 15 to 18 per cent of the mass entering the intake to the engine is not directly burned but used to cool, indicating the significance of cooling.

A consideration of a cooled turbine is the effect on the overall efficiency from the incurred losses. Whether it is beneficial to sacrifice some aerodynamic efficiency to reduce such losses is of significance. The losses arise from direct loss of turbine work due to the reduced mass flow, expansion of the gases not remaining adiabatic, loss in pressure and enthalpy from the mixing of spent cooling air with the main gas stream at the blade tips and work done by the blades to push the cooling air through the passages.

To understand cooling performance, a commonly used parameter is the ratio of temperature differences between the mean blade and coolant temperatures and that between the gas and coolant temperatures. Coolant temperature is controlled by conditions at the compressor bleed, and increases with the pressure ratio. Industrial gas turbines have the option of using a water-cooled heat exchanger to reduce the coolant and blade relative temperatures. Up to four stages of a turbine may be cooled, with air extracted from earlier stages of a compressor to cool the later turbine stages. The cooled turbine also offers benefits in the form of a higher blade loading coefficient (permitting use of fewer stages), a higher pitch/chord ratio (reducing the number of blades in a row) and a higher flow coefficient (implying a blade of smaller camber and consequent reduced surface area).

18.5 COOLANT PASSAGES WITHIN AIRFOIL

Cooling of the rotating airfoils represents unusual difficulties from engineering and manufacturing considerations. Serpentine cooling passages assure optimum cooling efficiency. The blades are made by the precision casting process, and are made of a Nickel-based single crystal super alloy because of its superior high temperature strength. The cooling holes are electro-discharge machined. The dovetail and shank portions are machined.

One layout of the cooling system consists of three independent channels located at the leading edge, mid-chord and trailing edge, with the coolant supplied at the root region through formed pockets. Cooling air impinges on the inner surface of the leading edge, and flows out in the form of a showerhead through four staggered rows of exit holes. A five-pass serpentine passage is used at the mid-chord channel to enhance heat transfer by convection. Two rows of film holes are placed on the pressure surface and three rows on the suction surface. The trailing edge channel is a three-pass serpentine, with air flowing out of ejection holes into the main gas path. Turbulence is promoted in the channels with ridges leaning in the direction of airflow to improve cooling effectiveness. The turbulent flow reduces velocity and improves heat transfer.

At elevated gas and metal temperatures, oxidation and creep impose limitations on the blade's capabilities. In the rotating blades the first stage gas temperature may range up to 1260° C (2300° F) and pressure to 1.11 MPa. Available cooling air temperature may run to 400° C (752° F).

The stationary nozzle guide vanes do not suffer from centrifugal stresses, but the high gas temperatures from the combustor directly impact the vanes at the turbine inlet. Unlike rotating airfoils, provision for admitting and extracting of the coolant is comparatively easy. Combined cycle power plants may opt for bleeding cooler steam from a convenient point in the heat recovery steam generator and return back hotter steam. The method entails extra plumbing with valves and sensors, but improves overall heat balance.

Cooling of stator vanes and annulus walls requires special attention. Cooling holes are machined using the electro-discharge machining process. A cover plate and coolant supply and recovery pipes are welded after machining of the cooling holes.

18.6 COOLANT DELIVERY TO ROTATING BLADE AND DISK

Providing the coolant to the rotating components of a turbine exposed to elevated temperatures presents interesting problems. Delivering and even recovering the coolant from stationary parts is in comparison easy.

One method calls for cooler air extracted from a gas turbine's compressor unit to be discharged through a nozzle of a tube that is placed close to the dovetail area. A cooling plate is located on the forward side of the rotating stage. Pockets are formed in the cooling plate to admit the air through the nozzle into holes in the rim area, and then passed into holes machined in the base of the blade's dovetail. The cool air then goes through the passages inside the airfoil through multiple circuits. After performing its task, the coolant escapes through holes on the suction and pressure surfaces of the airfoil and at its leading and trailing edges. The cooling plate also serves the dual purpose of retaining the blades and being ejected from the dovetail due to axially oriented forces produced as the gases flow over the airfoil.

The disk to which the airfoils are attached also requires cooling. High levels of circumferential stresses build up in the disk material due to the rotating field of centrifugal forces set up from the attached blades and also from the disk's own material density. The forces are proportional to the square of the running speed, with tangential stresses peaking inside the surface of the central bore. When metal temperatures are high, Young's modulus, yield and ultimate strength of the material of the disk rapidly deteriorate in the higher temperature regime. So, cooling the disk is essential in protecting its integrity. The coolant is brought in through a second tube located near the rim on the forward side of the disk, and the flow is directed towards the hub. The spinning surface than distributes the air over the entire area.

Note that circumferential stresses can reach high levels if a turbine rotor goes into over-speed condition. When the stress reaches the material's limit, the disk will burst with catastrophic consequences. Lowering the temperature aids in reducing the stresses, pointing to the need for cooling the disk when operating in an elevated temperature regime.

18.7 STEAM AND AIR COOLING

Most gas turbines use air to cool hot path components, but alternatives such as water and steam have also been selected. Single-crystal alloy turbine blades may have three independent cooling channels, employing impingement and film-cooling methods.

The effectiveness of cooling, flow characteristics and metal temperatures have been confirmed through hot wind tunnel tests. In a test on a three-stage turbine, cooling is provided for the first two stages. The first stage vanes use steam for cooling combined with some air ejection to obtain film cooling. Second and third stage vanes are cooled with air extracted from compressor mid-stages. Cooling air for the rotating blades is extracted from the compressor discharge, and then cooled externally before delivery to the blades.

A closed loop approach obtains optimum flow of the coolant while maintaining the right pressure differential between steam and gas to minimize thermal stress. A number of straight circular holes of 2 mm diameter are arranged in the vane, with the pitch selected to obtain a nearly uniform metal temperature. The inner convection cooling without impingement delivers enough effectiveness to permit a simple path for the coolant. The direction of flow takes into consideration heat flow from the outside. Fresh steam enters on the suction side to flow from the tip to the root region. Next, the steam is collected in the root area and directed towards the tip area along the pressure side and leading and trailing edges. In the final run it is collected to cool the outer end wall before flowing back to the recovery pipe. The enhanced features of the closed loop system control the thermal stresses in the structure because of the steep temperature gradient. Two other countermeasures permit further reduction in thermal stresses. A cooling film produced on the outer surface of the vane isolates the metal from hot gases in the vicinity, and thermal barrier coating on the vane reduces heat transfer because of its low heat conductivity.

The design of the vanes is similar to the air-cooled type, and employs the impingement technique with an inserted core. Steam enters the tip area to the core at the leading edge and collected in the root area, then flows to the core of the mid-chord region and recovered at the tip. The trailing edge is cooled with air using film cooling and blow-off holes in the area.

After the tests, each vane was disassembled and checked using the fluorescent penetrant inspection method. No cracks were found for the steam and air-cooled vanes, but the non-cooled vanes experienced cracks. The un-cooled vanes are estimated to have thermal stress higher by 30–40 per cent compared with the cooled vanes.

18.8 THERMAL BARRIER COAT

Thermal barrier coats represent perhaps the most promising and exciting development in super alloy coatings. Any mechanism by which the temperature limits can be raised by overcoming hot-section material restraints is of significant interest, and thermal barrier coatings offer this potential.

Coatings of ceramic or metallic or a combination of the two are applied on the substrate of a super alloy to preclude or inhibit direct interaction between the substrate and the potentially damaging environment in which the turbine airfoils operate. This damage can be either metal recession due to oxidation/corrosion or a reduction in the mechanical properties of the substrate due to diffusion of harmful species into the alloy at elevated temperature.

Coats used on turbine airfoils do not act as inert barriers. Rather, they provide protection by interaction with oxygen in the environment to form dense and tightly adherent oxide scales. The intent is to inhibit the diffusion of sulfur, nitrogen and other damaging elements. Hence, coatings tend to be rich in elements such as aluminum, chromium and silicon that readily participate in the formation of these protective scales.

The coat may be described as a multi-layer coating system. An inner layer covers the metallic substrate, with an insulating ceramic outer layer laid above it. The inner coat prevents the corrosive species in the gas stream from attacking the alloy. The ceramic layer insulates the metallic substrate from higher surface temperatures than it might otherwise be able to tolerate.

18.9 BOWING OF TURBINE INLET VANE

Attachment of the nozzle guide vanes to the annulus walls of the turbine are of special interest when considering thermal distortion of individual vanes. The fixity arrangement is generally in the form of machined plates at the two ends of vanes and configured to permit the vanes to be admitted into channels at the outer and inner case diameters. More than one vane may also be grouped to form a sector.

As the extremely hot gases from the combustor enter into the stationary stage, the direct impact causes the vanes to bow. The complex configuration of the vanes causes uneven thermal growth. At the two extremities the stiffness of the attachment design and also of the turbine casing limits the deformation. At the leading edge of the vanes also there is material available to cause some deformation. But at the trailing edge of the vane, there is so little material that substantial thermal deformation is inevitable. The differential growth pattern causes the trailing edge material in the proximity of the inner and outer casing region to see limited thermal growth, but the portion in the mid-section experiences far greater deformation. A direct consequence is bowing at the trailing edge of the vanes.

The profile of the airfoils is developed in order to optimize aerodynamic performance of the turbine unit. At the leading edge of the vanes the rounded shape provides extra material and stiffness. But at the trailing edge the prime concern is to ensure feeding of the hot gases into the succeeding rotating stage at the right angle of incidence and keeping formation of eddies at the leading edge to a minimum. So the thickness of the edge is also maintained to the least possible. Unfortunately, this also causes the trailing edge of the vanes to bow in its mid-section region. On a newer machine the effects may not be of concern. As the machine accumulates operating hours, thermal creep is bound to take its toll and the thermal bow becomes more prominent.

A direct effect of the bowing in the vanes is to misdirect the flow of gases into the rotating blades, especially in the mid-section region. The net result is to bring down the power developed in the turbine. A more efficient cooling in the form of holes built into the trailing edge generally helps to reduce the deformation. However, oxidation within the holes can restrict the passages and cause the flow of the coolant to be retarded.

18.10 FATIGUE DAMAGE AND LIFE LIMIT

Turbine components subjected to steady and alternating stresses accumulate a certain amount of damage with each load cycle. Depending on the total maximum stress the part experiences and the number of load cycles, the damage accumulation progresses till a point is reached when the endurance capability of the material is reached. The material then gives way, and cracks develop in the most heavily stressed area of the part. The failure mechanism sets in, the material finally gives up and the part fails.

Fatigue in the material is the primary cause of failure in turbine parts. Turbine and super alloy manufacturers spend considerable resources in establishing safe stress limits and associated number of loading cycles at given operating temperatures. In aviation applications the option to retire a part after the failure mechanism sets in is not there, the part must be replaced before cracks develop. At the same time, turbine users do not want to throw away a part before its useful life is exhausted, pointing to the need to accurately determine the number of load cycles the part is good for. Even in power generation and industrial gas turbines, failure of a turbine blade is out of question. The units operate at speeds high enough to cause tremendous amounts of kinetic energy to build up in the broken fragments. The dynamic energy may be large enough to cause the broken parts to penetrate the casing of the turbine and land far away, destroying almost everything in its path.

Depending on the source and magnitude of the dynamic stress the component is exposed to, the failure mechanism may be split into high-cycle fatigue and low-cycle fatigue categories. Rotating blades traverse through small fluctuations in pressure of the gas stream as they go past the stationary vanes. In a relatively short period of time the blades will accumulate such pressure, and hence stress, fluctuations running into millions of cycles. High-cycle material property data for the super alloy in the form of stress and number of cycles (or S/N) at the operating temperature must be available to ensure the blade will give satisfactory service. Generally, if the total maximum stress in the part is maintained below 50 per cent of the material's endurance strength, the part is considered acceptable for the intended application.

In contrast, low-cycle fatigue is set in when a part experiences periodical large variations in the dynamic stress. Thermal gradients during an engine's startup and shutdown phases produce high levels of thermal stress. As the operation stabilizes into a stable regime, the temperatures soak in and thermal stresses reduce to nearly zero. Centrifugal stresses also rise to the maximum as the engine reaches top speed, and then come down during deceleration and to nothing when the rotor comes to standstill. Data in the form of Goodman diagram and similar other techniques are employed to evaluate steady and alternating stresses. Material data based on fracture mechanics principles and rules based on Palmer-Minegren and related procedures are used for damage assessment at the end of each load cycle. Observations from fatigue failures of similar components also provide modification factors for the analysis in deciding the number of cycles before failure.

18.11 CONTROLLING SECONDARY FLOW LOSS

Secondary flow in a turbine cascade represents the complex three-dimensional flow pattern near the walls of a turbo-machine. The pattern is composed of discrete phenomena that tend to interact with each other. Development of vortex in various regions upset the usually laminar flow, and oftentimes strongly contribute to losses in aerodynamic efficiency.

Some dominant components of the secondary flow are vortices in the passage between the stationary and rotating airfoils, at the trailing edges and at the corners of the airfoils. Sometimes referred to as a horseshoe form of vortex is produced in the region where the airfoils meet the platform, and may manifest in the form of a series of vortices around the contour. Secondary losses constitute a considerable portion of the total losses in all turbine units, holding especially true of blades with a low aspect ratio.

The secondary losses may be controlled through modification at the leading edge profile of the rotating blades, close to the end wall. One method calls for interaction among the various vortices in order to neutralize one another. Thus, if a corner vortex were to swirl in one direction, causing the horseshoe vortex to spin in the opposite direction reduces the severity of both of them. The objective is to create a strong twist on the suction side branch of the horseshoe vortex, and cause it to interact with a vortex from the main passage of the opposite rotational direction.

The leading edge is locally modified with a bulb at the wall, with the flow going through a series of devices and sieves to straighten and break down turbulence. The bulb has to assume specific dimensions to obtain the desired level of interaction of the two vortices. A pronounced bulb aids on the suction side, and intensifies the desired horseshoe vortex locally. The action moves it away from the boundary layer on the suction side profile of the airfoil, and the counter-acting vortices reduce the losses at the end wall.

18.12 REVIEW QUESTIONS

18.12.1 Enumerate the problems encountered when hot and pressurized gases go through the turbine.

18.12.2 Why are piston engines not suitable for large aircrafts?

18.12.3 What are the prime characteristics of super alloys?

18.12.4 How does solidification of crystals along certain directions in turbine blades help withstand fatigue?

18.12.5 What are the merits and demerits of cooling hot parts in a turbine?

18.12.6 How is cool air inserted in rotating turbine blades?

18.12.7 What are the repercussions of bowing in turbine inlet vanes?

18.12.8 Why is cooling of a turbine disk required?

18.12.9 Under what circumstances does steam cooling of turbine blades improve the heat balance of a turbine?

18.12.10 How does thermal barrier coating provide protection?

18.12.11 How does low-cycle fatigue differ from high-cycle fatigue in a component?

18.13 EXAMPLE PROBLEMS

Example Problem # 18.13.1: A simply supported beam 48 in. long, 1.25 in. deep and 2.5 in. wide is impacted at the mid-span by a free falling sphere of weight 47.5 lbs from height h. Determine the height if it creates maximum momentary deflection of 0.878 in. in the beam. Assume beam material $E = 12 \times 10^6$ lbs/in^2.

Solution: When the load falls a distance h on the beam, it causes a peak instantaneous deflection δ in the beam where the ball strikes. The total work done by the ball equals $W(h + \delta)$. Assume a static equivalent load W_{eq} that produces an equal deflection δ at the same point on the beam, with the load applied gradually. The work done by this equivalent load is $\frac{1}{2}(W_{eq}\delta)$. At mid-span deflection $\delta = (W_{eq} l^3/48EI)$, where l is the length of the beam and EI represents the beam stiffness in flexure. Since the supports at the two ends of the beam are rigid, the work done by the falling and the stationary loads are the same, or

$$W(h + \delta) = \tfrac{1}{2}(W_{eq}\delta)$$

The static deflection δ maybe expressed in terms of the equivalent static load W_{eq}. For example, if the load is applied centrally on the beam,

$$W(h + \delta) = ½(W_{eq}\delta) = 47.5 \times (h + 0.878) = ½(W_{eq} \times 0.878)$$
$$h = (½W_{eq} + 0.878/2 - 47.5 \times 0.878)/47.5 = W_{eq}/95.0 - 0.869$$
$$EI = Ebd^3/12 = 12 \times 10^6 \times 2.5 \times 1.25^3/12 = 48.83 \times 10^5$$
$$\delta = (W_{eq}l^3/48EI) = 0.878 = W_{eq} \times (l^3/48EI) = W_{eq} \times 48^3/(48 \times 48.83 \times 10^5)$$

Or: $0.878 = W_{eq} \times (4.72 \times 10^{-4})$

So: $W_{eq} = 1860.72$ lb.

And: $h = W_{eq}/95.0 - 0.869 = 18.72$ in.

Example Problem # 18.13.2: A ball is shot upwards at a speed of 12.7 ft/sec from a height of 21.2 ft above the ground. Neglecting the resistance of air, calculate the time required to fall back to the ground.

Solution: Let v_1 = initial velocity, v_2 = velocity at maximum height and v_3 = velocity when the object reaches the ground. At start, it goes up with the gravity reducing its upward speed. At top height speed is zero, and gravity forces it to come down. Assume t_1 seconds is required to reach the maximum height of x_1 ft, and t_2 seconds is required to travel $(x_1 + 21.2)$ ft.

Now $v_2 = v_1 + gt_1$ and $v_3 = v_2 + gt_2$.

Velocity at highest point is zero. Hence, $0 = -12.7 + 32.2t_1$ or, $t_1 = 0.394$ seconds.

Also: $x_1 = -21.2t_1 + (32.2 \times t_1^2)/2 = -21.2 \times 0.394 + 16.1 \times (0.394)^2 = -5.86$ ft upwards.

Then: $(-x_1 + 21.2) = 27.06$ ft.

Also: $(x_1 + 21.2) = -0t_2 + (32.2 \times t_2^2)/2 = 27.06$, or $t_2 = 1.296$ seconds.

Thus, total time required to fall back to the ground $= t_1 + t_2 = 1.691$ seconds.

Example Problem # 18.13.3: The signal of a vibration meter reads amplitude of 0.724 V at 328 Hz and a phase angle of 49°. The output is amplitude of 0.184 ins. and phase angle of 81°. Calculate the gain, phase angle and transfer function of the instrument.

Solution: Gain = output/input = 0.184 in./0.724 V
= 0.254 in./V

Phase difference = (output angle) – (input angle) = (81 – 49)°
= 32°

Transfer function = 0.254∠32° in./V

Example Problem # 18.13.4: A strain gage sensor has maximum allowable stress = 110 × 10^3 lbs/in^2, and is used for a cantilever beam of depth b = 0.875 in., height h = 1.25 in. and length L = 87.5 in. If gage factor = 2.255 and nominal resistance = 127.5 Ω, calculate the maximum force that may be measured with the sensor.

Solution: Maximum force acceptable by the beam is from the maximum stress produced, gage factor and resistance are not needed:

$$F_{max} = (S_{max} \times bh^2)/6L$$
$$= (110 \times 10^3 \times 0.875 \times 1.25^2)/(6 \times 87.5)$$
$$= 286 \text{ lbs}$$

Example Problem # 18.13.5: The resistance of a push-pull circuit of an electric motor is 171 Ω, rated at 1.275 A. If the armature is centered, find the voltage of each coil and the change in current load.

Solution: Maximum voltage: $E_{max} = I \times R = 1.275 \text{ A} \times 171\ \Omega$

$= 218.03$ V

With armature centered: $E = E_{max}/2 = 109.01$ V

Variation in current is maximum if maximum voltage is applied on a coil, and no voltage on other coil.

Then: $\Delta I = E_{max}/R - 0 = (109.01 \text{ V})/(171\ \Omega) = 637.5 \text{ m}\dot{\text{A}}$

18.14 EXERCISE PROBLEMS

Exercise Problem # 18.14.1: Determine the maximum allowable stress of a strain gage sensor needs when it is used for a tension load on a coupon of diameter 1.25 in. if the load is 4475 lbs, gage factor is 1.875 and nominal resistance = 22 Ω.

Solution: Maximum force acceptable by the beam is from the maximum stress produced, gage factor and resistance are not needed:

Exercise Problem # 18.14.2: The resistance of an electric circuit is 3240 Ω, rated at 375 mA. Find the voltage of the coil when the current load is full.

Exercise Problem # 18.14.3: The gain of a measuring meter is 0.525 in./volt and phase difference is 78°. Calculate the output of an input signal 0.137 V and phase angle of 187°.

Exercise Problem # 18.14.4: A ball is shot upwards at a speed of 34.3 ft/sec. Neglecting resistance of air, calculate the time required to fall back to the ground.

Exercise Problem # 18.14.5: A cantilever beam 36 in. long, 2.0 in. deep and 1.0 in wide has a load of 410 lbs at the free end. If beam material $E = 30 \times 10^6$ lbs/in^2, calculate the radius of curvature of the beam.

REFERENCES & BIBLIOGRAPHY

Cheruvu, N. S., "Development of a Corrosion Resistant Directionally Solidified Material for Land Based Turbine Blades", ASME paper # 97-GT-425, New York, 1997.

Corman, J. C., "A Gas Turbine Combined Cycle Power Generation System for the Future", Proceedings, Yokohama Gas Turbine Congress", Japan, 1995.

Corten, H. T., Dolan, T. J., "Cumulative Fatigue Damage", Proceedings, International Conference on Fatigue of Metals, ASME and IME, page 235, New York, 1956.

Fleeting, R., Coats, R., "Blade Failures in the HP Turbine of RMS Queen Elizabeth II and Their Rectification", Transactions, Institute of Marine Engineers, volume 82, page 49, London, 1970.

Gatts, R. R., "Application of a Cumulative Damage Concept to Fatigue", Transactions, ASME, volume 83, Series D, page 529, 1961.

Harris, C. M., Crede, C. E., "Shock and Vibration Handbook", McGraw-Hill Book Co., 4th edition, New York, 1995.

Henry, D. L., "Theory of Fatigue Damage, Accumulation in Steel", Transactions, ASME, volume 77, page 913, New York, 1955.

Jay, R. L., MacBain, J. C., Burns, D. W., "Structural Response Due to Blade-Vane Interaction", ASME paper # 83-GT-133, New York, 1983.

Kool, G. A., Agema, K. S., Van Buijtenen, J. P., "Operational Experience with Internal Coatings in Aero and Industrial Gas Turbine Airfoils", Proceedings, ASME Turbo Expo, The Netherlands, paper # GT-2002-30591, New York, 2002.

Koul, A. K., Immarigeon, J. P., Dainty, R. V., Patnaik, P. C., "Degradation of High Performance Aero Engine Turbine Blades", Proceedings, ASM Materials Congress, pages 69–74, Pittsburgh, 1993.

Lucas, M. G., Ireland, P. T., Wang, Z., Jones T. V., "Fundamental Studies of Impingement Cooling Thermal Boundary Conditions", AGARD, CP-527, paper # 14, 1993.

Marco, S. M., Starkey, W. L., "A Concept of Fatigue Damage", Transactions ASME, volume 76, page 627, New York, 1954.

Manson, S. S., Frecke J. C., Ensign, C. R., "Application of a Double Linear Damage Rule to Cumulative Fatigue", Fatigue Crack Propagation, STP-415, ASTM, Philadelphia, page 384, 1967.

Marin, J., "Mechanical Behavior of Materials", Prentice-Hall, Englewood Cliffs, NJ, 1962.

McLean, M., "Directionally Solidified Materials for High Temperature Service", The Metals Society, London, page 153, 1983.

McQuiggan, G., "Design for High Reliability and Availability in Combustion Turbines", ASME paper # 96-GT-510, New York, 1996.

Miner, M. A., "Cumulative Damage in Fatigue", ASME Transactions, Journal of Applied Mechanics, volume 67, page A159, 1945.

Nomoto, H., Koga, A., Ito, S., Fukuyama, Y., Otoma, F., Shibuya, S., Sato, M., Kobayashi, Y., Matsuzaki, H., "The Advanced Cooling Technology for the 1500° C Gas Turbines: Steam Cooled Vanes and Air Cooled Blades", ASME paper # 96-GT-16, New York, 1996.

Palmgren, A. "Die Lebensdauer von Kugellagern", ZVDI, Volume 68, page 339, 1924.

Patnaik, P., Elder, J., Thamburaj, R., "Degradation of Aluminide Coated Directionally Solidified Super Alloy Turbine Blades", Superalloys Book, Reichmann et al editors, TMS AIME, pages 815–824, Warrendale, NJ, 1988.

Rangwala, A. S., "Dynamic Analysis of Turbo-machinery", McGraw-Hill, New York, 2004.

Rao, J. S., Pathak, A., Chawla, A., "Blade Life: A Comparison by Cumulative Damage Theory", ASME paper # 99-GT-287, New York, 1999.

Saravananamuttoo, H. I. H., Rogers, G. F. C., Cohen, H., "Gas Turbine Theory", Prentice Hall, Harlow, England, 2001.

Sauer, H., Muller, R., Vogeler, K., "Reduction of Secondary Flow Losses in Turbine Cascades by Leading Edge Modifications at the End Wall", ASME paper # 2000-GT-473, New York, 2000.

Shigley, J. E., Mitchell, L. D., "Mechanical Engineering Design", McGraw-Hill, New York, 1983.

Sims, C. T., Stoloff, N. S., Hagel, W. C., "Super Alloys II", John Wiley & Sons, New York, 1987.

Sondak, D. L., Dorney, D. J., "Simulation of Vortex Shedding in a Turbine Stage", ASME Journal of Turbo-machines, volume 121, pages 428–435, New York, 1999.

Strang, A., Lang, A., Pichoir, R., "Practical Implications of the Use of Aluminide Coatings for Corrosion Protection of Super Alloys in Gas Turbines", AGARD-CP-356, page 11, 1983.

Thomson, W. T., "Theory of Vibration with Applications", Prentice-Hall, Englewood Cliffs, NJ, 1988.

Tsuchiya, T., Furuse, Y., Yoshino, S., Chikami, R., Tsukuda, Y., Mori, M., "Development of Air-Cooled Ceramic Nozzles for a Power Generating Gas Turbine", ASME paper # 95-GT-105, New York, 1995.

Warnes, B. M., "Improved Pt-Aluminide Coatings Using CVD and Novel Platinum Electro-Plating", ASME paper # 98-GT-391, New York, 1998.

Wood, M. I., "Internal Damage Accumulation and Imminent Failure of an Industrial Gas Turbine Blade – Interpretation and Implications", ASME paper # 96-GT-510, New York, 1996.

19

CHAPTER

PERFORMANCE DETERIORATION

19.1 INTRODUCTION

Clearly a primary objective of turbine operators is to obtain peak efficiency and power output from the units. Large gaps between the rotor and the stator are the major culprit. Flow past the blade tips in the turbine causes highly energized gases to escape without doing useful work. But losses of even small magnitude such as from secondary flow in the cascades need to be investigated. Loss of metal around the leading edge of a vane at the inner and outer walls produces turbulence and contributes to flow losses. Vortices also tend to form in the contour area where the airfoil meets the end wall.

Centrifugal loads account for a large portion of radial variations in tip clearance. The rotating centrifugal force field due to an unbalanced disk causes the shaft to deflect from its true axis. The supporting bearings, housings and casing are distorted from the dynamic forces and bending moments. And if the spin speed were to coincide with the elastic shaft's natural frequency of vibration in the transverse direction, a state of resonance exists and the shaft and disk assembly will experience large dynamic motion.

Shaft misalignment is another factor responsible for flexural vibrations. At the coupling between an engine and the driven unit, thermal conditions may throw out the rotor from its aligned position prior to the engine's start. Radial and angular misalignment of the rotor can radically alter the flexural vibration signature of a rotating assembly, with consequent bending moments imposed on the shaft as also the outer case. This also has the adverse effect of putting unusual wear pattern in the bearings.

Differential thermal growth between the rotor and stator is inevitable due to the geometric variations in the structures. During startup and shutdown of a turbine unit, the varying growth pattern can cause the clearance at the tip of the blades to increase. On the other hand, shrinkage has the adverse effect of the tips rubbing against the case, which further opens the clearances.

Asymmetric clearance variations arise from non-uniform loads that usually act on the stationary structures of the engine. Distortion of casing arising from uneven heating can cause the casing or the shroud to open clearances along a portion of the segment, or segments, of the circumference while reducing them in other areas. This leads to the casing going out of roundness. Aviation gas turbines are susceptible to non-symmetric distortion of the casing under the influence of thrust and during aircraft maneuvers. Since the engine is not attached to the wing of the aircraft at its centerline, aerodynamic and thrust load reactions produce an applied moment, leading to a bent casing structure relative to the rotor.

19.2 SEALING CLEARANCE BETWEEN ROTOR AND STATOR

Control of radial clearances between blade tips and the casing is achieved by blowing cooler air on the external surface of the case in order to shrink the casing. The active clearance control system is employed for the high-pressure gas generator turbine of aviation engines. Cool air is directed on the outer surface of the casing from holes in tubes that are placed around it in the plane of the rotor stages. Temperature sensors and valves are employed to control the flow of cooling air during periods of varying temperature operation. Passive cooling systems for the low-pressure turbine merely blows cooler air on the shell.

Squeeze film dampers have the ability to attenuate vibrations in a rotating machine assembly, and in the process reduce clearance between the rotor and stator. Aircraft jet engines use rolling element bearings, which are inherently free of damping as witnessed by their alternative name of anti-friction bearings. Used in conjunction with a selected bearing, the damper introduces damping into the system to reduce synchronous response amplitudes. The method enhances stability and provides a margin of safety for blade loss conditions. Squeeze film dampers have also been developed and successfully employed for protection against sub-harmonic rotor instabilities operating on tilting pad bearings. Some high-pressure compressor manufacturers also use the dampers.

19.3 LABYRINTH AND HONEYCOMB SEALS

Labyrinth seals are extensively used in centrifugal compressors operating at high pressure and speed to control leakage. In a multi-stage centrifugal compressor, for example, the shaft seal labyrinth may be used to restrict leakage flow along the shaft to the backside of the preceding impeller. The eye-packing seal limits return flow leakage down the front of the impeller. A shaft seal is also provided to restrict leakage along the shaft to the preceding stage.

An in-line machine calls for the stages to be arranged in series, where the radial discharge from one impeller is made to turn around and feed into the eye of the succeeding stage. To alleviate axial force due to the pressure difference between the discharge and inlet pressures, a balance drum is used to counteract by directing the leakage flow through the drum into the inlet. Hence, the balance drum absorbs the full pressure differential developed in the compressor. The fluid within the seal has an average density proportional to the inlet and discharge pressures. The eye-packing and shaft seal configurations may be considered to be half-labyrinth seals, while the one at the balance drum is an interlocking or full labyrinth.

A honeycomb seal calls for a roughened stator surface to reduce leakage. It also offers the benefit of reducing circumferential flow velocity within the seal and the associated destabilizing cross-coupled stiffness coefficient. This seal type has been successfully used in compressors for the balance drum and as a turbine inter-stage seal for the high-pressure oxygen turbo-pump of the space shuttle main engine.

Honeycomb seals are also used in gas turbine expander section to seal the gap between rotating blade tips and the casing. Sectors of honeycomb backed with steel pads are inserted inside grooves in the casing in the plane of rotation. Knife-edged crests at the tip of the blades cut grooves into the honeycomb to provide effective sealing of the hot and pressurized gases and prevent them from escaping past the tips.

Brush seals use a biased pattern of wires in contact with a ceramic coating on the shaft, and sharply reduce leakage when compared with a labyrinth or honeycomb seal. Test measurements indicate the leakage rate of a single stage brush seal is a mere 5 per cent of a four-cavity tooth on a rotor labyrinth seal. And because of their compliant nature, damage due to transient impact between the rotor and stator components is held to a minimum. But it can prevent them from springing back to original shape after several hours of operation, and may need replacement.

19.4 DEFLECTIONS IN ROTATING AND STATIONARY STRUCTURES

The problem of sealing asymmetric gaps around the periphery is complicated due to a variety of loads acting on the rotating and stationary structures of aviation jet engines. Centrifugal, thermal, internal pressure and thrust loads constitute engine loads. At the same time, external aerodynamic, inertial and gyroscopic forces contribute to flight loads. The thin casing walls of the turbine lack enough rigidity and experience distortion from large loads. Linear and angular deflections in the components from the engine and flight loads may act in a specific direction or may be multi-directional. As a direct consequence, the deflections produced may also be symmetric or asymmetric with the engine centerline.

In the high-pressure section of a modern aircraft jet engine the blade tip seals are located inside the stator. The casing is lined with air seals to form a continuous shroud around the rotating blades. The arrangement provides for controlling the spillage of gas over the tip of the blades.

The clearance over the blade tips varies according to the power level produced by the engine. The differences around the 360° arise from the distortion occurring in both the stationary and rotating components due to the combination of induced engine power and flight loads acting on them.

19.5 UNBALANCE IN ROTOR

Centrifugal and thermal loads are responsible for the maximum radial variations in tip clearance. A disk of mass m, for instance, is mounted on a shaft of length l between two bearings and spinning at a fixed speed, ω. Assume the center of gravity of the disk to be located at a radial distance e away from the centerline of the shaft. Then a rotating centrifugal force $m\omega^2 e$ acts on the disk. The rotating force vector may be divided into horizontal and vertical components, causing the disk to execute vibratory motions in the two directions.

If the spin speed ω were to coincide with the elastic shaft's natural frequency of vibration in the transverse direction, a state of resonance is produced. The shaft and disk assembly will experience considerable vibrations. The dynamic motion will only be constrained by windage and with damping associated with the lubricant film in the bearings. If rolling elements bearings are used, even the latter source is absent.

Neglecting the effects of gravity, there are two forces acting on the disk, the elastic pull of the shaft tending to restore it to its original position and the centrifugal force on the center of gravity of the disk, causing it to travel in a circle when the system is fully symmetric about the axis of spin. The elastic restoring force depends on the bending stiffness of the shaft, and is proportional to deflection. This observation is also true of a shaft with many disks running in two or more bearings.

Bending in the shaft may arise from a number of operational factors, some of which arise internally in the machine and others from outside the machine. Dynamic balance of the machine's rotating assembly is an internal machine factor that causes bending in the shaft. A misaligned shaft is another cause of flexural vibrations. Misalignment may exist at the coupling, between the rotor and the stator and also between adjacent rotating and stationary parts.

Rotor unbalance and shaft bow are the most commonly encountered stimulus of flexural vibrations, and result in whirling or whipping about the shaft's true axis. Unstable operation of the machine in the form of chatter (or stick-slip) and large deflections and loads are other undesirable consequences.

19.6 SENSORS TO MEASURE TIP CLEARANCE

Sensors for measuring blade vibrations and tip clearances in rotating machinery rely on a number of technologies, including x-ray, inductive, capacitive, optical, eddy current, microwave and acoustic. Blade passage sensors typically are capable of measuring the clearance at the tip and the time of arrival. Accuracy level in the sensing devices should be of the order of 0.001 in. to avoid detraction from an active clearance control system's own tolerance stack up. Accuracy of measurement will also depend on the physical condition of the blade tip.

Measurement of clearances at the blade tip may be imperative in an active clearance control system if the objective is to control the dead space between the rotor's outer tip and the casing. To permit multiple measurement of clearance per blade of larger engines, sensor response should be in high-frequency range. High-temperature capable sensors and the associated communication system may need cooling, and must also be durable to withstand vibrations and impact while providing long service life. The sensors must be tolerant to moisture, dirt and combustion gases.

Non-contacting measurement techniques such as blade tip timing measure the clearance gap between the rotating and stationary components using optical probes mounted in the assembly casing. When vibrations of the external casing structure are not substantial, the time for the tip of a probe, the so-called blade arrival time, depends on rotor speed alone. But when the blade is vibrating, blade arrival time also depends on the amplitude and frequency of vibration. In principle, blade motion can be characterized from such data. As with other transducers such as strain gauges and accelerometers, whether a particular mode can be captured with a laser probe depends on the measurement location. The measurements are typically taken near the leading edge of the blade to maintain reasonable sensitivity to the motion of the first few bending and torsion modes.

19.7 VARYING ROTOR-STATOR CLEARANCE

Asymmetry in the mount system of wing-mounted aviation engines is primarily responsible for variation in clearances between the rotor and stator around the periphery. Large components with thin walls tend to lack stiffness even when assembled with mating components in accordance with guidelines. Bolted flange faces must be flat and parallel. Pre-torque of the bolts must be within prescribed limits. But rotor speeds in gas turbines are so high that bending, twisting and shearing loads will create large deflections. High operating temperatures approaching 2,500° F also produce deflections in the structures of aircraft engine that must be dealt with.

Tip clearance also varies with rotor speed as an aircraft takes off, cruises, decelerates and re-accelerates. Cold build engine clearances and response of the rotor and the casing vary as the engine speed and temperature change. Axi-symmetric clearance changes to the extent of 0.030 in. are oftentimes encountered during takeoff and re-acceleration in large commercial engines. Military engines place a greater emphasis on integrity and lesser concern on efficiency, so clearance variations are even more exaggerated.

A strenuous event in an aircraft flight takes place when an engine shuts down during flight. Relight in the combustor will then be attempted. As part of a new aircraft's acceptance testing, the engine is shutdown at altitude. This is followed by a dwell time during which the engine windmills due to the aircraft's forward momentum. The engine is then restarted and accelerated back to the same power setting at which the engine was shutdown. Since the engine is abruptly shutdown without going through deceleration, the casing cools even more than a normal shutdown, while the centrifugal forces are less because windmill speed is less than ground idle. Restart followed by acceleration leads to greater clearance closure than a takeoff condition due to lower temperatures of both the rotor and the casing.

19.8 EXCESS CLEARANCE AND BLADE AERODYNAMICS

The magnitude of clearance at the tip of the blades has a major impact on the flow pattern of air or gases in the region. Concurrent opening of the gap in both the high-pressure compressor and the high-pressure turbine will dramatically alter the specific fuel consumption, compressor stall margin and engine efficiency. Payload and flight range capability will also be seriously affected.

Close control of the clearances in the high-pressure turbine will improve service life, or time on the wing, of aviation engines. Together with related degradations that may occur in the components as more operating hours accumulate, the effects manifest in the form of increased exhaust gas temperature (EGT).

The EGT is a useful indicator in gauging the overall condition of an engine. The upper limit on the EGT is based on the maximum temperature level at which the high-pressure turbine can safely operate. The Federal Aviation Administration (FAA) certifies aircraft engines with a definite EGT limit. Crossing of the EGT limit requires removal of an engine from the wing for service maintenance. In this respect note that the cost of maintenance for a major overhaul of large commercial engines is prohibitive, and will likely keep climbing.

19.9 COMPRESSOR SURGE

A surging compressor is a major concern in its operation. Besides loss of performance, a stalling compressor can lead to structural damage. Some manifestations of a stall are pressure loss at the discharge end, reduced power output in the turbine and rising exhaust gas temperatures. Since aviation engines operate in a wide speed band, the risk of running into a stall is large.

Physical effects of a surge in an aircraft engine's compressor can be grave. A sudden stoppage in the airflow and accompanied by a loud noise emanating from the engine is common during such an event. Damage to the engine structure may be expected when repeated a number of times, particularly in the compressor and fan areas. The engine is usually able to recover from the stall and continue to proceed, albeit at a lower speed.

As the compression ratio increases, the danger of running into stall troubles rises correspondingly. The angle of attack as air flowing into the rotating blades is of significance. But the radial clearance between the rotating blade tips and the stationary outer casing in the compressor directly affects the ability of operation. The stall margin deteriorates, and stability of the flow is jeopardized since the flow is moving as the pressure increases. The flow reverses direction and the engine goes into a stall.

Inception of two forms of rotating stall has been observed in axial compressors. Pressure taps at discreet locations in the casing in the plane of a rotating stage may be used to obtain variations in pressure. One form of pressure signals is characterized by waves with length scale on the order of the compressor's circumference. The waves propagate at a speed of one-fourth to one-half the speed of the rotor. Another considerably different rotating stall is characterized by disturbances with a dominant length scale that is much shorter than the circumference, on the order of several blade pitches, and a propagation speed of 70 to 80 per cent of the rotor speed.

19.10 VORTEX FORMATION IN TIP CLEARANCE REGION

Aerodynamic rotating instability is responsible for intensifying secondary flow losses, tip clearance related noise and for setting up excitation of blade vibrations. This flow characteristic maybe due to a group of superimposed modes or stall in a part of the span. The instability appears in pressure or velocity frequency spectra in the form of a significant increase in amplitude within a frequency band that is well below the blade passing frequency, and is superimposed by lesser peaks with nearly constant spacing in the frequency. The pattern is marked in a group of consecutively numbered modes.

Primary parameters associated with the instabilities in axial compressors vary over a wide range of values. Tip clearance at the rotating blades is a major influencing factor in the initiation of the instability. Rotational speed of the instabilities in different machines varies between 25 and 90 per cent of rotor speed. Pressure difference across the blade row and axial component of flow velocity also play a substantial role.

Within the clearance volume, measurements can be made only for the biggest magnitude. Post processing of the data is by the ensemble-averaged method. An array of microphones distributed circumferentially in one axial plane determines higher order modes of the rotating instabilities. Simultaneously acquired signals permit correlation between various sensors.

Rotating instabilities propagate in the blade's tip region in the form of a wave front. Since the dominating mode order is about half the number of blades, the process repeats itself at every other blade at the same time. Direction of the propagation in the tip region is determined by correlating the signal from the pressure transducers on the pressure and suction sides of adjacent blades.

Pressure difference between the pressure and suction sides along the chord is nearly constant at the design point, hence the time averaged tip vortex intensity is not dependent on the axial position. But at an operating point near the stability limit the driving pressure difference at the leading edge initiates, indicating a propensity towards intensification of the vortex.

19.11 EFFECTS OF THERMAL RATCHETING ON TURBINE POWER OUTPUT

Operating clearances in an engine are determined to a considerable extent by the operating temperature in an engine. For instance, the takeoff pinch point would be lower for a hot engine takeoff since the rotating components will grow radially due to the larger thermal mass of the turbine rotor and blades as compared to the casing. An adequate period of cooling time must be provided before an aircraft can takeoff with hot engines.

Besides the thermal growth, substantial thermal stresses in the airfoils and in the rotor are induced when the takeoff is attempted with hot engines. Thermally induced growth is also responsible for another phenomenon known as 'bowed rotor'. Shortly after an engine is shutdown, heat buildup takes place in the upper portion of the engine from the heat retained in the components of the engine and lack of venting for this trapped heat. As a consequence, temperature gradients develop along the diameter of the rotors and cause the shafts to deflect, or bow. Depending on the prevailing wind in the vicinity of the engine and the consequent wind milling of the rotors, the maximum radial deflections from the chimney effect generally occur between 1 and 2 hours after the engine is shutdown. If the engine were to be restarted in this time period, the bowed rotor start will lead to considerably large amounts of unbalance in the two rotors, excessive vibrations in the engine, and rubbing between the blade tips and the casing. Bowed rotor starts are known to be severe enough to cause the low-pressure rotor of an aircraft engine to lockup due to the interference between the blade tips and the casing shroud.

Industrial and power generation steam and gas turbines display similar traits in regards to bowed rotor starts. As a rule, immediately following a shutdown, the rotor is then placed on turning gear to achieve better temperature distribution. Since the rotor is of a substantially heavier mass in this case, the cooling time required is also larger.

19.12 EXHAUST GAS TEMPERATURE AND TURBINE BLADE TIP CLEARANCE

A reliable indicator of the overall health of a gas turbine is the temperature of gases at the turbine's exhaust. When the temperature increases beyond a certain set limit, ground personnel in charge of engine maintenance are required to check for causes. If inspection of blade tips with a bore-scope and review of accelerometer signals and pressure and temperature readings indicate abnormalities, the decision may be made to take the engine off the wing based mostly because of the deteriorated exhaust gas temperature (EGT).

The prime reason behind increase in the EGT is excess clearance between the rotating blade tips and stator casing. The higher than normal clearance lets hot and pressurized gases to flow over the tip without performing work, and causes output from the turbine to fall. The torque sensor on the rotor detects the reduced output and sends the signal to the turbine's microprocessor. To account for the loss, the processor will order additional fuel to be delivered to the combustor. But the action also raises the temperature of the gases at the exhaust from the turbine. Concurrently, the turbine's efficiency will also fall. Also, the clearances at the blade tip will further increase from the higher operating temperature.

Overhaul of the turbine unit is perhaps the best solution to maintain efficiency and fuel costs within limits. If left unattended, the problem can only worsen as the engine accumulates extra operating hours. One solution is to replace the honeycomb shroud segments that line the turbine casing. Repairing of knife crested edges on the blade shrouds may also be done in conjunction.

19.13 REVIEW QUESTIONS

19.13.1 Discuss the role of misalignment and its impact on operating efficiency.

19.13.2 How do labyrinth seals function to seal leakage in rotating machines?

19.13.3 What are the merits and demerits of a brush seal?

19.13.4 What arc thc cffccts of a disk mountcd cccentrically on a shaft?

19.13.5 How does a flexible shaft in rotating machine affect its operation?

19.13.6 Discuss some technologies that aid measuring clearances at rotating blade tips.

19.13.7 Why are aviation engines susceptible to variations in tip clearance around the periphery?

19.13.8 How does clearance at the tip of the blades affect performance of a turbine engine?

19.13.9 Name the two leading factors that can cause a compressor to surge.

19.13.10 What are the causes behind bowing of a rotor in aviation and in power generation engines?

19.13.11 How does increased temperature of exhaust gases of a turbine affect its performance?

19.14 EXAMPLE PROBLEMS

Example Problem # 19.14.1: A material has $S_y = 112000$ lb/in², ultimate strength $S_{ut} = 147000$ lb/in², and endurance strength $S_e = 48700$ lb/in². The mean and alternating stress components are $\sigma_m = 38312$ lb/in² and $\sigma_{alt} = 13760$ lb/in². Find the factor of safety using the Gerber theory's parabolic relation.

Solution: The Gerber theory based on a parabolic relation is:

$$(S_{alt}/S_e) + (S_m/S_{ut})^2 = 1$$

Thus, $S_m = n\sigma_m = 38312n$ and $S_{alt} = n\sigma_{alt} = 13760n$. Substitution into the non-linear equation gives:

$$(13760n/48700) + (38312n/147000)^2 = 1$$

The factor of safety n is obtained by solving the equation, yielding $n = 2.285$.

Example Problem # 19.14.2: The rotor of a gas turbine runs at 5400 rpm and has mean blade radius = 28.45 in. The swirl velocity at inlet is 47.3 ft/min and 28.2 ft/min at exit. If 945 lbs/min of air and fuel is used and mechanical losses result in 89.2% efficiency, calculate the power developed in the turbine.

Solution: Tangential speed $U = \omega \times r = (2\pi \times 5400) \times 28.45/12 = 80399.7$ ft/min

From equation (7.4),

$$\begin{aligned} W &= (dm/dt) \times U \times \{\Delta V_w\} \\ &= 945 \times 80399.7 \times (47.3 + 28.2)/33000 = 173827.81 \text{ HP} \end{aligned}$$

So net power produced = 173827.81 × 0.892 = 155054 HP

Example Problem # 19.14.3: Calculate the specific work output of a simple-cycle gas turbine if the ratio of specific heats is 1.387, compression ratio is 29.3, inlet air temperature is 289 K and turbine inlet temperature is 1362 K.

Solution: From equation (8.2), $$\frac{W}{c_p T_1} = t\left(1 - \frac{1}{r^{(\gamma-1)/\gamma}}\right) - \left(r^{(\gamma-1)/\gamma} - 1\right)$$

Temperature ratio $t = 1362/289 = 4.713$

Then: $$\begin{aligned} \frac{W}{c_p T_1} &= 4.713\left(1 - \frac{1}{29.3^{(1.387-1)/1.387}}\right) - (29.3^{(1.387-1)1.387} - 1) \\ &= 1.31012 \end{aligned}$$

Example Problem # 19.14.4: A machine shaft has $D_i/D_o = 0.630$ and $D_o = 7.875$ in. If a bending moment of 429189 in-lbs acts on it as it transmits 27575 hp at 3600 rpm, determine the normal, shear and combined maximum shear stresses in the shaft.

Solution: $$\text{Torque} = 27575 \times 12 \times 33000/(2 \times \pi \times 3600) = 483001.59 \text{ in-lbs}$$

Normal stress is: $$\begin{aligned} \sigma_x &= \frac{32M}{\pi D_o^3[1-(D_i/D_o)^4]} \\ &= [32 \times 429189]/\{\pi \times 7.875^3 \times (1 - 0.630^4)\} \\ &= 10630.72 \text{ lbs/in}^2 \end{aligned}$$

Shear stress is: $$\tau = \frac{16T}{\pi D_o^3[1-(D_i/D_o)^4]}$$

$= [16 \times 483001.59]/\{\pi \times 7.875^3 \times (1 - 0.630^4)\}$

$= 5981.80$ lbs/in^2

Combined maximum shear stress: $\tau_{max} = \dfrac{16\sqrt{M^2 + T^2}}{\pi D_o^{\,3}[1 - (D_i / D_o)^4]}$

$= 16 \times (429189^2 + 483001.59^2)^{1/2}/\{\pi \times 7.875^3 \times (1 - 0.630^4)\}$

$= 8002.19$ lbs/in^2

Example Problem # 19.14.5: A pulley *B* on a shaft weighs 473 lbs, and transmits 48.75 HP at 900 rpm from a driver (Figure 19.1). The attached belt exerts downward force of 179 lbs. Determine the torque transmitted by the pulley, and calculate the reactions in the bearing.

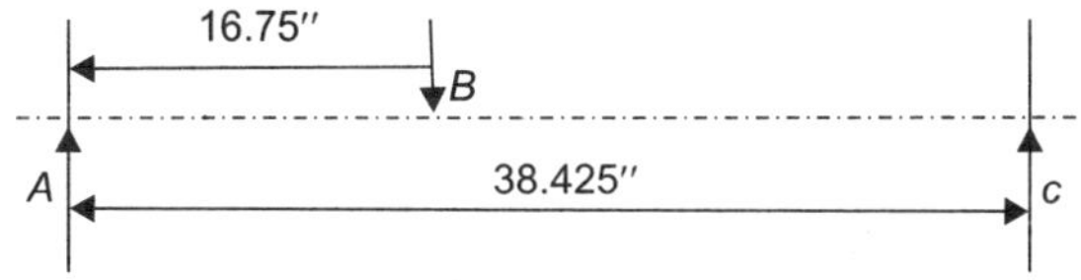

Fig. 19.1

Solution: Torque T $= (33000 \times \text{HP})/(2\pi \times \text{N})$

$= (33000 \times 48.75)/(2\pi \times 900)$

$= 284.63$ ft-lbs

Reactions at the bearings: $R_A + R_C = -179 - 473 = 652$ lbs ⇑

Sum of moments at *A*: $M_A = R_C \times 38.425 + (-179 - 473) \times 16.75 = 0$

Hence, $R_C = -(-179 - 473) \times 16.75/38.425 = 284.22$ lbs ⇑

And $R_A = 652 - 284.22 = 367.78$ lbs ⇑

19.15 EXERCISE PROBLEMS

Exercise Problem # 19.15.1: A turbine shaft carries bending moment of 3487 in-lbs and transmits 7546 in-lbs torque. If shaft diameter is 2.25 in., determine the normal, shear and combined maximum shear stresses in the shaft.

Exercise Problem # 19.15.2: Calculate the isentropic efficiency of a simple-cycle gas turbine if the ratio of specific heats is 1.387 and compression ratio is 38.275.

Exercise Problem # 19.15.3: Calculate the specific work output of a simple-cycle gas turbine if the ratio of specific heats is 1.347, compression ratio is 32.2, inlet air temperature is 289 K and turbine inlet temperature is 1312 K.

Exercise Problem # 19.15.4: A steam turbine spins at 1800 rpm and has mean blade radius = 16.425 in. The swirl velocity at inlet is 32.4 ft/min and 16.87 ft/min at exit. If 437 lbs/min of air and fuel is used and mechanical losses result in 89.5% efficiency, calculate the power developed in the turbine.

Exercise Problem # 19.15.5: A material has $S_y = 82500$ lb/in^2, ultimate strength $S_{ut} = 117000$ lb/in^2, and endurance strength $S_e = 29700$ lb/in^2. The mean and alternating stress components are $\sigma_m = 9856$ lb/in^2 and $\sigma_{alt} = 4532$ lb/in^2. Find the factor of safety using the Gerber theory's parabolic relation.

REFERENCES & BIBLIOGRAPHY

Cawley, J. D., "Oxidizing Seal for Turbine Tip Gas Path", US Patent 4,540,336, 1985.

Childs, D., "Turbo-machinery Rotor Dynamics", John Wiley & Sons, New York, 1993.

Childs, D., Kleynhans, G., "Experimental Rotor Dynamic and Leakage Results for Short (L/D = 1/6) Honeycomb and Smooth Annular Pressure Seals", Proceedings, Institute of Mechanical Engineers, 5th International Conference on Vibrations in Rotating Machinery, London, 1992.

Ciokajlo, J. J., "Blade Tip Clearance Control Apparatus Using Shroud Segment Annular Support Ring Thermal Expansion", US Patent 5,116,199, 1992.

Day, I. J., Breur, T., Escuret, J., Cherret, M., Wilson, A., "Stall Inception and the Prospects for Active Control in Four High Speed Compressors", ASME paper # 97-GT-281, New York, 1997.

European Regions Airline Association (ERA) "ERA Fast Facts", September, 2001.

Fasching, W. A., "CF6 Jet Engine Performance Improvement Summary Report", NASA CR-165612, 1982.

Ferguson, J., "Brushes as High Performance Gas Turbine Seals", ASME paper # 88-GT-182, New York, 1988.

Flotow, A., Mercadal, M., Tappert, P., "Health Monitoring and Prognostics of Blades and Disks with Blade Tip Sensors", Aerospace Conference Proceedings, IEEE, volume 6, pages 433–440, 2000.

Gaffin, W. O., "JT9D-70/59 Improved High Pressure Turbine Active Clearance Control System", NASA CR-159661, 1979.

Greathead, S., Bostow, P., "Investigations into Load Dependent Vibrations of High Pressure Rotor on Large Turbo-Generators", Proceedings, Institute of Mechanical Engineers, Conference on Vibrations in Rotating Machinery, Cambridge, pages 279–286, 1976.

Halila, E. E., Lenahan, D. T., Thomas, T. T., "Energy Efficient Engine, High Pressure Turbine Test Hardware Detailed Design Report", NASA CR-167955, 1982.

Heath, S., Imregun, M., "A Survey of Blade Tip Timing Measurement Techniques for Turbo-machinery Vibration", ASME paper # 97-GT-213, New York, 1997.

Haynes, J. M., Hendricks, G. J., Epstein, A. H., "Active Stabilization of Rotating Stall in a Three Stage Axial Compressor", ASME Journal of Turbo-machinery, volume 116, pages 226–239, New York, 1994.

Howard, W. D., Fasching, W. A., "CF6 Jet Engine Diagnostics Program: High Pressure Turbine Roundness/Clearance Investigation", NASA CR-165581, 1982.

Huber, F. W., Dietrich, D. J., "Clearance Control for the Turbine of a Gas Turbine Engine", US Patent 5,667,359, 1997.

Kawecki, E. J., "Thermal Response Turbine Shroud Study", Air Force Aero Propulsion Laboratory Technical Report AFAPL-TR-79-2087, 1979.

Kolthoff, P., "Device for Controlling Clearance Between Rotor and Shroud of a Turbine", US Patent 3,039,737, 1962.

Marquette, O. R., Childs, D. W., San Andres, L., "Eccentricity Effects on Rotor Dynamic Coefficients of Plain Annular Seals: Theory versus Experiment", ASME, Journal of Tribology, volume 119, pages 443–448, New York, 1997.

Olsson, W. J., Martin, R. L., "B747/JT9Dflight Loads and Their Effect on Engine Running Clearances and Performance Deterioration: Nacelle Aerodynamic and Inertial Loads (NAIL)/JT9D Jet Engine Diagnostics Programs", NASA CR-165573, 1982.

Rangwala, A. S., "Turbo-machinery Dynamics: Design and Operation", McGraw-Hill, New York, 2005.

Santiago, O., San Andres, L., Oliver, J., "Imbalance Response of a Rotor Supported on Open End Integral Squeeze Film Dampers", ASME paper # 98-GT-6, New York, 1998.

Shemeld, D., "A History of Development in Rotor Dynamics – A Manufacturers Viewpoint", Rotor Dynamic Instability Problems in High Performance Turbo-machinery, NASA Report # CP 2443, pages 1–18, 1986.

Soto, E. A., Childs, D. W., "Experimental Rotor Dynamic Coefficient Results for (a) a Labyrinth Seal With and Without Shunt Injection and (b) a Honeycomb Seal", ASME paper # 98-GT-8, New York, 1998.

Tseng, W., Hauser, A. A., "Blade Tip Clearance Control Apparatus with Shroud Segment Position Adjustment by Unison Ring Movement", US Patent 5,035,573, 1991.

Tryfonidis, M., Etchevers, O., Paduano, J. D., Hendricks, G. F., Epstein, A. H., "Pre-Stall Behavior of Several High Speed Compressors", ASME Journal of Turbo-machinery, Volume 117, pages 62–80, New York, 1995.

Wardle, R. L., "Floating Blade Shrouds", US Patent 2,927,724, 1960.

Wiseman, M. W., Guo, T., "An Investigation of Life Extending Control Techniques for Gas Turbine Engines", Proceedings of the American Control Conference, Arlington, VA, June 25–27, 2001.

20
CHAPTER

SYSTEMIC FAULTS

20.1 INTRODUCTION

Failures of unimaginable proportions and consequent destruction in turbo-machinery occur under certain circumstances that are totally external to the engine's own system. A foreign object ingested into the fan of an aviation engine has the potential to bring down an entire airplane. An abrupt shedding of a large electrical load can cause a power generation turbine to speed up and cause the rotating disks to burst. Faults in the lube oil processing system can cause oil flow stoppage to the bearings, leading to rotating blades grinding against the casing.

A flock of birds striking an engine during the takeoff mode when the engine is required to develop full thrust is a prime example. In spite of engine manufacturers taking adequate steps, broken fan and compressor blades may penetrate the casing and hit the passenger cabin. Fire may break out in the combustor section due to the sudden loss of air and engulf the aircraft's wing and the fuselage. Debris in the form of broken tire treads and metal objects lying on the runway also hold hazards.

In tandem with failure of the speed controller and malfunctioning steam admittance valves in the steam chest of a high-pressure turbine, a sudden loss of power demand on the electrical generator can cause the rotor to run away to failure. Accessories such as clogged oil filter, pressure or flow restrictions in the lubrication pump or cooling system will prevent the bearings from holding the rotor within the clearance region.

Sophisticated control mechanisms notwithstanding, the operating speed of modern turbines is so high it is not possible to avoid running the engines into cataclysmic breakdowns. A failure in what may at first sight appear to be a minor component can lead to rapid failures in a string of components of the sub-assembly. And where remedial action commands from the microprocessor-based controller are not possible, the end consequences may very well prove to be disastrous.

20.2 FAULTS ARISING FROM EXTERNAL CONDITIONS

The basic principle of air and gases flowing over the stationary and rotating airfoils in a turbine engine develops power. But the same principle is also responsible for the cause of aerodynamic interference and failure of the blades due to unacceptable blade passing frequencies. This is a leading factor in failures arising from internal causes.

In contrast, there are other factors lying outside the engine system that at times are responsible for failures. Strictly beyond the control of operating personnel, the failures may even be practically instantaneous and catastrophic.

Ingestion of foreign objects such as birds and other debris is a major cause of accidents in the aviation industry. If one engine of a multi-engine aircraft is affected, the aircraft has the option to continue for at least a short duration and land at a nearby airfield. But when the birds are ingested into more than one engine at the same time, it would be difficult for the flight to continue to operate, in the process jeopardizing the safety of passengers and crew.

Reliability of lubrication oil is an important requirement for continuity of service of turbine engines. Oftentimes, however, the oil delivered may not be at the required pressure and flow rate due to a ruptured oil line fitting, problems in suction or discharge pumps, filtration or cooling systems. Within minutes the journal will rub against the bearing surface, and metal chips loosen and grind both surfaces. The tightly controlled clearances in the bearings and between the rotating blade tips and stator will be lost and cause widespread damage throughout the engine. Aviation engines with rolling-element bearing supports also rapidly deform, permitting the rotor assemblies to shift both axially and radially. By far the loss of lubricant oil to the bearings is a prime cause of failures in most turbines.

Land-based and aviation turbine engines occasionally go into an over-speed condition. Failure in the speed control mechanism coupled with malfunctioning valves leads to the problem. But the rotating blades and disks can only be stressed up to their endurance limit. Since circumferential stress varies with the second power of speed, going into a speed range beyond a fixed limit can cause a disk to burst, in the process destroying an entire facility. Loss of even a partial section of a rotating airfoil has the potential to cause severe vibrations in the rotor and in the stator. The broken pieces will also damage stator and rotor parts in the vicinity.

20.3 TIME LAG BETWEEN SENSOR INDICATION AND COMMAND EXECUTION

A typical control system based on the feedback principle offers a number of advantages in obtaining optimum performance from a turbine engine. The control mechanism also plays a central role in preventing breakdown of a component and possible spread and development of a difficult situation into a full-blown failure of the entire engine. A reduction in speed or power developed generally goes a long way to alleviate a faulty condition.

But the control system based on the feed back approach has certain drawbacks that on occasion cannot prevent the failure mechanism from spreading after initiation. The feed back principle relies on the output of a modified input signal to further modify the input signal to achieve a more acceptable result in the output. As an illustration, if the power output of the turbine does not meet the control room's demand, the microprocessor will order extra fuel to be injected in the combustor to produce additional power.

However, a time delay is involved in the measurement of the output signal, it's processing in the controller, ordering the corrective action and its execution. The time delay is admittedly small and may be measured in milliseconds. The operating speed of modern turbines is high enough, and the time lag has the potential to prevent a critical component from further disintegrating after the failure mechanism ensues. The failure may then spread to adjacent parts and grow into an outright failure and engulf large parts of the overall system.

A large time delay causes a system to be less sensitive. In conjunction with time lag, stability of a control system is also of great significance. The two objectives of stability and sensitivity contradict each other. A sensitive system will react promptly to an unacceptable situation. An overly sensitive system, however, will cause the corrective action to overshoot the necessary output requirement, and further cause it to take the remedial action in the opposite direction. The net result will be lack of stability in operation. Thus, a control system must aim for both acceptable levels of time lag and stability to obtain a proper combination of stable operation and time lag.

20.4 DEFICIENT CONTROL VALVES

A major element of any automatic control system is valves. The valves regulate flow of pneumatic air or hydraulic oil to the actuator to carry out a certain command ordered by the microprocessor. A large array of valves is used in turbo-machinery to ensure safety, control operational parameters such as flow and pressure and perform a host of other tasks. Pressure release valves, for instance, are required to actuate when the steam pressure exceeds a set limit. The safety mechanism then permits some amount of steam to be released and to reduce the steam pressure.

The valves depend on mechanical elements like a spring, stem, valve disk and manual, electric or pneumatic device to operate. Rubber diaphragms are frequently backed with a steel disk to obtain leak-proof seal. Some or all of the parts are assembled inside the valve body.

The elements of the valve may be exposed to air, moisture or other chemicals that lead to corrosion. In particular, salt water and sea air are major culprits behind corroded components that tend to inhibit the operation of the valve. Suspended oily sand and ferric oxide particles have the ability to scratch and erode metallic parts. After exposure over prolonged periods of time, the valve loses its capacity to provide the requisite control. The deficiencies may be serious enough to make the valve lose much of its functionality.

Deficient valves are a major cause of failures and result in accidents in many turbine installations. Over time the valves degrade to an extent that prevents them from providing the controlling capability. In this respect, routine preventive maintenance is of significance. Periodic cleaning, repairing and replacement of broken parts should be of value in avoiding major failures.

20.5 FLOW BLOCKAGE IN COMPRESSOR

Air flowing into the compressor sometimes may get partially blocked when dry vegetation and other items get ingested into the intake during stormy weather. In spite of a bank of filtration elements installed at the intake of a land-based turbine, twigs, branches and leaves get past the filters. The foreign material may get entangled in the stationary vanes to partially block the flow of air.

Even if the offending element is small in size and affect just one stator vane, the blocked passage will misdirect the flow of air into the arriving rotating blade, leading to turbulence at its leading edge. The turbulence will also affect the adjacent blades by increasing the flow angle to the next blade and retarding the angle of attack to the preceding blade. A direct consequence is to induce a rotating stall in the compressor. Since the rotor is spinning, the stalled flow will be rotating in the direction opposite to that of the rotor.

When the ingested object is small in size, the associated problem may not be severe, and even permit continuous operation of the turbine engine as the object gradually disintegrates and the stall dissipates.

A larger and harder object, on the other hand, may leave some dent marks on the airfoils. The disturbed geometry of the smooth airfoil surface may then also act as an obstruction and have its own rotating stall characteristics. The stall generally manifests in the form of reduced pressure at the compressor discharge while also affecting temperature. Vibration signals at the bearings will also increase, with the dynamic activity increasing with the severity of the stall.

20.6 FAULT IN SPEED CONTROL MECHANISM

Power generation turbines are required to operate at a near-fixed speed condition. Industrial gas turbines may operate in a narrow band of speed range. To ensure the turbine unit does not step into a higher regime, a speed control mechanism is required. The mechanical form consists of two masses spinning around the shaft axis and connected with lever arms, collar and retainer springs. At steady operation of the turbine, centrifugal force acting on the masses keeps them in equilibrium position. When rotor speed increases, the centrifugal force correspondingly goes up and causes the masses to shift away from the axis of rotation.

The motion is then transmitted through mechanical elements or electrical signals to a servo-mechanism to either increase or decrease the flow of fuel or steam into the turbine. In steam turbines the inlet valve is placed inside the steam chest section of the high-pressure unit. Gas turbine valves may be placed in the vicinity of either liquid or gas fuel manifold. Rotor speed is reduced as the control valve reduces the fuel or steam flow, which will attempt to bring down the speed.

In the case of over-speed condition in a turbine, loss of control may be attributed to failure of the speed controller and fuel or steam admittance valves. Oxidation can corrode parts such as springs, levers and mechanical actuators as the engine accumulates operational hours. The consequent loss of material increases the risk of the parts failing when stressed, and prevents them from performing their assigned task. So when the driven unit experiences loss of load, fuel in the case of a gas turbine or steam in a steam turbine keeps flowing unabated. If the speed is left unchecked in the control room, the result is the rotor accelerates to a point where the components burst.

20.7 LUBRICATION SYSTEM FAILURES

Problems in the lubrication system arise from a number of causes. When the oil cooling system is not operating at its peak, temperature of the oil above 260° F will lead to deteriorated viscosity and load carrying capability. If the magnetic chip detector or oil filter is clogged, debris in the form of carbon particles, sand and metal chips is likely to flow into the annular space between the journal and the bearing. Where the pressure or flow of the oil from the pump is not sufficient, the bearing will starve and cause the oil film to break. As a consequence, friction losses will mount, scoring will damage the mating surfaces and temperature in the bearing will rise.

The efficiency of the cooling system must not be gauged by the temperature differential between the incoming and outgoing oil. The temperature of the oil coming into the bearing housing is far more significant. Excess oil temperature rapidly diminishes the viscosity of the lubricant, with detrimental impact on the performance of the bearing. At the other end of the temperature scale, the viscosity becomes very high. Thus, a turbine unit operating in the winter in colder climates may need to be warmed since high viscosity oil is too thick and causes problems in delivering the oil.

Accumulated debris in the oil reservoir requires removal and cleaning. The oil also tends to get exhausted after a period due to the mixing of steam and other fluids getting into the bearing sump.

Periodically scouring the reservoir tank, replenishing and replacing the lubricant oil are essential if the turbine is to perform at its best. Proper running of the oil pump, valves and tube fittings is also important.

20.8 FRACTIONAL FREQUENCY VIBRATION AND WHIP IN JOURNAL BEARING

Unstable operation is sometimes encountered in journal bearings. The increased dynamic activity may be attributed to a host of operating factors, including imposed load, running speed, oil viscosity and bearing clearance. Changes in any of these parameters are oftentimes responsible for the bearing vibrations to shift both in terms of amplitude and frequency. Shaft vibrations increase considerably when the phenomenon occurs from hydrodynamic effects within the oil film, giving rise to the expression, 'bearing whip'. The frequency mostly is slightly below half the running speed of the shaft.

Turbulence in the fluid film and temperature are two aspects that play a major role in bearing operation. The Reynolds number determines whether the bearing operates in the laminar, transition or turbulent regime. Defined by the lubricant's density and viscosity, journal diameter, radial clearance and shaft speed, laminar flow prevails when the Reynolds number is below 750. Turbulence sets in when Reynolds number is in excess of 1500, and the transition occurs between these two values. Both the laminar and turbulent operation may coexist since viscosity and film thickness vary throughout the film.

Generally, the turbulence raises the load capacity by increasing the film thickness but it also boosts power losses, and the consequent temperature gain lowers the operating viscosity. This in turn lowers the load capacity. It may be surmised that the effects of turbulence on stability are deleterious because of the increased film thickness. An important bearing parameter is the Sommerfeld number, and is determined by the lubricant's viscosity, shaft running speed, journal radius, clearance and static pressure on the bearing's projected surface.

20.9 SHAFT FAILURE DUE TO EXCESSIVE TORSIONAL VIBRATION

Elastic twisting of a shaft is superimposed on its regular spinning motion, making it difficult to distinguish the torsional oscillations. Also, the support bearings are there to permit the shaft to freely rotate without interference. The type and size of the bearings and their location does not affect the elastic twists and de-twists, making them nearly impervious to detection from outside. Thus, when a rotating machinery goes through a torsional natural frequency of vibration during a run up or down in speed, the vibrations remain unnoticed.

If now a dynamic torsional excitation force were to act with a frequency that coincides with one of the natural frequencies of the rotating system, a state of resonance prevails. In addition to stresses due to the regular torque load, dynamic torsional stresses will be produced in the shaft. However, there is little damping available in the system, mostly attributable to the molecular rubbing in the material and perhaps from wind effects. So the dynamic stresses at the resonant speed will mount rapidly.

The situation has been observed to occur in power generation steam and gas turbine rotors that operate at a fixed speed of 3600 rpm for 60 Hz electrical power. When the natural frequency of vibration in the combined turbine and generator rotor is close to 60 Hz or its second multiple 120 Hz, failures in the form of cracks develop in the fillet in shoulder areas where shaft radii change. Where operation continues over prolonged periods, the possibility of ruinous failures in the shaft cannot be ruled out.

At first sight it is difficult to understand the interconnection between the mechanical elastic vibrations at 60 Hz or its second multiple and the electrical signal from the generator at the same frequency. A possible explanation may be based on the frequency of the current of the main bus bars of a regional power grid. Several generators may be on the power grid, contributing large amounts of electrical power. If one errant generator experiences periodic twists and de-twists during its regular spinning at 3600 rpm, it disrupts production of a smooth sinusoidal electrical signal. Variations in the phase angle from the dynamic oscillations will interfere with the generation of the sinusoidal current. At the same time the regional bus bar will force the errant rotor to turn smoothly without the intermittent oscillations, thus setting up the condition for resonance.

20.10 FOREIGN OBJECT DAMAGE

Aviation-based jet engine manufacturers are mandated to ensure broken fragments of fan, compressor and turbine blades do not penetrate the outer casing. Some manufacturers opt for steel fan case to prevent separated blade sections from piercing and exiting radially. With an aluminum fan case, bundles of Kevlar, a synthetic fiber with capacity to withstand penetration from a partial or full fan blade acting as a missile, have also been employed successfully to reduce weight.

The section where the blade separates is also of interest since it determines the centrifugal force and dynamic energy with which it is released. Compressor and turbine blades are relatively shorter and easier to contain after a separation incident. Stationary components will then cause the parts to break into smaller pieces, and the flow of air and gases will direct the debris axially through the fan or turbine exhaust duct.

However, if simultaneously more than one blade were to break, the possibility of failure to contain within the casing cannot be ruled out. Associated with the breakage is the abrupt reduction in the flow of air to the combustor. The usual flow of fuel cannot be burned, and flames of unburned fuel will leap out through the turbine's exhaust. Within the combustor itself fires may start and spread outside the engine.

20.11 ICE AND WATER INGESTION

Aircraft takeoffs and landings in less than ideal weather conditions are all too common. Sheets of water and ice may get ingested into the inlet to the engine. Rainwater drops may not impact upon the structural integrity of the intake cowling or the fan blades. Sheets of ice, on the other hand, may cause dents if its temperature is low enough and is in a compacted form.

If in a near blizzard condition an aircraft is required to depart, the swirling motion of air all around the engine may cause problems. Immediately before the start of the takeoff roll, it is necessary to point the aircraft nose and engines into the air stream so that air is admitted freely. The fan than does not have to produce extra suction, reserving all the power for the imminent takeoff of the aircraft. Occasionally, however, the combustor may be robbed of the oxygen required to burn all of the fuel injected for the takeoff condition, and cause the engine to stall.

Engine manufacturers carry out a crosswind test to check the ability of the engine to start and develop thrust. Large fans are placed perpendicular to the engine axis near the inlet mouth to produce the crosswind effect. The engine is then started following a standardized procedure. The engine is required to light and continue to develop power.

Water ingestion test calls for a spray of water to form a mist blown directly into the inlet. A slab of ice of appropriate dimensions and maintained at a fixed temperature is also flung into the inlet mouth. Visual inspection after the test must ensure nicks, cuts and dents are not produced on the parts.

20.12 BIRD INGESTION LEADING TO FAN BLADE FAILURE

A powerful suction is created at the inlet of aviation jet engines. After compression and combustion, the turbine unit produces power to drive the fan, which in turn breathes in the air and builds up thrust to propel the aircraft to move forward. Maximum thrust, and hence also suction at the inlet, is required when the aircraft gets off the ground and is in the climb mode. Besides air, however, foreign objects in the path of the flow of air sometimes get sucked into the engine's inlet. When a flock of birds directly crosses the flight path, the risk of one or more birds striking the fan is greatly enhanced. Figure 20.1 illustrates such a situation.

Fig. 20.1 Aircraft in Vicinity of Birds

Observations from ornithologists indicate high population growth of Canada geese, brown pelicans, sand-hill cranes and wild turkeys. Wildlife related strikes have increased in recent years in many parts of the world. The birds pose a risk to airliners because birds can get sucked into jet engines and cause catastrophic failures.

In 2009, 4 Canada geese weighing 8 pounds on average struck a US Airways Airbus A320 jet flying over the Hudson River in New York. The birds crashed into both engines less than two minutes after takeoff from LaGuardia Airport, and robbed them of almost all thrust. The cockpit crew was forced to land the aircraft in the middle of the river. All 155 people on board survived what authorities referred to as a 'miracle on the Hudson'.

20.13 LOSS OF CONTROL DURING ABORTED FLIGHT LANDING

A significant string of events occurs in an aircraft's engines when the control tower orders the pilot to abort a scheduled landing of the flight. As the airplane comes in for the landing with the altitude decreasing, the engine's low and high-pressure rotors are running at slightly above their flight idle condition. The engines develop just enough power to overcome the overall aerodynamic drag. The combustor is then receiving fuel at the set rate, and the turbine metal temperatures are at the appropriate level.

As soon as the command to abort is received, the pilot will push the engine throttle to obtain more thrust so that the aircraft can get into the climb mode. If the instruction were to be received when the aircraft is close to the ground, the ascent will need to be rapid. The abrupt demand for additional power will cause a far higher flow of fuel into the combustor than for the flight idle condition. But the air mass flow from the compressor is still that at the low power setting. The result is not all of the fuel is burned, causing flames from the unburned fuel to leap out of the turbine's exhaust and the exhaust gas temperature to soar.

Still another factor is the amount of cooling air for the hot path components. Lack of adequate coolant during the rapid shift from idle to takeoff conditions will lead to gas and metal temperatures climbing up and consequent burning of the airfoils. Related other effects will be due to rubbing of blade tips against the stator case and opening up of the clearances.

Following the incident, ground personnel will opt to carry out a bore scope inspection of the full engines and possible removal of the engines from the wing for repairs.

20.14 DAMAGE IN TURBINE FROM FLIGHT TAKEOFF ABORT

A somewhat similar sequence of events as in an aborted landing is likely to occur when an aircraft is ordered to stop from taking off. Opinion is divided as to which event, an aborted takeoff or an aborted landing, is more stressful and inflict damage to the engines.

As the takeoff roll initiates, throttle setting of the engines is at full power to enable the aircraft to accelerate, gain speed and break off from earth's gravity. An abrupt nearly shut down fuel flow will cause gas temperatures to plunge, producing excessive thermal gradients in the turbine's flow path components. As a consequence, thermal fatigue related damage will be high on the components.

Differential thermal distortion will occur between the rotor and stator masses. The thermally related shrinking of the two modules does not keep pace with each other as the turbine rotors spin. As a consequence, the sudden reduction in the power will also alter the available clearances at the blade tips. Damage at the casing shrouds and at the blade tips may be severe. Bore scope inspection of the engine should not be ruled out after the incident.

20.15 SYSTEM RELATED POWER GENERATION TURBINE TRIP

As much as possible, power generation and other industrial turbine users are generally advised to avoid the unit from going into tripping condition and cause the turbine to shut down suddenly. Sometimes, however, situations may arise that cannot prevent the trip. A trip in land-based and marine turbines can occur from a variety of reasons, such as failure in the lubricant oil system or generator cooling system, loss of airfoil segments, or the exhaust gas temperature going out of prescribed limits.

Protection for the entire turbine is then necessary. The only way would be for the turbine engine's microprocessor to order operations to cease once the signal from the sensors for lube oil pressure, coolant flow, rotor vibrations or exhaust gas temperature indicate the readings are not acceptable.

But the consequences of the tripping of operations will be felt in the combustor and turbine modules of the engine. Substantial thermal gradients in the turbine's airfoils will inflict severe thermal fatigue related damage during a sharp shutdown. It is estimated the thermal fatigue damage in the rotating airfoils following a trip is four times the damage experienced following a normal shutdown. Stator vane airfoils at the inlet to the turbine will also see considerable damage in the form of thermal distortion and bowing at the leading edge. The bow will prevent the turbine from developing full power due to the disturbed aerodynamic flow pattern. Premature retirement and replacement of the airfoils may be necessitated, especially of the first stage airfoils.

20.16 REVIEW QUESTIONS

20.16.1 Name the important factors both inside and outside a turbine engine that can lead to stoppage of a unit.

20.16.2 What is the role of the speed controller in a turbine? How does its failure affect operations?

20.16.3 Name some factors that can cause a lubrication oil system to stop delivering oil to the bearings.

20.16.4 How does turbulence in the fluid film of a journal bearing affect its operation?

20.16.5 Under what conditions do torsional vibrations in a spinning shaft create a problem?

20.16.6 Why is it difficult for an airplane to commence operations in stormy weather?

20.16.7 Why is it necessary for an aircraft to point in the direction of prevailing winds prior to starting the engine?

20.16.8 How does ingestion of extraneous material into an aviation engine's inlet affect its operation?

20.16.9 Describe the major effects on an aircraft's engines when the pilot is ordered to abort a scheduled landing?

20.16.10 Why are torsional elastic vibrations in a rotating shaft difficult to detect?

20.16.11 Name some factors that can cause a power generation turbine to come to a sudden stoppage.

20.17 EXAMPLE PROBLEMS

Example Problem # 20.17.1: If C_6H_{12} combines with O_2, calculate the number of O_2 molecules required and molecules of CO_2 and of H_2O for stoichiometric combustion.

Solution: The general rule for stoichiometric combustion is:

$$C_xH_y + nO_2 \Rightarrow aCO_2 + bH_2O$$

where $a = x$, $b = y/2$ and $n = x + y/4$

For C_6H_{12}, $a = 6$, $b = 12/2 = 6$ and $n = x + y/4 = 6 + 3 = 9$

Thus, for stoichiometric combustion,

$$C_6H_{12} + 9O_2 \Rightarrow 6CO_2 + 6H_2O$$

Example Problem # 20.17.2: A disk of 14.875 in. radius and 3.625 in. width is attached at one end of a hollow shaft of inner radius 1.127 in. and thickness 0.173 in. The disk material has density ρ = 0.249 lbs/in³. The shaft is supported in two roller bearings placed 3.127 in. and 12.428 in. from the disk centerline. Disregarding weight of the shaft, calculate the reactions at the bearings due to the weight of the disk.

Solution: Weight of the disk: W = Volume × density = $\pi R^2 l\rho$

$$= \pi \times 14.875^2 \times 3.625 \times 0.249 = 627.44 \text{ lbs}$$

Label bearing support near the disk B and the far bearing A. Assume x as the distance along the shaft, with $x_A = 0$, $x_B = 9.301$ in., and $x_{Disk} = 12.428$ in.

For static equilibrium, $(R_A + R_B - 627.44) = 0$, or $R_A = 627.44 - R_B$.

Also: $\Sigma M = 0$. Taking moments about A: $627.44 \times 12.428 - R_B \times 9.301 = 0$

Hence: $R_B = 838.38$ lb., and $R_A = 627.44 - 838.38 = 210.94$ lb. ↓

Example Problem # 20.17.3: A steel alloy has yield strength of 88.25×10^3 lbs/in^2 in simple tension. If in a two-dimensional stress system the principal stresses are 22.4×10^3 lbs/in^2 compression and 18.7×10^3 lbs/in^2 tension, calculate the factor of safety from the maximum shear stress theory and Miscs-Hencky theory.

Solution: The stresses are $\sigma_1 = -22.4$, $\sigma_2 = 0$ and $\sigma_3 = 18.7$ ksi

By the maximum shear stress theory the equivalent single tensile stress is:

$$s = \sigma_1 - \sigma_3 = -22.4 - (18.7) = -41.1 \times 10^3 \text{ lbs/in}^2.$$

So the factor of safety is 88.25/41.1 = 2.147

The Mises-Hencky theory for combined stresses is:

$$\begin{aligned} 2s^2 &= (\sigma_1 - \sigma_2)^2 + (\sigma_2 - \sigma_3)^2 + (\sigma_3 - \sigma_1)^2 \\ &= (-22.4 - 0)^2 + (0 - 18.7)^2 + (18.7 + 22.4)^2 \\ &= 2540.66 \end{aligned}$$

Hence: $s = 35.64 \times 10^3$ lbs/in^2

The factor of safety $= 88.25/35.64 = 2.476$

Example Problem # 20.17.4: A torque load of 11925 in-lbs is applied to a 2.75 in. diameter and 15.875 in. long test coupon. Calculate the shear stress induced from the load and the angle of twist if material shear modulus is 11.5×10^6 lbs/in^2.

Solution: For torsion: $(T/I_p) = (\tau_s/R) = (G\theta/L)$, where T = torque, I_p = cross-section polar inertia, τ_s = shear stress, R = radius, G = shear modulus, θ = angle of twist and L = length of specimen.

Shear stress $\tau_s = (TR/I_p) = (11925 \times 2.75/2)/[(\pi/32) \times 2.75^4]$
$= 2921.80$ lb./in^2

Angle of twist $\theta = (TL/GI_p) = (11925 \times 15.875)/[11.5 \times 10^6 \times (\pi/32) \times 2.75^4]$
$= 0.002933$ radians $= 0.168°$

Example Problem # 20.17.5: A gas turbine rotor has maximum design cold clearance at the blade tip of 0.0045 in. Static weight deflection is 0.00225 in. and expected dynamic deflection from unbalance is 0.0014 in. Determine the maximum and minimum clearances on that stage.

Solution: If cold design clearance and dynamic deflection are uniform around the periphery, deadweight deflection is maximum at the base.

Maximum clearance at top of rotor = 0.0045 + 0.00225 – 0.0014
= 0.00535 in.

Minimum clearance at base of rotor = 0.0045 – 0.00225 – 0.0014
= 0.00085 in.

20.18 EXERCISE PROBLEMS

Exercise Problem # 20.18.1: Air pressure outside an aircraft is measured 9.745 ins. of mercury, and pressure at the nozzle exhaust is 78.24 ins. of mercury. Calculate area of the exhaust nozzle if thrust developed due to the pressure differential is 12765 lbs.

Exercise Problem # 20.18.2: A ball is shot vertically up at a speed of 40.5 ft/sec. Calculate the time required for it to return back to ground.

Exercise Problem # 20.18.3: A machine rotor has maximum design cold clearance at the blade tip of 0.00685 in. Static weight deflection is 0.00375 in. and expected dynamic deflection from unbalance is 0.00220 in. If dynamic deflection varies linearly with unbalance, what percentage of unbalance increase will cause the tips to rub at the base?

Exercise Problem # 20.18.4: A ball weighing 75.0 lbs is shot vertically up at a speed of 38.5 ft/sec. Calculate its velocity as it comes back to the ground.

Exercise Problem # 20.18.5: A torque load of 118.75 in-lbs is applied to a 0.75 in. diameter and 4.25 in. long test coupon. Shaft material shear modulus is 11.5×10^6 lbs/in^2. Calculate the angle of twist in degrees.

REFERENCES & BIBLIOGRAPHY

"AFBMA Standards for Ball and Roller Bearings", Revision # 4, June, 1972.

Allaire, P. E., Li, D. F., Choy, K. C., "Transient Unbalance Response of Four Multi-Lobe Journal Bearings", Journal of Lubrication Technology, ASME, New York, 1980.

Allison Gas Turbine Division, TM # 55-2840-231-23, 1981.

American Petroleum Institute, "Centrifugal Compressors for Petroleum, Chemicals and Gas Service Industries", API Standard, 617, Sixth Edition, 1995.

ASME Report – "Pressure/Viscosity in Rolling Element Bearings", Volume II, New York, 1954.

Bailey, J. K., Galbato, A. T., "Evaluating Bearings for High Speed Operation", Machine Design, October, 1981.

Bently, D. E., "Monitoring Rolling Element Bearings", Orbit 3, No. 3, pages 2–15, 1982.

Forster, N. H., "High Temperature Lubrication of Rolling Contacts with Lubricants Delivered from the Vapor Phase and as Oil Mists", Ph. D. Thesis, University of Dayton, 1996.

Greathead, S., Bostow, P., "Investigations into Load Dependent Vibrations of High Pressure Rotor on Large Turbo-Generators", Proceedings, Institute of Mechanical Engineers, Conference on Vibrations in Rotating Machinery, Cambridge, pages 279–286, 1976.

Hagg, A. C., Sankey, G. O., "Some Dynamic Properties of Oil Film Journal Bearings with Reference to Unbalance Vibration of Rotors", Journal of Applied Mechanics, volume 23(2), pages 302–306, New York, 1956.

Harris, T. A., "Rolling Bearing Analysis", John Wiley & Sons, New York, 1984.

Monk, R., "Vibration Measurement Gives Early Warning of Mechanical Fault", Process Engineering, pages 135–137, November, 1972.

Pinkus, O., Sternlicht, B., "Theory of Hydrodynamic Lubrication", McGraw-Hill, New York, 1961.

Raimondi, A. A., Boyd, J., "A Solution for the Finite Journal Bearing and Its Application to Design and Analysis, Parts I, II and III", Transactions of ASLE, volume 1, number 1, pages 159–209 in Lubrication Science and Technology, Pergamon, New York, 1958.

Vazquez, J. A., Barrett, L. E., Flack, R. D., "Flexible Bearing Supports Using Experimental Data", ASME paper # 00-GT-404, New York, 2000.

INDEX

B

C

D

E

F

J

K

L

M

N

R

S

T

U